AF389614

Human Enhancement Technologies and Our Merger with Machines

Human Enhancement Technologies and Our Merger with Machines

Editors

Woodrow Barfield
Sayoko Blodgett-Ford

MDPI • Basel • Beijing • Wuhan • Barcelona • Belgrade • Manchester • Tokyo • Cluj • Tianjin

Editors
Woodrow Barfield
Visiting Professor, University of Turin
Affiliate, Whitaker Institute, NUI, Galway

Sayoko Blodgett-Ford
Boston College Law School
USA

Editorial Office
MDPI
St. Alban-Anlage 66
4052 Basel, Switzerland

This is a reprint of articles from the Special Issue published online in the open access journal *Philosophies* (ISSN 2409-9287) (available at: https://www.mdpi.com/journal/philosophies/special_issues/human_enhancement_technologies).

For citation purposes, cite each article independently as indicated on the article page online and as indicated below:

LastName, A.A.; LastName, B.B.; LastName, C.C. Article Title. *Journal Name* **Year**, *Volume Number*, Page Range.

ISBN 978-3-0365-0904-4 (Hbk)
ISBN 978-3-0365-0905-1 (PDF)

Cover image courtesy of N. M. Ford.

Contents

About the Editors

Woodrow Barfield (currently, Visiting Professor of Law, University of Turin, Italy; affiliate, Whitaker Institute, NUI, Galway): Woodrow (Woody) Barfield, Ph.D., J.D., LLM is a speaker, lecturer, and author who served as a Professor of Engineering at the University of Washington where he received the National Science Foundation Presidential Young Investigator Award and where he directed the Sensory Engineering Laboratory. His research revolves around the design and use of wearable computers, augmented and virtual reality systems, and law and technology. He served as the Senior Editor of *Presence: Teleoperators and Virtual Environments* and recently published "Musings on Presence Twenty-Five Years after 'Being There'" about his work on *Presence* within virtual environments. He is currently Associate Editor of the *Virtual Reality Journal* and is the Senior Editor of the first and second editions of *Fundamentals of Wearable Computers and Augmented Reality*. His book "Cyber Humans: Our Future with Machines" talks about the possibility of a future in which humans may eventually merge with the technology/machines we are creating, including artificial intelligence. He edited "The Cambridge Handbook of the Law of Algorithms", co-edited "The Research Handbook on Law and AI" (with Ugo Pagallo), co-authored "Advanced Introduction to Law and Artificial Intelligence" (with Ugo Pagallo), and co-edited "Virtual Environments and Advanced Interface Design", "Human Factors in Intelligent Transportation Systems", and the "Research Handbook on the Law of Virtual and Augmented Reality" (with Marc Blitz). Woody earned his PhD in Engineering from Purdue University and his JD degree and LLM in intellectual property law and policy, serving as an editor of the *Journal of Law, Technology & Arts*. He has authored numerous publications and presented at national and international conferences, including keynote addresses and articles in engineering and law review journals.

Sayoko Blodgett-Ford (Adjunct Professor, Boston College Law School, Newton Centre, MA 02459, USA, and Principal, GTC Law Group PC & Affiliates, Westwood, MA 02090, USA): Sayoko's practice focuses on intellectual property, contracts and privacy/data protection in mergers and acquisitions. She has extensive expertise in connection with technology M&A transactions. Sayoko is an Adjunct at Boston College Law School, where she teaches Artificial Intelligence and the Law and Mobile Apps & Big Data-Legal Contributions. She has also taught Privacy Law (2017–2019). She has also served as Lecturer in Law at the University of Hawai'i Law School, where she has taught Internet Law & Policy, Intellectual Property, Contracts Drafting, High-Growth Entrepreneurship and Administrative Law, among other subjects. She served as a Court Appointed Arbitrator for the Hawai'i State District Court—First Circuit. She previously served as acting general counsel of Tetris Online, Inc. and as Senior Manager of the Intellectual Property Group at Nintendo of America Inc. in Redmond, Washington. At Nintendo, Sayoko was responsible for domestic and international intellectual property clearance, defense, registration, and enforcement. Her particular areas of focus included patent, trademark, copyright, trade secret and advertising, as well as privacy law compliance and video game ratings compliance. Sayoko has also taught advertising law in the University of Washington IP LLM program. Prior to Nintendo, Sayoko practiced at Foley Hoag LLP and was a law clerk for Judge Douglas P. Woodlock, U.S. District Court for the District of Massachusetts. Sayoko holds a B.S. in Physics from the College of William and Mary, an M.S. in Physics from the University of Maryland, and a J.D. from Yale Law School. Her publications include Human Enhancements and Voting: Towards a Declaration of Rights and

Responsibilities of Beings, in *Philosophies* https://doi.org/10.3390/philosophies6010005; A Real Right to Privacy for Artificial Intelligence (book chapter) in Research Handbook on the Law of Artificial Intelligence (Publisher: Edward Elgar) (Woodrow Barfield and Ugo Pagallo, editors) (2018) https://www.e-elgar.com/shop/research-handbook-on-the-law-of-artificial-intelligence; Advertising Legal Issues in Virtual and Augmented Reality (book chapter) in Research Handbook on the Law of Virtual and Augmented Reality (Publisher: Edward Elgar) (Woodrow Barfield and Marc Blitz, editors) (2018) (co-authors: Woodrow Barfield and Alexander Williams); and Data Privacy Legal Issues in Virtual and Augmented Reality (book chapter) in Research Handbook on the Law of Virtual and Augmented Reality (Publisher: Edward Elgar) (Woodrow Barfield and Marc Blitz, editors) (2018) (co-author Mirjam Supponen) https://www.e-elgar.com/shop/research-handbook-on-the-law-of-virtual-and-augmented-reality.

Preface to "Human Enhancement Technologies and Our Merger with Machines"

This book represents a collection of papers that comprised two Special Issues which were published in the online journal *Philosophies*. Eight of the papers are from the Special Issue on "Human Enhancement Technologies and Our Merger with Machines", which was edited by Woodrow Barfield and Sayoko Blodgett-Ford. Three of the papers are from the Special Issue on "Cyberphenomenology: Technominds Revolution", which was edited by Professor Jordi Vallverdu (who kindly consented to the use of papers collected for his Special Issue in this edited book). All the papers discuss issues related to emerging technologies designed to enhance human bodies and minds from the perspective of ethics, law, and policy. Considering the papers which comprise this book, we do not focus solely on the integration of technology within human bodies, such as in neuroprosthetics; instead, we emphasize a wider scope of the topic and also include a discussion of genetic, biological, and pharmacological enhancements to humans. Diverse approaches are leading to a future world where humans may be permanently or temporarily enhanced with technology through the use of artificial parts, by manipulating (or reprogramming) human DNA, and other enhancement techniques that will surely emerge in the coming decades (and combinations thereof). Such techniques could include precision medicine, more sophisticated genome sequencing and gene editing (CRISPR), cellular implants, and computationally powerful wearable devices that can be implanted in the human body, including the brain. Other enhancements for humans could result from nanoscale drugs (including antibiotic "smart bombs") to target specific strains of bacteria and to repair cells. And, in the future, we may be able to implant devices such as bionic eyes and bionic kidneys within the human body, or even artificially grown and regenerated human organs. Whatever the form of the enhancement technology, important ethical, legal, and policy issues arise. The collection of papers in this book continues the discussion on this important topic which has already garnered interest within the philosophy community. Here, we present some unique issues, and we hope to motivate further discussion on this topic.

Should the reader think we are operating in the realm of science fiction, only discussing issues of ethics, law, and policy which may be relevant for a distant future, consider that in 2018, in research funded in the US by the Defense Advanced Research Projects Agency (DARPA), a person with a brain chip piloted a swarm of drones using signals from the brain. More recently, a research team from Columbia University tested the convergence of neural networks. This was achieved by combining brain implants, artificial intelligence, and a speech synthesizer to translate brain activity into recognizable robotic produced words. The implications of this neuromorphic technology for human enhancement are tremendous, including allowing paralyzed people the ability to communicate more easily and the potential to read human thoughts via cognitive imaging. In the view of the editors, we are on the cusp of significantly upgrading (or at least modifying) the human ecosystem. We are already witnessing the merging of artificial circuitries with biological intelligence, retrieved in the form of electric, magnetic, and mechanical transductions. These and other technologies that will be integrated into or combined with the human body all give rise to the ethical, legal, and policy issues considered in this book. They also require a continuing effort to re-evaluate current laws and, if appropriate, modify or develop new laws that address enhancement technology. A legal, ethical, and policy response to current and future human enhancements should strive to protect the rights of all parties involved, and to recognize responsibilities of humans to other conscious and living beings, regardless of what they look like or what abilities they may have (or lack). In the final paper, a potential ethical approach is outlined in which rights and responsibilities should be respected even if enhanced humans are perceived by non-enhanced (or less-enhanced) humans as "no longer human" at all.

Woodrow Barfield, Sayoko Blodgett-Ford
Editors

Editorial

Introduction to Special Issue "Human Enhancement Technologies and Our Merger with Machines"

Woodrow Barfield [1,*] and Sayoko Blodgett-Ford [2,3,*]

1   Whitaker Institute, National University of Ireland, H91 TK33 Galway, Ireland
2   Boston College Law School, Boston College, Newton Centre, MA 02459, USA
3   GTC Law Group PC & Affiliates, Westwood, MA 02090, USA
*   Correspondence: Jbar5377@gmail.com (W.B.); blodgesa@bc.edu (S.B.-F.)

Citation: Barfield, W.; Blodgett-Ford, S. Introduction to Special Issue "Human Enhancement Technologies and Our Merger with Machines". *Philosophies* **2021**, *6*, 9. https://doi.org/10.3390/philosophies6010009

Received: 22 January 2021
Accepted: 25 January 2021
Published: 29 January 2021

Publisher's Note: MDPI stays neutral with regard to jurisdictional claims in published maps and institutional affiliations.

We are pleased to introduce the authors and papers which form the Special Issue "Human Enhancement Technologies and Our Merger with Machines". We should first note that humans are already becoming equipped with technology that ranges from artificial limbs, heart pacers, retinal and neuroprostheses, cochlear implants, and other emerging technologies used to enhance human capabilities. We had such current and future "mechanical enhancements" in mind when we first considered editing a Special Issue. However, as we developed the Special Issue further, based on the authors submissions in response to the call for papers, we realized that it was essential to develop a broad definition of "enhancement" to include chemical enhancements, genetic enhancements, pharmaceutical enhancements, and beyond, as all such areas are currently under development and relate to the topic of the Special Issue. In our view, a Special Issue on human enhancement technologies is not only timely but covers a topic that invokes many interesting and challenging ethical, legal, practical, and philosophical issues. When inviting scholars to contribute to the Special Issue, we sought authors from different academic disciplines such that we would produce a Special Issue with different perspectives and points of view. We think a cross-disciplinary approach to human enhancement technologies is important given the complexity and depth of the challenges that need to be addressed. Thus, the Special Issue consists of eight papers contributed by a diverse group of scholars from law, technology, and social science that focus on issues which involve a future in which people will be equipped with enhancement technologies for medical and nonmedical reasons.

The paper by Dunagan, Grove, and Halbert, "The Neuropolitics of Brain Science and Its Implications for Human Enhancement and Intellectual Property Law", focuses on how human enhancements directed at the brain may change what we know about human creativity and innovation. The authors make the points that the possibility of direct brain-to-brain communication, the use of cognitive-enhancing drugs to enhance intelligence and creativity, and the extended connections between brains and the larger technological world can lead to a linkage between intellectual property (IP) law and policy. To address IP issues resulting from human enhancements, the authors suggest that new conceptualizations of the brain are needed to challenge the notion of the autonomous individual and to address creativity and originality as possibly occurring beyond that of an individual creation. Fiorella Battaglia, in "Agency, Responsibility, Selves, and the Mechanical Mind", also focuses on the subject of mental activities and human enhancement, discussing moral issues that arise not only when neural technology directly influences and affects people's lives, but also when the impact of its interventions in creating an "extended mind" allows us to conceptualize the mind in new and unexpected ways, using a "mechanical mind" approach. By reviewing theories of consciousness, theories of subjectivity, and third person perspective on the brain, the paper identifies both a major area of transformation in philosophy of action understood in terms of additional epistemic devices and as a result "self-objectification" as a new area of concern, which she characterizes in terms of "alienation" following Ernst Kapp's philosophy of technology.

Allen Coin, Megan Mulder, and Veljko Dublievic, in "Ethical Aspects of BCI Technology: What Is the State of the Art?", address uses of enhancement technology to discuss how brain activity can be interpreted through both invasive and noninvasive monitoring devices, allowing for novel, therapeutic solutions for individuals with disabilities and for other nonmedical applications. However, the authors note that, to date, a number of ethical issues have been identified from the use of brain–computer interface (BCI) technology. Based on a review and analysis of the literature, the authors identify key areas of BCI ethics that they argue warrant further research efforts, arguing that the physical and psychological effects of BCI technology and, particularly, BCI ethics should be addressed in future work. From more of a technological view of human enhancement, Kevin Warwick in "Superhuman Enhancements via Implants: Beyond the Human Mind" takes a practical look, based on his unique perspective of one of the world's first cyborgs, at some of the possible enhancements for humans through the use of implants, particularly in the brain or nervous system. An interesting point is made by Warwick that some cognitive enhancements may not turn out to be practically useful, whereas others may turn out to be mere steps on the way to the construction of superhumans. Warwick's paper focuses on the latter issue—enhancements that take recipients beyond the human norm rather than any implantations employed merely for therapy. Warwick takes the perspective that humans are essentially their brains, and that bodies serve as interfaces between brains and the environment. Speculating about the future of human enhancement, Warwick considers the possibility of building an "Interplanetary Creature" having an intelligence and, possibly, a consciousness of its own.

Chia Wei Fahn, in "Marketing the Prosthesis: Supercrip and Superhuman Narratives in Contemporary Cultural Representations", examines prosthetic technology in the context of posthumanism and disability studies, with a focus on how the disabled body is empowered through prosthetic enhancement and cultural representations. Fahn cross-examines empowering marketing images and phrases embedded in cinema and media that emphasize how disability becomes a superability with prosthetic enhancement. She discusses questions such as: Why is it that the disabled have yet to reap the rewards or prosthetic technology? How are disabled bodies, biotechnology, and posthuman possibilities commodified and commercialized? Most importantly, what impact will human enhancements have on our society? Presenting a different but important type of enhancement technology, Marcelo de Araujo in "The Ethics of Genetic Cognitive Enhancement: Gene Editing or Embryo Selection?" discusses how recent research with human embryos in different parts of the world has sparked a new debate on the ethics of genetic human enhancement. While focusing on gene-editing technologies, especially CRISPR, Araujo notes that less attention has been given to the prospect of pursuing genetic human enhancement by means of in vitro fertilization (IVF) in conjunction with in vitro gametogenesis, genome-wide association studies, and embryo selection. Thus, he examines the different ethical implications of the quest for cognitive enhancement by means of gene-editing on the one hand, and embryo selection on the other. He argues that the philosophical debate on the ethics of enhancement should take into consideration public attitudes to research on human genomics and human enhancement technologies.

S.J. Blodgett-Ford, in "Human Enhancements and Voting: Towards a Declaration of Rights and Responsibilities of Beings", explores the ethics of "voting" in the context of human enhancements. The author discusses "voting" for enhanced humans who fall into two broad categories—those with *moderate* enhancements, where the "humanity" of the enhanced person is not seriously in question, and those with *extreme* enhancements, where humans who lacked similar enhancements would likely question the "humanity" of the enhanced human (people a hundred years ago might think enhanced people of today are beyond humanity). The question of who has the right to vote and how they can vote is explored in the context of voting in the United States at the federal level. The author points out that for voting, there is risk that existing patterns of discrimination will continue substantially "as is" for humans with physical and mental enhancements, who are viewed as "still human". For the "beyond human" category of enhancements, the article argues that the established rules and practice of voting are likely to be directly challenged by extreme physical and/or mental enhancements, and that for both moderate and extreme enhancements, humans who are not enhanced may be disenfranchised if certain enhancements become prevalent among the voting population. Extending the discussion, the author borrows from the Universal Declaration of Human Rights (UDHR) of 1948 and applies an extended ethical theory of engagement to advocate that voting rights and responsibilities should be reframed from a foundational working hypothesis that all living and all conscious "beings" (including all enhanced and nonenhanced humans) are of value and should have a right to vote as a cornerstone of a new Declaration of Rights and Responsibilities of Beings.

Given the use of enhancements technologies to create "super soldiers", Sahar Latheef and Adam Henschke in "Can a Soldier Say No to an Enhancing Intervention?" discuss how technological advancements have provided militaries with the possibility to enhance human performance and to provide soldiers with better warfighting capabilities. Though these technologies hold significant potential, their use is not without cost to the individual. The paper explores the complexities associated with using human cognitive enhancements in the military, focusing on how the *purpose* and *context* of these technologies could potentially undermine a soldier's ability to say no to these interventions. The authors explore situations that could potentially compel a soldier to accept such technologies and how this acceptance could impact rights to individual autonomy and informed consent within the military. The authors argue that though in some situations, a soldier may be compelled to accept enhancements, with their right to say no diminished, it is not a blanket rule, and safeguards ought to be in place to ensure that autonomy and informed consent are not overridden.

To conclude, due to the importance of the ethical and other philosophical, legal, social, and scientific issues involved in human enhancement technologies (broadly defined), we hope that the Special Issue generates more discussion and research on the topics covered in the eight papers and beyond. In our view, a cross-disciplinary approach within the fields of philosophy, engineering, computer science, law, and social science is critical to approach the involved ethical issues as increasingly more powerful technology is attached to and implanted within the human body, and as other modifications become possible.

Funding: This research received no external funding.

Acknowledgments: The editors gratefully acknowledge the efforts of the contributors to the special edition and the quality of their papers.

Conflicts of Interest: The authors declare no conflict of interest.

 *philosophies*

Article

Cyborgs and Enhancement Technology

Woodrow Barfield [1] and Alexander Williams [2,*]

1   Professor Emeritus, University of Washington, Seattle, Washington, DC 98105, USA; Jbar5377@gmail.com
2   140 BPW Club Rd., Apt E16, Carrboro, NC 27510, USA
*   Correspondence: zander.alex@gmail.com; Tel.: +1-919-548-1393

Academic Editor: Jordi Vallverdú
Received: 12 October 2016; Accepted: 2 January 2017; Published: 16 January 2017

Abstract: As we move deeper into the twenty-first century there is a major trend to enhance the body with "cyborg technology". In fact, due to medical necessity, there are currently millions of people worldwide equipped with prosthetic devices to restore lost functions, and there is a growing DIY movement to self-enhance the body to create new senses or to enhance current senses to "beyond normal" levels of performance. From prosthetic limbs, artificial heart pacers and defibrillators, implants creating brain–computer interfaces, cochlear implants, retinal prosthesis, magnets as implants, exoskeletons, and a host of other enhancement technologies, the human body is becoming more mechanical and computational and thus less biological. This trend will continue to accelerate as the body becomes transformed into an information processing technology, which ultimately will challenge one's sense of identity and what it means to be human. This paper reviews "cyborg enhancement technologies", with an emphasis placed on technological enhancements to the brain and the creation of new senses—the benefits of which may allow information to be directly implanted into the brain, memories to be edited, wireless brain-to-brain (i.e., thought-to-thought) communication, and a broad range of sensory information to be explored and experienced. The paper concludes with musings on the future direction of cyborgs and the meaning and implications of becoming more cyborg and less human in an age of rapid advances in the design and use of computing technologies.

Keywords: cyborg; enhancement technology; prosthesis; brain–computer interface; new senses; identity

---

1. Cyborgs and Prostheses

The human body is in the process of experiencing a rapid transformation from a completely biological entity created based on instructions provided by human DNA to a body becoming far more "computational and technological" [1]. While this paper focuses on the theme of "human" enhancement technology, we also review some computational enhancements to animal subjects because such studies provide examples of the future direction of enhancement technology and in some cases these very technologies will be implemented into the human body and likely within one or two decades. Generally, body-worn and implantable technology serves to identify cyborgs as a constellation within which the identities of the members of cyborg groups "negotiate" their individual significance. We describe "cyborg culture" or "cyborg being" as a particular way of life, or set of beliefs, which expresses certain meanings in the context of cyborg technologies; particularly in the case of many self-imposed cyborgs that "way of life" is to become transhuman [2,3]. Broderick describes a transhuman as a person who explores all available and future methods for self enhancement that eventually leads toward the radical change of posthuman—which is to ultimately become nearly unlimited in physical and psychological capability (i.e., to go beyond human) [4].

Using a semiotic framework, cyborg enhancement technologies can be viewed as signs which are subject to the criteria of ideological evaluation [5] which for self-enhanced cyborgs is a culture of "technologically savvy" and to some extent nonconformists, and, as noted, transhumanists. In general, we use the term "cyborg technology" to refer to technology integrated into the human body which not only restores lost function but enhances the anatomical, physiological, and information processing abilities of the body [6]. With this definition in mind a person with a heart pacer is a cyborg as is a person with an artificial arm controlled by thought. In terms of scope and content, the focus of the paper is not on drug enhancements to amplify human performance or methods of genetic engineering to enhance the body, nor does the paper focus on mobile consumer products such as smartphones or tablets which some refer to as a cyborg enhancement. Instead the paper focuses more so on the body itself—which we theorize is becoming an information processing technology based on the implantation of computing technology directly within the body. Finally, we use the term "cyborg prosthesis" to refer to artificial enhancements to the body providing computational capability, one example is an artificial hippocampus another is a brain–computer interface.

Table 1 provides an overview of cyborg technologies and enhancements designed to augment human abilities and is organized around: (1) technology which "externally interfaces" with the body; (2) implants within the body; and (3) technology which modifies in some way brain activities. The last category may include devices like Google Glass and other types of "eye-worn" technology, that while not directly implanted within the body, do in fact help to augment the world with information and thus enhance the information processing abilities of humans. Further, many refer to people wearing such devices as "cyborgs" therefore the following table includes a brief section—"Computing Attachment as Enhancement", to more fully represent the range of technologies available that help create what to some is the "common view" of a cyborg. And, to a lesser extent, enhancements to aid mobility in the form of exoskeletons are included in Table 1 to provide a more complete range of cyborg technologies that are emerging now. Additionally, there are currently a large number of enhancement technologies that are available either as commercial products or as emerging technologies, to review them all would be beyond the scope of this paper, therefore Table 1 is provided mainly to motivate discussion on the topic and to provide some organizing principles and categories to frame the debate on our future as cyborgs. Finally, two examples in Table 1 are of animal studies, again to show the direction of cyborg technology and to give the reader a more complete overview of the cyborg future which awaits us.

Similar to our Table 1, Kevin Warwick in this special edition on *Cyberphenomenology: Technominds Revolution* [7] presented a four-case description of enhancement (or cyborg) technologies. Case 1 represents technology positioned close to the human body, but not integrated into the body; case 2 is technology implanted into the body but not the brain/nervous system (whether for therapy or enhancement); case 3 represents technology linked directly to the brain/nervous system for therapeutic purposes; and case 4 is technology linked to the brain/nervous system to create "beyond normal" levels of performance. We present Warwick's classification as an alternative method for parsing distinctions between cyborg enhancements keeping in mind the fluidity of some of these, and our, categories—namely that Warwick's case 3 technology may only be a matter of a software rewrite away from a case 4 technology and that a prosthesis in our table may also have direct neural links.

Table 1. Overview of Cyborg Enhancement Technologies.

Enhancement Type/Category	Description	Significant Example
I. General External Enhancements to the Body		
Prostheses to Replace or Restore Lost Functions Prostheses are becoming more controllable through the use of control theory principles, and are integrally connected to the body, upgradable, and under some circumstances controlled by thought via a brain–computer interface (which may or may not be wireless).		
Limb Prostheses to Restore Mobility	Artificial limb replacement with multiple degrees of freedom, more and more controllable by thought	• DEKA Arm's, among other, myoelectric and brain-controlled prosthesis [8]. See also 'modifying the brain' in part III of this table.
Retinal Prosthesis to Restore Vision	Rectify visual sense degradation; provide enhancement to visual sense	• Implantable Miniature Telescope for treatment of AMD (age-related macular degeneration) [9] • Argus II Retinal Prosthesis System, an implanted device to treat adults with severe retinitis pigmentosa. The System has three parts: a small electronic device implanted in and around the eye, a tiny video camera attached to a pair of glasses, and a video processing unit that is worn or carried by the patient [10]
Cochlear Implant to Restore Hearing	Improve auditory sensitivity, the implant consists of an external portion that sits behind the ear and a second portion that is surgically placed under the skin	• Med-Els SYNCHRONY Cochlear Implant [11]
Computing Attachment as Enhancement Increasing our computational resources through technology directly integrated with our bodies allows us to scale our capabilities, senses, and interaction with our environment and with external technology. Insomuch as wearable computing integrates with our senses and responds to our thoughts, it represents a significant move towards becoming a cyborg.		
Computing Device Worn by the Body	Extraneous computing directly integrated with prosthetic part	• Jerry Jalava's USB Fingertip [12]
	Direct-interface wearable computing, such devices allow information to be projected into the world whenever and wherever it is needed	• Steve Mann's Eyetap Wearable Computer [13] • Google Glass [14]
Computing Grafted onto the Body	Attached computing device providing sensory input	• Neil Harbisson's "Eyeborg" auditory-augmented vision, allows color to be heard [15]
	Attached computing not directly integrated with the brain but accessible by the user and others with wireless capability	• Rob Spence's eye camera records and transmits images [16]

Table 1. *Cont.*

Enhancement Type/Category	Description	Significant Example
Epidermal Enhancement	Epidermal printed circuits on the surface of the skin	• Biostamp digital tattoo interacts with smartphones [17,18]
	Attached via surface	• Cyborg Nest's magnetic north sensor attached by surface-to-surface "barbells" [19]
II. Enhancement Technology Implanted Within Body		
Passive Implant Cyborg technology implanted within the body, such technology might not interact with the body through a feedback loop but be worn by the body, either collecting or storing information.		
Radio Frequency (RF) or Wi-Fi Subcutaneous Technology	Programmable storage/transmitter implanted under the skin	• RFID chips for location and medical information [20] • Anthony Antonellis "Net Art Tattoo" sends pictures to smartphones [21]
	Interactive implanted chips/LEDs	• LED tattoos' programmable lights [22]
Active/Sensor Implants Implants with closed-loop feedback coupled with computational capabilities providing medical information, technological interaction, and extra-sensory input.		
Biometric Sensors	Closed-loop measurement systems	• Heart pacemaker & defibrillator monitor and correct heart function [23] • Inflammation treatment implants: a nerve stimulator that interfaces between the immune and nervous system to treat a broad range of inflammation-related diseases, from diabetes to congestive heart failure [24]
	Open-loop measurement systems	• Implantable sensors measure and transmit glucose levels [25] • Tim Cannon's "Circadia" measures temperature [26]
Non-Medical Functional Implants	Extra-sensory detection	• Moon Ribas' seismic sensor vibrates to earthquakes [27]
	Functional computational implants	• Tooth implanted microphone/speaker [28] • Brain activated wireless controller [29]

Table 1. *Cont.*

Enhancement Type/Category	Description	Significant Example
Interfacing with Nervous System This class of implants are more thoroughly integrated with the body and provide higher levels of integration with the wearer. Through this integration, the feedback loops their systems create can be considered artificial extensions of our own body's.		
Direct Nervous System Interfacing	Nerve to nerve and nerve to machine communication	• Kevin Warwick's proof-of-concept research allowing him to control a robot arm and to create artificial sensation [30] (Warwick also experimented with BrainGate technology which is a neural interface allowing movement of an external device using thought)
Recreating Sensation	Computer generated sensation transmitted to nerves	• "Bionic" fingertip creates sensation of roughness in amputee [31]
III. Brain Enhancement or Modification		
Neuron Control Technologies that directly interface with the brain are the height of cyborg integration. This first class deals with interfaces with the least specificity, generally used to suppress large groups of neuron clusters affected by disease.		
Suppressing Neuron Activity	Implants to control neuron groups	• Deep brain stimulation for treatment of movement disorders [32]
	External brain stimulation	• Transcranial direct-current stimulation for treatment of depression (and others) [33]
Reading the Mind To interface with the brain, technology is required to observe neuron activity and technology is required to affect specific neuron groups. Neuron activity is first measured, then translated by a computer, and finally sent as some form of output, the most compelling of which are affective of other neuron groups—that is, a direct mind link. Telepathy, new sensations, and expanded senses are all resultant technologies from this area of cyborg enhancement.		
Interacting with Technology	Linking thoughts of movement with limbs	• Battelle Memorial Institute partially restores motor control in paralyzed hand via brain chip [34] • Similar techniques can be used to control a robotic arm [35]
Modifying the Brain	Linking thoughts between subjects	• Electroencephalogram linked minds coordinated in virtual game [36]
	Linking sensory areas between subjects	• Miguel Nicolelis directly linked senses between two animal subjects [37]

Table 1. *Cont.*

Enhancement Type/Category	Description	Significant Example
Influencing Memory The specificity required to read and create neuron activity in relation to senses and thought can also be applied to memory, the recursive core of the human self. Cyborg technologies that influence memory can create and dismantle identity as well as cure degenerative disease, assist in learning, and expand knowledge bases.		
Memory Encoding	Aid in memory creation	• Theodore Berger's artificial hippocampus [38]
	Aid in memory retrieval	• DARPA Restoring Active Memory program [39]
Memory Content	Memory modification	• MIT's Ramirez & Liu creating false memories in lab mice [40]
IV. Exoskeletons and Mobility Aids		
Prostheses of Heightened Function While not technically separate in cyborg classification from 'normal' prostheses, these prostheses tend to be more non-anthropomorphic, have reduced thought control functions, and have more specific design specifications intended to enhance certain abilities.		
Sports Prostheses	Provide performance greater than the biological analogues'	• Ossur's Cheetah Xtend [41] • Hugh Herr's climbing prosthetic [42]
Exoskeletons Technology designed around existing limbs to increase mobility. These enhancements can greatly increase our natural capabilities or restore lost functionality. All are closed-loop feedback systems with the body, and, in addition, the powered exoskeletons contain computational systems which increase their level of cyborg enhancement.		
Unpowered	Mechanical extensions of limbs towards performance goal	• Powerskip aids in dramatically increasing jump height [43]
Powered Load Reinforcement	Power-assisted leg exoskeletons designed to take loads off wearer	• Hugh Herr MIT's load bearing leg exoskeleton [44]
Powered Mobility Assist	Powered and computer controlled leg exoskeletons for walking	• Homayoon Kazerooni's exoskeleton [45]

1.1. Medical Necessity Creates Cyborgs

With Table 1 as background for the discussion which follows, recently people's bodies have become enhanced by use of technology with computational capabilities (that is, have become more "cyborg"), based on medical necessity; for example, debilitating disease affecting the central nervous system in the case of Parkinson's patients, or due to accidents or injuries (see [46] for additional examples of cyborgs). One example of a current cyborg is Jerry Jalava who, after suffering injuries sustained from a motorcycle accident, embedded a 2 GB USB drive on the tip of his prosthetic finger, essentially converting his finger into a hard drive; however, unlike other cyborg technologies, the USB drive isn't permanently fused to his finger, instead its inside a rubber tip that fits directly onto the nub of his prosthesis [12]. In contrast, another DIY cyborg, Tim Cannon, has integrated technology directly into his body by implanting a computer chip in his arm that can record and transmit biometrical data [26]. The above devices compute and provide information to the wearer, both characteristics of cyborg technology and of being a cyborg.

Considering cyborg enhancements (Table 1), as indicated by Dietrich and Laerhoven [47], interesting questions are raised related to how technology mediates the relation of person to the "world and self" as reflected in Verbeek's work, which is a postphenomenological approach that technology only bears meaning in a use context (e.g., how cyborg technology is actually used), and specifically the concept of embodied interaction [48]. In fact, the concept of "embodiment" is at the center of phenomenology, which rejects the Cartesian separation between mind and body on which many traditional philosophical approaches are based. In place of the Cartesian model, phenomenology explores our experiences as embodied actors interacting in the world, participating in it, and acting through it, in the absorbed and unreflective manner of normal experience. In terms of our identity resulting from the use of cyborg enhancements (see [3] and [48]), Locke's discussion of personal identity is relevant [49] to the technology presented in this paper. To Locke, personal identity is a matter of psychological continuity, a person psychologically evolves from "an adventure" (that is, becoming cybernetically enhanced) to a new evolved identity (say a transhuman), afterwards, the person's desires, intentions, experiential memories, and character traits may reflect the reality of a new cyborg identity.

Considering the above discussion, an interesting question is how one's perception of their body, that is, their embodiment, is affected by cyborg technologies—for example, are cyborg parts considered an extension of the body, or as separate from the body creating a new sense of identity for an individual? And will one's sense of identity change with the use of neuroprosthetic devices that allow memories to be edited, stored, and transferred? Surely one's sense of identity will be radically altered if one's experiences and memories become artificial and not necessarily tied to actual experiences. Additionally, in the coming cyborg age, will enhancements to human abilities, for example, in the form of telephoto vision or the ability to detect magnetic fields, change not only our functionality but our sense of experiencing the world?

On the point of increasing the computational capabilities of the body, for Canadian filmmaker Rob Spence, loss of vision was the motivating factor for converting him into a cyborg [16]. After an accident left him partially blind, he decided to create his own electronic eye in the form of a camera, which can be used to record everything he sees just by looking around. Even more interesting, though, the eye-camera has wireless capability; the system could allow another person to access his video feed and view the world through his artificial right eye. Unlike with a biological eye, Spence can upgrade the hardware and software of his cyborg enhancement. In our view the ability to upgrade the body is a major benefit of becoming a cyborg (and is likewise a fundamental characteristic of a cyborg) and essentially allows people to transcend human abilities resulting from evolution. It would be easy to imagine fundamentally new ways of seeing, experiencing, and feeling the world through these enhancements.

Given that necessity spawns invention, people paralyzed from spinal cord injuries are beginning to receive brain implant technology which may allow them to move again. How does the technology work? Generally, the "cyborg technology" bypasses the patient's severed spine by sending a signal from the brain directly to technology placed on the patient's muscles [36,45,46]. In the procedure, the surgeons first map the exact spot in the patient's motor cortex that control the muscles in a particular part of the body, then implant a tiny computer chip at that location. The next step is to "teach the chip" how to read the patient's thoughts. This is done by placing the patient inside an MRI machine where the patient watches a video of a hand moving in specific ways and at the same time imagines moving his own hand that way. The implanted chip reads the brain signals, decodes them, and translates them into electrical signals where they are transmitted to the muscles of the patient's forearm. Next, the patient is "plugged into" technology by running a cable from his skull to a computer and then to electrodes on his arm. Effectively, when the patient focuses his mind on moving his hand, it moves. This aspect of cyborg technology—creating a feedback loop between the body and technology—is not only a characteristic of what it means to be a cyborg but a potential "game changer" in connecting our senses and mind to external technology (especially to control the technology using thought), and, given appropriately powerful new technologies, may even influence our sense of experiencing that world. However, this experimental and developing cyborg technology, still needs improvement before it will become common treatment for paralyzed patients and accessible to other populations (for different reasons than medical necessity); for example, it needs to be wireless so there is not a cable plugged into the skull and researchers need to figure out a way to send a signal from the body back to the brain (that is, close the feedback loop) so the patient can sense when his body is moving [6].

As another example of an implantable device which is used due to medical necessity, Setpoint, a technology company, is developing computing therapies to reduce systemic inflammation by stimulating the vagus nerve using an implantable pulse generator [24]. This device works by activating the body's natural inflammatory reflex to dampen inflammation and improve clinical signs and symptoms. Thus far, the company is developing an implanted neuromodulation device to treat rheumatoid arthritis, a disease currently afflicting over two million people in the U.S. alone. Each advance in cyborg devices spurred by medical necessity is leading to advances in technology which make the body more computational, with closed-loop feedback and upgradeable technology, and in some cases controllable by thought—these are all characteristics of the future direction of cyborg technologies.

1.2. Enhancements, Thought Control, and Communication

Even with the brain's tremendous complexity (estimated to be 85–100 billion neurons, with 100 trillion synaptic connections) as shown in the table above, progress is being made towards the integration of the human brain with machines and sensors—this idea will ultimately allow the brain to be "cognitively enhanced" and to have additional computational capabilities [6]. For example, researchers at the Rehabilitation Institute of Chicago, have developed a thought-controlled bionic leg which uses neuro-signals from the upper leg muscles to control a prosthetic knee and ankle [50]. The prosthesis uses pattern recognition software contained in an on-board computer to interpret electrical signals from the upper leg as well as mechanical signals from the bionic leg. When the person equipped with the prosthesis thinks about moving his leg, the thought triggers brain signals that travel down his spinal cord, and ultimately, through peripheral nerves, are read by electrodes in the bionic leg, which then moves in response to the proceeding thought.

Among other things, what's interesting about the human enhancement movement is that it's not just major research centers that are developing thought controlled prosthesis and other enhancement technologies, hackers are beginning to enter the fray which will increase the speed at which the body will become computational (from a digital sense) and will challenge our sense of identity as a new technologically enhanced person. Take body hacker and inventor Shiva Nathan, a teenager, who after being inspired to help a family member who lost both arms below the elbow, created a robotic arm which can be controlled by thought [51]. The technology uses a commercially available MindWave Mobile headset to read EEG waves and uses Bluetooth to send the data to a computer which then translates them into limited finger and hand movements. In addition, in Sweden, researchers at Chalmers University of Technology are developing a thought-controlled prosthesis for amputees in the form of an implantable robotic arm. And in the U.S., the FDA has approved a thought-controlled prosthetic limb invented by Dean Kamen that provides multiple degrees of freedom, is the same size and weight as a natural human arm, and works by detecting electrical activity caused by the contraction of muscles close to where the prosthesis is attached [8]. The electrical signals, initially generated by thought are sent to a computer processor in the prosthetic arm, which triggers a specific movement in the prosthesis. In FDA tests, the artificial arm/hand has successfully assisted people with household tasks such as using keys and locks and preparing food [8].

Researchers at Brown University and *Cyberkinetics* in Massachusetts, are devising a microchip that is implanted in the motor cortex just beneath a person's skull that will be able to intercept nerve signals and reroute them to a computer, which will then wirelessly send a command to any of various electronic devices, including computers, stereos and electric wheelchairs. In this case a person's sense of identity will expand to accommodate feedback not only from the body's sensors, but from sensors on external devices. And consider a German team that has designed a microvibration device and a wireless low-frequency receiver that can be implanted in a person's tooth [28]. The vibrator acts as microphone and speaker, sending sound waves along the jawbone to a person's eardrum. Given that our sense of identity in the world is derived partially through mind-world interactions, developments extending our body's reach and methods of influence upon the world may create a new, or at least significantly different, human phenomenology.

Further, there is also research on brain-to-brain communication, including major efforts in this area from the Defense Advanced Research Projects Agency (DARPA) in the U.S. But in a university research laboratory, University of Washington researchers have created a system that represents a noninvasive human-to-human brain interface, allowing one person to send a brain signal via the Internet to control the hand motions of another person at a different location [52]. The system uses electrical brain recordings and a form of magnetic stimulation, in which one person wearing a cap with electrodes is hooked up to an electroencephalography machine (which reads electrical activity in the brain) that sends a signal to another person with a cap equipped with the stimulation site for a transcranial magnetic stimulation coil which is placed directly over the person's left motor cortex, (which controls hand movement). As a proof-of-concept study, Professor Rao looked at a computer screen while playing a simple video game with his mind. When he was supposed to fire a cannon at a target, he imagined moving his right hand, causing a cursor to hit the "fire" button. Almost instantaneously, another person who wore noise-canceling earbuds and wasn't looking at a computer screen, involuntarily moved his right index finger to push the space bar on the keyboard in front of him, as if firing the cannon. The technologies used by the researchers for recording and stimulating the brain are both well-known. Electroencephalography, or EEG, is routinely used by clinicians and researchers to record brain activity noninvasively from the scalp. Transcranial magnetic stimulation is a noninvasive way of delivering stimulation to the brain to elicit a response. Its effect depends on where the coil is placed; in this case, it was placed directly over the brain region that controls a person's right hand. By activating these neurons, in a proof-of-concept study, Rao and his team concluded that the stimulation convinced the brain that it needed to move the right hand [53].

1.3. Computational Skin

If we can design artificial limbs controlled by thought and if we can implant technology into the body, can we enhance the skin, the largest sense organ, with computational capabilities? If so, this would be a major step in our cyborg future. Based on recent advances in technology, the answer is yes, but first a digression into popular culture. Enhancing the body's surface such that it is transformed into a "computational device" represents a change in our very self-identity as our skin is perceived as the barrier between our internal self and the external world—the *surface* of a person's identity, if you will. We theorize that any change to that visual biological-self model has the potential to increase our capabilities and interactiveness with the world, but also to potentially shift the normal of human "appearance".

On the point of popular culture and our cyborg future, a recent study showed that nearly forty percent of Americans under the age of forty have at least one tattoo (see generally [54,55]); however, like any trade, the tattoo industry must innovate to expand and gain new clients. In an analog world, one way to innovate is to make the switch to digital technology. Rather than being passive as are current tattoos, digital tattoos are active, they *do* things, and they are getting smart [17]. Digital tattoos have the potential to do more than serve the function of art or self-expression, even though these are laudable goals, they will indeed become digital devices as useful as smartphones—and may even monitor our health.

It is possible now to use a type of ink in a tattoo that responds to electromagnetic fields, which raises a host of new opportunities for cyborgs. In fact, Nokia patented a ferromagnetic ink technology that can interact with a device through magnetism. The basic idea is to enrich tattoo ink with metallic compounds that are first demagnetized (by exposing the metal to high temperatures) before the ink is embedded in a person's skin. Once the tattoo has healed, the ink is re-magnetized with permanent magnets. The resulting tattoo is then sensitive to magnetic pulses, which can be emitted by a device such as a cellular phone. Interestingly, a digital tattoo would allow a person's ringing phone to result in a haptic sensation experienced by the body; that is, the person would experience the phone ringing literally through the tattoo; an interesting computational capability for cyborgs and an interesting change in our use of technology and of our body.

If the tattoo consists of putting electronics on the surface of the skin, many more possibilities for body hacking exist. For computational skin, materials scientist and University of Illinois Professor John Rogers is developing flexible electronics that stick to the skin to operate as a temporary tattoo [18]. These so-called "epidural electronics" (or Biostamp) are a thin electronic mesh that stretches with the skin and monitors temperature, hydration and strain, as well as monitoring a person's body's vital signs [17,18]. The latest prototype of the Biostamp is applied directly to the skin using a rubber stamp. The stamp lasts up to two weeks before the skin's natural exfoliation causes it to come away. Rogers is currently working on ways to get the electronics to communicate with other devices like smartphones so that they can start building apps (eventually such devices will communicate with devices that are implanted within the body). Developing sensors worn by or implanted within the body that communicate with and controls external devices is a new capability for humans, essentially extending the "reach of the body" beyond that of the body's physical boundaries. Google, isn't far behind in developing digital tattoos, as the company's Advanced Technology and Projects Group patented the idea of a digital tattoo consisting of various sensors and gages, such as strain gauges for tracking strain in multiple directions (how the user is flexing), EEG and EMG (electrical impulses in the skeletal structure or nerves), ECG (heart activity), and temperature.

Considering another digital tattoo designed for a medical monitoring purpose, University of Pennsylvania's Brian Litt, a neurologist and bioengineer, is implanting LED displays under the skin for medical and bio-computation purposes [55]. These tattoos consist of silicon electronics less than 250 nanometers thick, built onto water soluble, biocompatible silk substrates. When injected with saline, the silk substrates conform to fit the surrounding tissue and eventually dissolve completely, leaving only the silicon circuitry. The electronics can be used to power LEDs that act as photonic tattoos. Litt is perfecting a form of this technology that could be used to build wearable medical devices—say, a tattoo that gives diabetics information about their blood sugar level. These examples highlight the use of cyborg devices to compute data, monitor the body, and eventually form closed-loop feedback systems with the body. Additionally, they demonstrate our increasing tendency to electively distance ourselves from our natural biology and technologically modify our very human form.

1.4. Body Hackers and Implantable Sensors

The body hacking movement, especially about implantable sensors within the body, gained momentum from the work of Professor Kevin Warwick starting in 1998 at the University of Reading [30]. Professor Warwick was one of the first people to hack his body when he participated in a series of proof-of-concept studies which first involved implanting a sensor into his shoulder (see his paper, this special edition). Warwick's "cyborg application" consisted of the use of an RFID implant which allowed Professor Warwick to switch on lights and open doors as he entered rooms (thus Warwick was-able-to link his body directly to external devices). Later, others extended Warwick's seminal work using RFID devices and other implantable sensors. For example, in an extension of Professor Warwick's early work, Dr. John Halakha of Harvard Medical School, chose to be implanted with an RFID chip used to access medical information [56]. His implant stores information which can direct anyone with the appropriate reader to a website containing his individual medical data. He believes that implantable chips such as these can be valuable in situations where patients arrive at the hospital unconscious or unresponsive.

Another person with an RFID implant, Meghan Trainor has a less pragmatic but highly creative application for her implant [57]. Trainor received the implant as part of her master's thesis for NYU's Interactive Telecommunications Program. Her implant serves as part of an interactive art exhibit; RFID tags are embedded in sculptures which can be manipulated to play sounds stored in an audio database. Trainor can use the implant in her arm to further manipulate these sounds. Additionally, body hacker Anthony Antonellis implanted an RFID chip into his hand which can be wirelessly accessed by a smartphone [21]. While the chip holds only about 1 KB to 2 KB of data, it allows Antonellis to access and display an animated GIF on his phone that is stored on the implant. Since the RFID chip can transfer and receive data, Antonellis can swap out 1KB files as he pleases. Antonellis views the implant as a "net art tattoo", something for which quick response codes (QR, or matrix barcode), are commonly used. Similarly, Karl Marc, a tattoo artist from Paris designed an animated tattoo that makes use of a QR code and a smartphone [58]. The code basically activates software on the phone that makes the tattoo move when seen through the phone's camera. The use of cyborg technology to transform the body into electronic art is surely a new mode of interacting with the world and a hint of what is to come in the future.

1.5. Vision Enhancements

Given that the eye is a major sensory organ in terms of providing information about the world, there is extensive current research oriented towards creating an artificial eye with "telephoto capabilities" (and research to detect energy beyond the range of our sensors). Who would benefit from such technology? Clearly the cyborg movement would benefit from providing the visual system enhanced computational abilities, but so too would the millions of people worldwide who have the advanced form of age-related macular degeneration (AMD), a disease which affects the region of the retina responsible for central, detailed vision. For such people an implantable telescope could help restore the essential visual modality [9]. In fact, in 2010, the U.S. Federal Drug Administration (FDA) approved an implantable miniature telescope (IMT), which works like the telephoto lens of a camera [9]. The IMT technology reduces the impact of the central vision blind spot due to end-stage AMD and projects the objects the patient is looking at onto the healthy area of the light-sensing retina not degenerated by the disease.

The tiny telescope is implanted behind the iris, the colored, muscular ring around the pupil and represents a tantalizing vision of our cyborg future consisting of enhanced sensory modalities. And of course, since our sense of identity is derived, among others, from sensory information—"hacking" the visual modality could potentially alter the information we use to perceive and make sense of our position in the world.

Some people appear intent on changing their senses and, by extension, their identity by becoming transhuman. For example, Neil Harbisson, who was born with a rare condition (achromatopsia) that allows him to see only in black and white and shades of grey, has become a cyborg due to necessity [15]. After viewing a talk on cybernetics, in the spirit of a hacker, Neil wondered if he could turn color into sound, based on the idea that a specific frequency of light could be made equivalent to a specific sound wave. To become a cyborg, Neil had a sound conducting chip implanted in his head, along with a flexible shaft with a digital camera on it, attached to his skull [15]. With his latest software upgrade, Neil says he is able to hear ultraviolet and infrared frequencies, can have phone calls delivered to his head, and has a Bluetooth connection which allows him to connect his "Eyeborg" to the Internet. Using "cyborg technology" Neil has created a new way of perceiving the world and has thus expanded the boundaries of human experience and interaction with the world.

2. Brain Enhancements and Neuroprosthesis

Through the Restoring Active Memory (RAM) program, the U.S. defense research institute, DARPA, is funding research to accelerate the development of technologies able to address the public health challenge of helping service members, and others, overcome memory deficits by developing new neuroprosthetics to bridge gaps in the injured brain [59]. The end goal of RAM is to develop and test a wireless, fully implantable neural-interface medical device for human clinical use. A number of additional and significant advances, however, will be targeted on the way to achieving that goal; such advances may be milestones for our cyborg future.

To start, DARPA is supporting the development of multi-scale computational models with high spatial and temporal resolution that describe how neurons code declarative memories—those well-defined parcels of knowledge that can be consciously recalled and described in words, such as events, times, and places [39,59]. Researchers will also explore new methods for analysis and decoding of neural signals to understand how targeted stimulation might be applied to help the brain reestablish an ability to encode new memories following brain injury. "Encoding" refers to the process by which newly learned information is attended to and processed by the brain when first encountered. Building on this foundational work, researchers will attempt to integrate the computational models developed under RAM into new, implantable, closed-loop systems able to deliver targeted neural stimulation that may ultimately help restore memory function [59]. Interestingly, RAM and related DARPA neuroscience efforts are monitored by members of an independent Ethical, Legal, and Social Implications (ELSI) panel [60]. Communications with ELSI panelists supplement the oversight provided by institutional review boards that govern human clinical studies and animal use. Given that cyborg technology can be used for multiple purposes, this panel provides the oversight needed to monitor developments in the field.

Additional progress is being made in other areas of brain–computer interface design, Figure 1 provides a broad overview. For example, scientists have used brain scanners to detect and reconstruct the faces that people are thinking of. In one study, Yale scientists hooked participants up to an fMRI brain scanner—which determines activity in different parts of the brain by measuring blood flow—and showed them images of faces in two sets [61]. The first set established a statistical relation between the images of the faces and corresponding areas of brain activity while the second set attempted to recreate those faces from observation of brain activity alone. Alan Cowen and Professor Marvin Chun were, in fact, able to recreate images of these faces to some degree of likeness [61]. One can imagine in the future that a witness to a crime might reconstruct a suspect's face based on "extracting" the image from his mind (of course, this will raise privacy issues). However, Yale researchers pointed out that an important limitation of the technology as it exists now, is that this sort of technology can only read active parts of the brain, not passive memories.

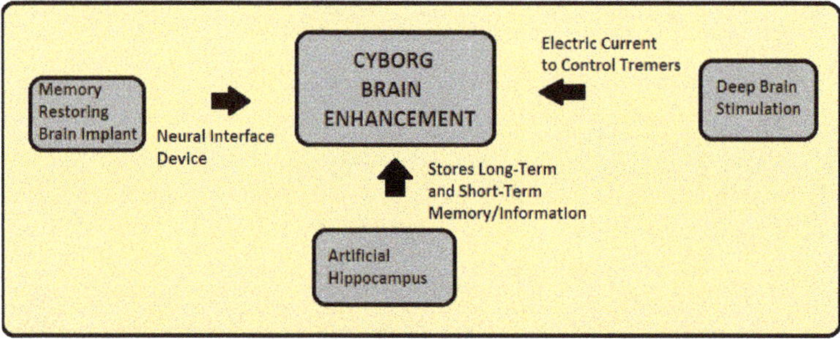

Figure 1. Some basic technologies of cyborg brain enhancement.

A major advance in cyborg technology is the development of a hippocampus prosthesis, which we view as a type of cognitive prosthesis (a prosthesis implanted into the nervous system in order to improve or replace the function of damaged brain tissue) [62]. In some cases, prosthetic devices replace the normal function of a damaged body part; this can be simply a structural replacement (e.g., reconstructive surgery) or a rudimentary, functional replacement. As an important cyborg technology, University of Southern California researchers are testing the usefulness of an artificial hippocampus which will mimic the brain's memory center [62]. The device may one-day help those with brain damage, epilepsy, and Alzheimer's disease. Additionally, the same device could, with additional developments, allow one's brain to be directly connected to the Internet which could potentially project our self-identity to the emerging Internet of Things. To create the neuroprosthesis, lead researcher Theodore Berger and his team used principles of nonlinear systems theory to develop and apply methods for quantifying the dynamics of hippocampal neurons. In this approach, properties of neurons are assessed experimentally by applying a random interval train of electrical impulses as an input and electrophysiologically recording the evoked output of the target neuron during stimulation [62]. The input train consists of a series of impulses, with interimpulse intervals varying according to a Poisson process. Thus, the input is "broadband" and stimulates the neuron over most of its operating range; that is, the statistical properties of the random train are highly consistent with the known physiological properties of hippocampal neurons. Nonlinear response properties are expressed in terms of the relation between progressively higher-order temporal properties of a sequence of input events and the probability of neuronal output, and are modeled as the kernels of a functional power series [38]. This example highlights the complexity of the engineering behind cyborg technologies designed for the brain and the importance of algorithms for our cyborg future.

Another "cyborg" brain technology is deep brain stimulation (DBS) which consists of a surgical procedure used to treat several disabling neurological symptoms—most commonly the debilitating motor symptoms of Parkinson's disease (PD), such as tremor, rigidity, stiffness, slowed movement, and walking problems [32]. The procedure is also used to treat essential tremor and dystonia. At present, the procedure is used only for individuals whose symptoms cannot be adequately controlled with medications. DBS uses a surgically implanted, battery-operated medical device called an implantable pulse generator (IPG)—similar to a heart pacemaker and approximately the size of a stopwatch to deliver electrical stimulation to specific areas in the brain that control movement, thus blocking the abnormal nerve signals that cause PD symptoms [32]. Before the procedure, a neurosurgeon uses magnetic resonance imaging (MRI) or computed tomography (CT) scanning to identify and locate the exact target within the brain for surgical intervention. Some surgeons may use microelectrode recording—which involves a small wire that monitors the activity of nerve cells in the target area—to more specifically identify the precise brain area that will be stimulated. Generally, these areas are the thalamus, subthalamic nucleus, and globus pallidus. The lead (also called an electrode)—a thin, insulated wire—is inserted through a small opening in the skull and implanted in the brain. The tip of the electrode is positioned within the specific brain area. The extension is an insulated wire that is passed under the skin of the head, neck, and shoulder, connecting the lead to the implantable pulse generator. The IPG (the "battery pack") is usually implanted under the skin near the collarbone. Once the system is in place, electrical impulses are sent from the IPG up along the extension wire and the lead and into the brain. These impulses block abnormal electrical signals and alleviate PD motor symptoms.

Additionally, the work by Professor Potter and his team [63] involving embodied networks of cultured neurons in simulation and robotic studies is relevant for our cyborg future. A "cultured neuronal network" is a cell culture of neurons that is used as a model to study the central nervous system, especially the brain. For future cyborgs, cultured neuronal networks may be connected to an input/output device such as a multi-electrode array, thus allowing two-way communication between the person and the network. Interestingly cultured neurons are often connected via computer to a real or simulated robotic component, creating a hybrot or animat, respectively [64]. Hochberg and Donoghue [65] with colleagues have created brain–computer interface technology to demonstrate that people with paralysis can control external devices by translating neuronal activity directly into control signals for assistive devices (specifically a robotic arm) [66].

Extending his original work with RFID sensors, Professor Warwick had a BrainGate interface implanted into his nervous system to link his body to technology external to his body. Most notably, Professor Warwick could control an electric wheelchair and an artificial hand, using the neural interface [30]. In addition to being able to measure the signals transmitted along the nerve fibers in Professor Warwick's left arm, the implant was also able to create artificial sensation by stimulating the nerves in his arm using individual electrodes. This bi-directional functionality was demonstrated with the aid of another person and a second, less complex implant connecting to her nervous system. Based on Warwick's results, this was an early proof-of-concept display of electronic communication between the nervous systems of two humans.

3. Towards "New Senses"

With regard to modifying and enhancing the body, can a new sense be created? In our view, new senses will certainly be developed if by "new sense" one meant to enhance a current sense in such a way that sensory information beyond the range of its sensory receptor(s) can be experienced. Substituting one sense for another is a well-researched topic and represents another way to modify the body and create a cyborg future. Increasing and/or extending the range of our senses may be desirable given we see and hear across certain frequencies, and that the eyes and ears can only detect information within a given distance to the sensory receptors. Given this explanation, Neil Harbisson already has an extra sense, thusly new states of identity for Neil are already being created. In the future, by hacking and modifying the bodies already existing senses, we may develop enhanced vision and greater sensitivity to olfactory, gustatory, or haptic information—we may even combine senses. As we create new senses for humans by the use of cyborg implants, without considering post-human levels of modification, will one's identity change? In the sense of Locke, would it even be possible for our identity not to change?

While EEG and fMRI technologies are leading to significant advances in the use of brain scans for lie detection, other research in neuroscience is more directly related to the topic of telepathic communication, a totally new mode of communication. How will telepathic communication impact one's sense of their individuality if brains are telepathically networked together? Professor Miguel Nicolelis from Duke University has developed important technology for the brain in this area that we believe is leading to a cyborg future for humanity [37]. His research is oriented toward brain-to-brain communication, brain machine interfaces and neuroprosthesis in human patients and non-human primates. Based on his studies, Dr. Nicolelis was one of the first to propose and demonstrate that animals and human subjects can utilize their electrical brain activity to directly control neuroprosthetic devices via brain–machine interfaces. As early as 2012 Professor Nicolelis speculated about the possibility that two brains could exchange information [37], and later, Nicolelis reported that his research team at Duke University Medical Center had achieved a back-and-forth exchange between two rodent brains. To test his brain interface technology, his team trained two animals to press one of two levers when an LED turned on in exchange for a drink of water. Microelectrodes were placed in each of the two animals' cortices and when one rat pressed the correct lever, a sample of cortical activity from that rat's brain was wired to the second animal's brain located in a chamber where the "it's-time-to-drink" LED was absent [37]. As evidence that information was exchanged between the two brains, the rat on the receiving end of the prosthesis proceeded to press the correct lever (to receive a drink) that had been messaged over the brain link. Summarizing the results—Nicolelis and his team provided proof-of-concept technology and preliminary results that telepathy may be possible as a future form of communication.

Related to Professor Nicolelis's work, results from studies with human subjects show that telepathy may in fact be a viable technology for the general public within a few decades (or less!). For example, using EEG technology, researchers at the University of Southampton, England, reportedly demonstrated communication from person-to-person using thought [67]. More recently, as described earlier in this paper, at the University of Washington, researchers demonstrated a working brain-to-brain interface with human subjects also using EEG technology [68]. According to the researchers, the next step is to determine *what* kind of information can be sent between people's brains.

In a study which has importance for our cyborg future, Duke University neuroscientist Miguel Nicolelis, and his team report that they have created a "sixth sense" through a brain implant in which infrared light is detected by lab rats [37]. Even though the infrared light can't be seen, lab rats are able to detect it via electrodes in the part of the brain responsible for the rat's sense of touch—so remarkably, the rats reportedly feel the light, not see it. In order to give the rats their "sixth sense", Duke researchers placed electrodes in the rat's brains that were attached to an infrared detector [37]. The electrodes were then attached to the part of the animals' brains responsible for processing information about touch. The rats soon began to detect the source of the 'contact' and move towards the signal. In addition to these important findings, the Duke scientists found that creating the infrared-detecting sixth sense did not stop the rats from being able to process touch signals, despite the electrodes (providing input for the infrared detection system) being placed in the tactile cortex. Sixth sense or not, in our view, the study by Nicolelis and his team is another step toward integrating brain–computer technology into the human body; and thus contributing to a cyborg future that will alter our senses and change our sense of identity as mere products of biology [6,37].

Additionally, in the military domain, DARPA, through funding, is trying to build "thought helmets" to enable telepathic communication using brain–computer interfaces to give soldiers extra senses, such as night vision, and the ability to "see" magnetic fields caused by landmines [39]. Finally, as another example, to create a "sixth sense", some DIY cyborgs have implanted magnets in their fingertips [69]. A cyborg with a magnet implanted in their finger, can sense magnetic fields that would otherwise be completely undetectable. The implant allows those who have received it the ability to not only sense magnetic fields, but to pick up tiny metal objects with their fingertips, and determine whether metals are ferrous. How extra senses will affect our sense of identity as a human being will be a fascinating topic of discussion in the near future.

4. Modifying Memory

As a future cyborg technology, neuroscientists foresee a future world where minds can be programmed in order to create artificial memories. In terms of challenges to one's sense of identity in the world, cyborg technologies which can edit memories [6], or add new memories to one's repertoire of experiences, has the potential to fundamentally change our self-identity, world view, and more [53]. Based on recent advances in brain-to-brain communication, some scientists argue that memories may be implanted into a person's mind, and that memories from one mind can be transferred to another. In fact, scientists have already successfully implanted a false memory into the brain of a mouse. To create a memory interface, MIT scientists Steve Ramirez and Xu Liu tagged brain cells in one mouse associated with a specific memory and then tweaked that memory to make the mouse believe an event had happened (to that mouse) when it hadn't; other laboratories are producing similar results [40]. While implanting a memory in humans equipped with a neuroprosthetic device won't happen in the immediate future, Ramirez et al., have shown that in principle, it should be possible to isolate a human memory and activate it [40]. In fact, Michael J. Kahana, who serves as director of the University of Pennsylvania's Computational Memory Lab commented on the MIT study, "We would have every reason to expect this would happen in humans as it happened in mice" (see [70]). Clearly, improvements in neuroprosthetic technologies are occurring rapidly and moving humanity toward a cyborg future.

5. Conclusions and Future Directions

Barfield [6] proposed that the capabilities of "cyborg technology" consists of several characteristics: (1) the technology is upgradeable allowing software and hardware improvements to be applied to the body in ever shorter cycle times (see [71]); (2) the technology offers the body additional computational capabilities, thus transforming the body into an information processing technology; (3) cyborg technology is integrated with the body through closed-loop feedback systems; and (4) is becoming more and more controllable by thought [6]. If these trends of innovation and research continue, we will soon be faced with a very new sort of human with very different sorts of capabilities [6]. How the law and public policy relates to technologically enhanced people is addressed in another paper by the authors [72], but surely, major changes in public policy and law will need to be debated and enacted to account for people with superior and quite different abilities.

In light of our impending cyborg future, how we view ourselves as individuals and as humans is certain to become subject to upheaval and change. As prostheses become more advanced, and additional capabilities more integrated within the body and brain, it would not be difficult to imagine an individual electing to replace their basic biological parts with the upgraded cyborg version—but how will this capability affect our sense of identity? Add that to the growing trend of body hacking and modification and it does not seem unreasonable to assume that in the near future we humans may look very different than we do today [73]. Our sense of humanness, insomuch as it is rooted in our biology, will quickly erode as these enhancement technologies develop and grow in use. The body of a human may have little to do with the destiny of their birth; as we replace our bodies with the customizable and the upgradable, so we replace the old world of biological phenotypes with a new, creative world of our making.

The mind is also on the verge of transformative changes. Brain implants that repair damaged memories might one day lead to the creation of new memories or telepathic communication (see [74]). Brand new and exotic experiences could be purchased and uploaded. Entire lives could be lived in an instant. Couple this with the potential for memory modification and we would be left with a sense of self and identity decoupled from memories derived from interacting with the world. Psychologically continuous "selves" would no longer indicate distinct persons. Expanded consciousness's through brain implants and computer interfacing could make tracing identity through biological continuity equally problematic. If we are no longer what we remember ourselves to be and have expanded far beyond our biology, then who are we? If we can modify our bodies and our memories, then we can modify our senses and our very ways of being in the world. The core phenomenology of being human will change, perhaps to the point of unrecognizability. What then could we say of human nature if all that we hold to be consistent and true is subject to modification—or even attack? Certainly, new philosophies on identity will be required in parallel with new social structures and technological advancements.

The phenomena and minutia of our existence have forever been locked to the biology of our brains, but as dynamic and varied as brains are, they are limited by their finite physicality. The human of the near future could be nearly unlimited in their cognitive capabilities. How could the man who sees in radio and feels the solar wind relate to the old human? However the future human manifests, the new human could very possibly be beyond our current understanding. The first steps in that journey have already been made.

Author Contributions: Both authors contributed equally to this paper.

Conflicts of Interest: The authors declare no conflict of interest.

References

1. LaGuardia, D. *Trash Culture: Essays in Popular Criticism*; Xlibris Publishing: Bloomington, IN, USA, 2008.
2. Masci, D. *Human Enhancement: The Scientific and Ethical Dimensions of Striving for Perfection*; Pew Research Center: Washington, DC, USA, 2016.
3. Haraway, D.J. A Manifesto for Cyborgs: Science, Technology, and Socialist-Feminism in the 1980's. *Soc. Rev.* **1985**, *15*, 65–107.
4. Broderick, D. Trans and Post. In *The Transhumanist Reader*, 1st ed.; Moor, M.M., Vita-More, N., Eds.; Wiley-Blackwell: London, UK, 2013; pp. 430–437.
5. Hebdige, D. From Culture to Hegemon. In *Subculture: The Meaning of Style*; Routledge Press: London, UK, 1979.
6. Barfield, W. *Cyber Humans: Our Future with Machines*; Springer: New York, NY, USA, 2016.

7.   Warwick, K. Homo Technologicus: Threat or Opportunity? *Philosophies* **2016**, *1*, 199–208. [CrossRef]
8.   Kamen, D. DEKA Prosthesis. Available online: http://www.dekaresearch.com/founder.shtml (accessed on 10 September 2016).
9.   Lipshitz, I. Intraocular Telescopic Implants for Age-related Macular Degeneration Eyes. *Eur. Ophthalmic Rev.* **2015**, *9*, 159–160. [CrossRef]
10.   Cruz, L.; Coley, B.; Dorn, J.; Merlini, F.; Filley, E.; Christopher, P.; Chen, F.K.; Wuyyuru, V.; Sahel, J.; Stanga, P.; et al. The Argus II Epiretinal Prosthesis System Allows Letter and Word Reading and Long-term Function in Patients with Profound Vision Loss. *Br. J. Ophthalmol.* **2013**. [CrossRef] [PubMed]
11.   FDA Approves MED-EL's SYNCHRONY Cochlear Implant. Available online: http://www.businesswire.com/news/home/20150123005073/en/FDA-Approves-MED-EL%E2%80%99s-SYNCHRONY-Cochlear-Implant (accessed on 20 September 2016).
12.   Yu, J. USB Prosthetic Finger Gives New Meaning to Thumbdrives. Available online: http://www.cnet.com/news/usb-prosthetic-finger-gives-new-meaning-to-thumbdrives/ (accessed on 8 October 2016).
13.   Mann, S.; Fung, J.; Aimone, C.; Sehgal, A.; Chen, D. Designing EyeTap Digital Eyeglasses for Continuous Lifelong Capture and Sharing of Personal Experiences. In Proceedings of the ALT.CHI 2005, Portland, OR, USA, 2–7 April 2005; ACM Press: New York, NY, USA, 2005.
14.   Google Glass. Available online: https://en.wikipedia.org/wiki/Google_Glass (accessed on 29 September 2016).
15.   Eyeborg Project. Available online: http://eyeborgproject.com/ (accessed on 12 September 2016).
16.   Hornyak, T. Eyeborg: Man Replaces False Eye with Bionic Camera. IEEE Spectrum, 2010. Available online: http://spectrum.ieee.org/automaton/biomedical/bionics/061110-eyeborg-bionic-eye (accessed on 15 September 2016).
17.   Estes, A.C. The Freaky, Bioelectric Future of Tattoos. 2014. Available online: http://gizmodo.com/the-freaky-bioelectric-future-of-tattoos-1494169250 (accessed on 1 December 2016).
18.   Ahlberg, L. Off the Shelf, on the Skin: Stick-on Electronic Patches for Health Monitoring. 2014. Available online: http://news.illinois.edu/news/14/0403microfluidics_JohnRogers.html (accessed on 20 September 2016).
19.   Sajej, N. Your First Step to Becoming a Cyborg: Getting This Pierced in You. Motherboard. Available online: http://motherboard.vice.com/read/cyborg-implant-magnetic-north (accessed on 20 September 2016).
20.   Wicks, A.V. Radio Frequency Identification Applications in Hospital Environments. *Hosp. Top.* **2006**, *84*, 3–8. [CrossRef] [PubMed]
21.   Antonellis, A. Net Art Implant (and Video). Available online: http://www.anthonyantonellis.com/news-post/item/670-net-art-implant (accessed on 9 October 2016).
22.   Neifer, A. Biohackers are Implanting LED Lights Under their Skin. Motherboard, 2015. Available online: http://motherboard.vice.com/read/biohackers-are-implanting-led-lights-under-their-skin (accessed on 18 September 2016).
23.   Graham-Rowe, D. New Pacemaker Needs No Wires. *MIT Technol. Rev.* 2011. Available online: https://www.technologyreview.com/s/426164/new-pacemaker-needs-no-wires/ (accessed on 18 September 2016).
24.   Andersson, U.; Tracey, K.J. A New Approach to Rheumatoid Arthritis: Treating Inflammation with Computerized Nerve Stimulation. *Cerebrum* **2012**, *2012*, 3. [PubMed]
25.   Implantable Sensor Measures Blood Sugar Levels. 2010. Available online: http://archive.azcentral.com/health/news/articles/2010/07/28/20100728implantable-sensor-measures-blood-sugar-levels.html (accessed on 25 September 2016).
26.   Hoppenstedt, M. The DIY Cyborg. Describing Tim Cannon's Computing Implant. Available online: http://motherboard.vice.com/blog/the-diy-cyborg (accessed on 5 October 2016).
27.   Ribas, M. QUARTZ. Available online: http://qz.com/677218/this-woman-a-self-described-cyborg-can-sense-every-earthquake-in-real-time/ (accessed on 2 October 2016).
28.   Frucci, A. Tiny Bluetooth Microphone Goes in a Hole Drilled in Your Teeth. 2008. Available online: http://gizmodo.com/374120/tiny-bluetooth-microphone-goes-in-a-hole-drilled-in-your-teeth (accessed on 24 September 2016).
29.   Liao, L.D.; Chen, C.Y.; Wang, I.J.; Chen, S.F.; Li, S.Y.; Chen, B.W.; Chang, J.-Y.; Lin, C.-T. Gaming Control Using a Wearable and Wireless EEG-based Brain-computer Interface Device with Novel Dry Foam-based Sensors. *J. Neuroeng Rehabil.* **2012**, *9*, 5. [CrossRef] [PubMed]

30. Warwick, K.; Gasson, M.N.; Hutt, B.; Goodhew, I.; Kyberd, P.; Andrews, B.; Teddy, P.; Shad, A. The Application of Implant Technology for Cybernetic Systems. *Arch. Neurol.* **2003**, *60*, 1369–1373. [CrossRef] [PubMed]

31. Amputee Feels Texture with a Bionic Fingertip: The Future of Prosthetic Touch Resolution: Mimicking Touch. Available online: https://www.sciencedaily.com/releases/2016/03/160308084937.htm (accessed on 10 September 2016).

32. Kringelbach, M.L.; Jenkinson, N.; Owen, S.L.; Aziz, T.Z. Translational Principles of Deep Brain Stimulation. *Nat. Rev. Neurosci.* **2007**, *8*, 623. [CrossRef] [PubMed]

33. Kalu, U.G.; Sexton, C.E.; Loo, C.K.; Ebmeier, K.P. Transcranial Direct Current Stimulation in the Treatment of Major Depression: A Meta-Analysis. *Psychol Med.* **2012**, *42*, 1791–1800. [CrossRef] [PubMed]

34. Bouton, C.E.; Shaikhouni, A.; Annetta, N.V.; Bockbrader, M.A.; Friedenberg, D.A.; Nielson, D.M.; Sharma, G.; Sederberg, P.B.; Glenn, B.C.; Mysiw, W.G.; et al. Restoring Cortical Control of Functional Movement in a Human with Quadriplegia. *Nature* **2016**, *533*, 247–250. [CrossRef] [PubMed]

35. NIH. In *Parlyzed individuals use though-controlled robotic arm to reach and grasp.* Available online: https://www.nih.gov/news-events/news-releases/paralyzed-individuals-use-thought-controlled-robotic-arm-reach-grasp (accessed on 20 December 2016).

36. Makeig, S.; Gramann, K.; Jung, T.-P.; Sejnowski, T.J.; Poizner, H. Linking Brain, Mind and Behavior. *Int. J. Psychophysiol.* **2009**, *73*, 95–100. [CrossRef] [PubMed]

37. Nicolelis, M. *Beyond Boundaries: The New Neuroscience of Connecting Brains with Machines—And How It Will Change Our Lives*; St. Martin's Press: New York, NY, USA, 2012.

38. Berger, T.; Glanzman, D.L. *Toward Replacement Parts for the Brain: Implantable Biomimetic Electronics as Neural Prostheses*; MIT Press: Cambridge, MA, USA, 2005.

39. DARPA. Restoring Active Memory (RAM). Available online: http://www.darpa.mil/program/restoring-active-memory (accessed on 9 October 2016).

40. Ramirez, S.; Ryan, T.J.; Tonegawa, S. Creating a False Memory in the Hippocampus. *Science* **2013**, *341*, 387–391. [CrossRef] [PubMed]

41. Cheetah Xtend. Available online: https://www.ossur.com/prosthetic-solutions/products/sport-solutions/cheetah-xtend (accessed on 12 December 2016).

42. Humpheries, C. The Body Electric, MIT Technology Review. Available online: https://www.technologyreview.com/s/531541/the-body-electric/ (accessed on 8 December 2016).

43. Powerskip. Available online: http://www.powerskip.de/mainpage.html (accessed on 10 December 2016).

44. Herr, H. Biomechanical Walking Mechanisms Underlying the Metabolic Reduction Caused by an Autonomous Exoskeleton. *J. NeuroEng. Rehabil.* **2016**, *13*, 4. [CrossRef]

45. Kazerooni, H. Human Augmentation and Exoskeleton Systems in Berkeley. *Int. J. Humanoid Res.* **2007**, *4*, 575–605. [CrossRef]

46. Lanxon, N. Practical Transhumanism: Five Living Cyborgs. 2012. Available online: http://www.wired.co.uk/article/cyborgs (accessed on 12 September 2016).

47. Dietrich, M.; Laerhoven, K.V. An Interdisciplinary Approach on the Mediating Character of Technologies for Recognizing Human Activity. *Philosophies* **2016**, *1*, 55–67. [CrossRef]

48. Verbeeks, P.P. *What Things Do: Philosophical Reflections on Technology, Agency, and Design*, 2nd ed.; Pennsylvania State University Press: University Park, PA, USA, 2005.

49. Dunn, J. *Locke: A Very Short Introduction*; Oxford University Press: Oxford, UK, 2005.

50. Hargrove, L.J.; Simon, A.M.; Young, A.J.; Lipschutz, R.D.; Finucane, S.B.; Smith, D.G.; Kuiken, T.A. Robotic Leg Control with EMG Decoding in an Amputee with Nerve Transfers. *N. Engl. J. Med.* **2013**, *369*, 1237–1242. [CrossRef] [PubMed]

51. Cirincione, M. Teen Designs Robotic Prosthetics Using Supplies from RadioShack and Home Depot. Available online: http://www.usnews.com/news/the-next-generation-of-stem/articles/2015/05/12/teen-designs-robotic-prosthetics-using-supplies-from-radioshack-and-home-depot (accessed on 10 December 2016).

52. Rao, R.P.N.; Stocco, A.; Bryan, M.; Sarma, D.; Youngquist, T.M.; Wu, J.; Prat, C.S. A Direct Brain-to-Brain Interface in Humans. *PLoS ONE* **2014**, *9*, e111332. [CrossRef] [PubMed]

53. Agar, N. *Truly Human Enhancement: A Philosophical Defense of Limits*; MIT Press: Cambridge, MA, USA, 2014.

54. Hoover, R. PEW: Majority of Americans Leery of Artificial 'Human Enhancements'. Available online: http://cnsnews.com/news/article/rachel-hoover/pew-majority-americans-leery-artificial-enhancement-humans (accessed on 14 September 2016).

55. Bourzac, K. Implantable Silicon-Silk Electronics, MIT Technology Review. Available online: https://www.technologyreview.com/s/416104/implantable-silicon-silk-electronics/.com/s/531541/the-body-electric/ (accessed on 6 December 2016).
56. Halamaka, J.D. A Chip in My Shoulder. 2007. Available online: http://geekdoctor.blogspot.com/2007/12/chip-in-my-shoulder.html (accessed on 19 September 2016).
57. Trainor, M. Meghan Trainor in Musiques & Cultures Digitale: Volume 6. Available online: https://depts.washington.edu/open3dp/2011/02/meghan-trainor-in-musiques-cultures-digitale-volume-6/ (accessed on 29 December 2016).
58. QR Code Tattoos. Describing the Work of Karl Marc. Available online: http://www.qrscanner.us/qr-tatoos.html (accessed on 5 January 2017).
59. Strickland, E. DARPA Project Starts Building Human Memory Prosthetics. Available online: http://spectrum.ieee.org/biomedical/bionics/darpa-project-starts-building-human-memory-prosthetics (accessed on 4 January 2017).
60. NIH. ELSI Research Program: The Ethical, Legal and Social Implications (ELSI) Research Program. Available online: https://www.genome.gov/10001618/the-elsi-research-program/ (accessed on 9 December 2016).
61. Cowen, A.S.; Chun, M.M.; Kuhl, B.A. Neural Portraits of Perceptions: Reconstructing Face Images from Evoked Brain Activity. *Neuroimage* **2014**, *94*, 12–22. [CrossRef] [PubMed]
62. Berger, T.W.; Baudry, M.; Brinton, R.D.; Liaw, J.-S.; Marmarelis, V.Z.; Park, Y.; Sheu, B.J.; Tanguay, A.R., Jr. Brain-implantable Biomimetic Electronics as the Next Era in Neural Prosthetics. *Proc. IEEE* **2001**, *89*, 993–1012. [CrossRef]
63. Potter, S.M. Distributed Processing in Cultured Neuronal Networks. *Prog. Brain Res.* **2001**, *130*, 1–14.
64. Bakkum, D.J.; Dhkolnik, A.C.; Ben-Ary, G.; Gamblen, P.; DeMsrse, B.; Potter, S.M. Removing Some "A" from AI: Embodied Cultured Networks. In *Embodied Artificial Intelligence*, 2004 ed.; Ida, F., Pfeifer, R., Steels, L., Kuniyoshi, Y., Eds.; Springer: New York, NY, USA, 2008; pp. 130–145.
65. Hochberg, L.R.; Donoghue, J.P. Sensors for Brain-Computer Interfaces. *IEEE Eng. Med. Biol. Mag.* **2006**, *25*, 32–38. [CrossRef] [PubMed]
66. Hochberg, L.R.; Bacher, D.; Jarosiewicz, B.; Masse, N.Y.; Simeral, J.D.; Vogel, J.; Haddadin, S.; Liu, J.; Cash, S.S.; van der Smagt, P.; et al. Reach and Grasp by People with Tetraplegia Using a Neurally Controlled Robotic Arm. *Nature.* **2012**, *485*, 372–375. [CrossRef] [PubMed]
67. University of Southampton. Brain-Computer Interface Allows Person-to-Person Communication through Power of Thought. *ScienceDaily.* 6 October 2009. Available online: www.sciencedaily.com/releases/2009/10/091006102637.htm (accessed on 3 October 2016).
68. Direct Brain Interface between Humans. *ScienceDaily.* 5 November 2014. Available online: https://www.sciencedaily.com/releases/2014/11/141105154507.htm (accessed on 27 November 2016).
69. Berg, D. I Have a Magnet in My Finger, Gizmodo. 2012. Available online: http://gizmodo.com/5895555/i-have-a-magnet-implant-in-my-finger (accessed on 8 December 2016).
70. Kim, M. MIT Scientists Implant a False Memory into a Mouse's Brain. *The Washington Post.* 25 July 2013. Available online: http://www.washingtonpost.com/national/health-science/inception-mit-scientists-implant-a-false-memory-into-a-mouses-brain/2013/07/25/47bdee7a-f49a-11e2-a2f1-a7acf9bd5d3a_story.html (accessed on 28 November 2016).
71. Wander, J.D.; Rao, R.P.N. Brain-computer Interfaces: A Powerful Tool for Scientific Inquiry. *Curr. Opin. Neurobiol.* **2014**, *25*, 70–75. [CrossRef] [PubMed]
72. Barfield, W.; Williams, A. Law, Cyborgs, and Technologically Enhanced Brains. **2017**, unpublished manuscript.
73. Harbisson, N. The Man Who Hears Colour. *BBC News.* 11 November 2014. Available online: http://www.bbc.com/news/technology-29992577 (accessed on 5 December 2016).
74. Brown University. Controlling Movement through Thought Alone. 2016. Available online: http://www.brown.edu/Administration/News_Bureau/2006-07/06-002.html (accessed on 8 December 2016).

 *philosophies*

Essay

The Process of Evolution, Human Enhancement Technology, and Cyborgs

Woodrow Barfield

Professor Emeritus, Chapel Hill, NC 27516, USA; Jbar5377@gmail.com

Received: 10 January 2019; Accepted: 19 February 2019; Published: 22 February 2019

Abstract: The human body is a remarkable example of the process of evolution which ultimately created a sentient being with cognitive, motor, and information-processing abilities. The body can also be thought of as an amazing feat of engineering, and specifically as an example of molecular nanotechnology, positioning trillions of cells throughout the body, and creating the billions of unique individuals that have existed since the beginning of humanity. On the other hand, from an engineering perspective, there are numerous limitations associated with the human body and the process of evolution to effect changes in the body is exceedingly slow. For example, our skeletal structure is only so strong, our body is subject to disease, and we are programmed by our DNA to age. Further, it took millions of years for *Homo sapiens* to evolve and hundreds of thousands of years for hominids to invent the most basic technology. To allow humans to go beyond the capabilities that evolution provided *Homo sapiens*, current research is leading to technologies that could significantly enhance the cognitive and motor abilities of humans and eventually create the conditions in which humans and technology could merge to form a cybernetic being. Much of this technology is being developed from three fronts: due to medical necessity, an interest within the military to create a cyborg soldier, and the desire among some people to self-enhance their body with technology. This article discusses the processes of biological evolution which led to the current anatomical, physiological, and cognitive capabilities of humans and concludes with a discussion of emerging technologies which are directed primarily at enhancing the cognitive functions performed by the brain. This article also discusses a timeframe in which the body will become increasingly equipped with technology directly controlled by the brain, then as a major paradigm shift in human evolution, humans will merge with the technology itself.

Keywords: evolution; neuroprosthesis; cultural technology; human enhancement; engineering

---

1. Introduction

In a discussion of technologically enhanced humans in the 21st century, the emergence of *Homo sapiens* several hundred thousand years ago is a good starting point for that discussion. Through evolution, humans first evolved to live as hunter-gatherers on the savannah plains of Africa [1]. The forces of evolution operating over millions of years provided our early human ancestors the skeletal and muscular structure for bipedal locomotion, sensors to detect visual, auditory, haptic, olfactory, and gustatory stimuli, and information-processing abilities to survive in the face of numerous challenges. One of the main evolutionary adaptations of humans compared to other species is the capabilities of our brain and particularly the cerebral cortex. For example, the average human brain has an estimated 85–100 billion neurons and contains many more glial cells which serve to support and protect the neurons [2]. Each neuron may be connected to up to 10,000–12,500 other neurons, passing signals to each other via as many as 100 trillion synaptic connections, equivalent by some estimates to a computer with a 1 trillion bit per second processor [3].

Comparing the brain (electro-chemical) to computers (digital), synapses are roughly similar to transistors, in that they are binary, open or closed, letting a signal pass or blocking it. So, given a median estimate of 12,500 synapses/neurons and taking an estimate of 22 billion cortical neurons, our brain, at least at the level of the cerebral cortex, has something on the order of 275 *trillion* transistors that may be used for cognition, information processing, and memory storage and retrieval [2]. Additionally, recent evidence points to the idea that there is actually subcellular computing going on within neurons, moving our brains from the paradigm of a single computer to something more like an Internet of the brain, with billions of simpler nodes all working together in a massively parallel network [4].

Interestingly, while we have rough estimates of the brain's computing capacity, we do not have an accurate measure of the brain's overall ability to compute; but the brain does seem to operate at least at the level of petaflop computing (and likely more). For example, as a back-of-the-napkin calculation, with 100 billion neurons connected to say 12,500 other neurons and postulating that the strength of a synapse can be described using one byte (8 bits), multiplying this out produces 1.25 petabytes of computing power. This is, of course, a very rough estimate of the brain's computing capacity done just to illustrate the point that the brain has tremendous complexity and ability to compute. There are definitely other factors in the brain's ability to compute such as the behavior of support cells, cell shapes, protein synthesis, ion channeling, and the biochemistry of the brain itself that will surely factor in when calculating a more accurate measure of the computational capacity of the brain. And that a brain can compute, at least at the petaflop level or beyond, is, of course, the result of the process of evolution operating over a period of millions of years.

No matter what the ultimate computing power of the brain is, given the magnitude of 85–100 trillion synapses to describe the complexity of the brain, it is relevant to ask—will our technology ever exceed our innate capabilities derived from the process of evolution? If so, then it may be desirable to enhance the body with technology in order to keep pace with the rate at which technology is advancing and becoming smarter ("smartness" in the sense of "human smartness") [5]. This is basically a "cyborg oriented" approach to thinking about human enhancement and evolution. Under this approach, even though much of the current technology integrated into the body is for medical purposes, this century, able-bodied people will become increasingly enhanced with technology, which, among others, will include artificial intelligence embedded in the technology implanted within their body [5,6].

An interesting question is why would "able-bodied" people agree to technological enhancements, especially those implanted under the skin? One reason is derived from the "grinder movement" which represents people who embrace a hacker ethic to improve their own body by self-implanting "cyborg devices" under their skin. Implanting a magnet under the fingertip in order to directly experience a magnetic field is one example, and arguably creates a new sense [5]. Another reason to enhance the body with technology is expressed by transhumanists, who argue that if a technology such as artificial intelligence reaches and then surpasses human levels of general intelligence, humans will no longer be the most intelligent being on the planet; thus, the argument goes, we need to merge with technology to remain relevant and to move beyond the capabilities provided by our biological evolution [5]. Commenting on this possibility, inventor and futurist Ray Kurzweil has predicted that there will be low-cost computers with the same computational capabilities as the brain by around 2023 [6]. Given continuing advances in computing power and in artificial intelligence, the timescale for humans to consider the possibility of a superior intelligence is quickly approaching and a prevailing idea among some scientists, inventors, and futurists is that we need to merge with the intelligent technology that we are creating as the next step of evolution [5–7]. Essentially, to merge with technology means to have so much technology integrated into the body through closed-loop feedback systems that the human is considered more of a technological being than a biological being [5]. Such a person could, as is possible now, be equipped with artificial arms and legs (prosthetic devices), or technology performing the functions of our internal organs (e.g., heart pacer), and most importantly for our future to merge with technology, functions performed by the brain itself using technology implanted within the brain (e.g., neuroprosthesis, see Tables 1 and 2).

Of course, the software to create artificial general intelligence currently lags behind hardware developments (note that some supercomputers now operate at the exaflop, i.e., 10^{18} level), but still, the rate of improvements in a machine's ability to learn indicates that computers with human-like intelligence could occur this century and be embedded within a neuroprosthesis implanted within the brain [6]. If that happens, commentators predict that humans may then connect their neocortex to the cloud (e.g., using a neuroprosthesis), thus accessing the trillions of bits of information available through the cloud and benefiting from an algorithm's ability to learn and to solve problems currently beyond a human's understanding [8]. This paper reviews some of the technologies which could lead to that outcome.

Evolution, Technology, and Human Enhancement

While the above capabilities of the brain are remarkable, we should consider that for millennia the innate computational capabilities of the human brain have remained relatively fixed; and even though evolution affecting the human body is still occurring, in many ways we are today very similar anatomically, physiologically, and as information processors to our early ancestors of a few hundred thousand years ago. That is, the process of evolution created a sentient being with the ability to survive in the environment that *Homo sapiens* evolved to successfully compete in. In the 21st century, we are not much different from that being even though the technology we use today is vastly superior. In contrast, emerging technologies in the form of exoskeletons, prosthetic devices for limbs controlled by the brain, and neuroprosthetic devices implanted within the brain are beginning to create technologically enhanced people with abilities beyond those provided to humans through the forces of evolution [1,5,7]. The advent of technological enhancements to humans combined with the capabilities of the human body provided through the process of evolution brings up the interesting point that our biology and the technology integrated into the body are evolving under vastly different time scales. This has implications for the future direction of our species and raises moral and ethical issues associated with altering the speed of the evolutionary processes which ultimately created *Homo sapiens* [9].

Comparing the rate of biological evolution to the speed at which technology evolves, consider the sense of vision. The light-sensitive protein opsin is critical for the visual sense; from an evolutionary timescale, the opsin lineage arose over 700 million years ago [10]. Fast forward over a hundred million years later, the first fossils of eyes were recorded from the lower Cambrian period (about 540 million years ago). It is thought that before the Cambrian explosion, animals may have sensed light, but did not use it for fast locomotion or navigation by vision. Compare these timeframes for developing the human visual system, to the development of "human-made" technology to aid vision [11]. Once technology created by humans produced the first "vision aid", the speed of technological development has operated on a timescale orders of magnitude faster than biological evolution. For example, around 1284, Salvino D'Armate invented the first wearable eye glasses and just 500 years later ("just" as in the planet is 4.7 billion years old and anatomically modern humans evolved a few hundred thousand years ago), in the mid 1780s, bifocal eyeglasses were invented by Benjamin Franklin. Fast forward a few more centuries and within the last ten to fifteen years, the progress in creating technology to enhance, or even replace the human visual system has vastly accelerated. For example, eye surgeons at the Massachusetts Eye and Ear Infirmary, are working on a miniature telescope to be implanted into an eye that is designed to help people with vision loss from end-stage macular degeneration [12]. And what is designed based on medical necessity today may someday take on a completely different application, that of enhancing the visual system of normal-sighted people to allow them to detect electromagnetic energy outside the range of our evolutionary adopted eyes, to zoom in or out of a scene with telephoto lens, to augment the world with information downloaded from the cloud, and even to wirelessly connect the visual sense of one person to that of another [5].

From an engineering perspective, and particularly as described by control theory, one can conclude that evolution applies positive feedback in adopting the human to the ambient environment in that the more capable methods resulting from one stage of evolutionary progress are the impetus used to create the next stage [6]. That is, the process of evolution operates through modification by descent, which allows nature to introduce small variations in an existing form over a long period of time; from this general process, the evolution into humankind with a prefrontal cortex took millions of years [1].

For both biological and technological evolution, an important issue is the magnitude of the exponent describing growth (or the improvement in the organism or technology). Consider a standard equation expressing exponential growth, $f_{(x)} = 2^x$. The steepness of the exponential function, which in this discussion refers to the rate at which biological or technological evolution occurs, is determined by the magnitude of the exponent. For biological versus technological evolution, the value for the exponent is different, meaning that biological and technological evolution proceed with far different timescales. This difference has relevance for our technological future and our continuing integration of technology with the body to the extent that we may eventually merge with and become the technology. Some commentators argue that we have always been human–technology combinations, and to some extant I agree with this observation. For example, the development of the first tools allowed human cognition to be extended beyond the body to the tool in order to manipulate it for some task. However, in this paper, the discussion is more on the migration of "smart technology" from the external world to either the surface of the body or implanted within the body; the result being that humans will be viewed more as a technological being than biological.

Additionally, given the rate at which technology and particularly artificial intelligence is improving, once a technological Singularity is reached in which artificial intelligence is smarter than humans, humans may be "left behind" as the most intelligent beings on the planet [5,7]. In response, some argue that the solution to exponential growth in technology, and particularly computing technology, is to become enhanced with technology ourselves, and deeper into the future, to ultimately become the technology [5,7]. In that context, this article reviews some of the emerging enhancement technologies directed primarily at the functions performed by the brain and discusses a timeframe in which humans will continue to use technology as a tool, which I term the "standard model" of technology use, then become so enhanced with technology (including technology implanted within the body) that we may view humans as an example of technology, which I term the "cyborg-machine" model of technology [5].

The rate at which technological evolution occurs, as modeled by the law of accelerating returns and specifically by Moore's law which states that the number of transistors on an integrated circuit doubles about every two years—both of which in the last few decades have accurately predicted rapid advancements in the technologies which are being integrated into the body—make a strong argument for accelerating evolution through the use of technology [6]. To accelerate evolution using technology is to enhance the body to beyond normal levels or performance, or even to create new senses; but basically, through technology implanted within the body, to provide humans with vastly more computational resources than provided by a 100-trillion-synapse brain (many of which are not directly involved in cognition). More fundamentally, the importance of the law of accelerating returns for human enhancement is that it describes, among others, how technological change is exponential, meaning that advances in one stage of technology development help spur an even faster and more profound technological capability in the next stage of technology development [6]. This law, describing the rate at which technology advances, has been an accurate predictor of the technological enhancement of humans over the last few decades and provides the motivation to argue for a future merger of humans with technology [5].

2. Two Categories of Technology for Enhancing the Body

Humans are users and builders of technology, and historically, technology created by humans has been designed primarily as external devices used as tools allowing humans to explore and manipulate the environment [1]. By definition, technology is the branch of knowledge that deals with the creation and use of "technical means" and their interrelation with life, society, and the environment. Considering the long history of humans as designers and users of technology, it is only recently that technology has become implanted within the body in order to repair, replace, or enhance the functions of the body including those provided by the brain [5]. This development, of course, is the outcome of millions of years of biological evolution which created a species which was capable of building such complex technologies. Considering tools as an early example of human-made technology, the ability to make and use tools dates back millions of years in the sapiens family tree [13]. For example, the creation of stone tools occurred some 2.6 million years ago by our early hominid ancestors in Africa which represented the first known manufacture of stone tools—sharp flakes created by knapping, or striking a hard stone against quartz, flint, or obsidian [1]. However, this advancement in tools represented the major extent of technology developments over a period of eons.

In the role of technology to create and then enhance the human body, I would like to emphasize two types of technology which have emerged from my thinking on this topic in the last few decades, these include "biological" and "cultural" technology (Table 1). While I discuss each in turn, I do not mean to suggest that they evolved independently from each other; human intelligence, for example, has to some extent been "artificially enhanced" since humans developed the first tools; this is due in part to the interaction between the cerebral cortex and technology. This view is consistent with that expressed by Andy Clark in his books, "Natural-Born Cyborgs: and the Future of Human Intelligence," and "Mindware: An Introduction to the Philosophy of Cognitive Science." Basically, his view on the "extended mind" is that human intelligence has always been artificial, made possible by the technology's that humans designed which he proposes extended the reach of the mind into the environment external to the body. For example, he argues that a notebook and pencil plays the role of a biological memory, which physically exists and operates external to the body, thus becoming a part of an extended mind which cognitively interacts with the notebook. While agreeing with Andy Clark's observations, it is worth noting that this paper does not focus on whether the use of technology extends the mind to beyond the body. Instead, the focus here, when describing biological and cultural technology, is simply to emphasize that the processes of biological evolution created modern humans with a given anatomy and physiology, and prefrontal cortex for cognition, and that human ingenuity has led to the types of non-biological technology being developed that we may eventually merge with. Additionally, the focus here is on technology that will be integrated into the body, including the brain, and thus the focus is not on the extension of cognition into the external world, which of course is an end-product of evolution made possible by the *Homo sapiens* prefrontal cortex.

With that caveat in mind, biological technology is that technology resulting from the processes of evolution which created a bipedal, dexterous, and sentient human being. As an illustration of biology as technology, consider the example of the musculoskeletal system. From an engineering perspective, the human body may be viewed as a machine formed of many different parts that allow motion to occur at the joints formed by the parts of the human body [14]. The process of evolution which created a bipedal human through many iterations of design, among others, took into account the forces which act on the musculoskeletal system and the various effects of the forces on the body; such forces can be modeled using principles of biomechanics, an engineering discipline which concerns the interrelations of the skeleton system, muscles, and joints.

In addition, the forces of evolution acting over extreme time periods ultimately created a human musculoskeletal system that uses levers and ligaments which surround the joints to form hinges, and muscles which provide the forces for moving the levers about the joints [14]. Additionally, the geometric description of the musculoskeletal system can be described by kinematics, which considers the geometry of the motion of objects, including displacement, velocity, and acceleration (without taking into account the forces that produce the motion). Considering joint mechanics and structure, as well as the effects that forces produce on the body, indicates that evolution led to a complex engineered body suitable for surviving successfully as a hunter-gatherer. From this example, we can see that basic engineering principles on levers, hinges, and so on were used by nature to engineer *Homo sapiens*. For this reason, "biology" (e.g., cellular computing, human anatomy) may be thought of as a type of technology which developed over geologic time periods which ultimately led to the hominid family tree that itself eventually led a few hundred thousand years ago to the emergence of *Homo sapiens*; the "tools" of biological evolution are those of nature primarily operating at the molecular level and based on a DNA blueprint. So, one can conclude that historically, human technology has built upon the achievements of nature [13]. Millions of years after the process of biological evolution started with the first single cell organisms, humans discovered the principles which now guide technology design and use, in part, by observing and then reverse engineering nature.

Referring to "human-made" technology, I use the term "cultural technology" to refer to the technology which results from human creativity and ingenuity, that is, the intellectual output of humans. This is the way of thinking about technology that most people are familiar with; technology is what humans invent and then build. Examples include hand-held tools developed by our early ancestors to more recent neuroprosthetic devices implanted within the brain. Both biological and in many cases cultural technology can be thought of as directed towards the human body; biological technology operating under the slow process of evolution, cultural technology operating at a vastly faster pace. For example, over a period of only a few decades computing technology has dramatically improved in terms of speed and processing power as described by Moore's law for transistors; and more generally, technology has continuously improved as predicted by the law of accelerating returns as explained in some detail by Ray Kurzweil [6]. However, it should be noted that biological and cultural technologies are not independent. What evolution created through millions of years of trial-and-error is now being improved upon in a dramatically faster time period by tool-building *Homo sapiens*. In summary, *Homo sapiens* are an end-product of biological evolution (though still evolving), and as I have described to this point, a form of biological technology supplemented by cultural technology. Further, I have postulated that biological humans may merge with what I described as cultural technology, thus becoming less biological and more "digital-technological".

Table 1. Comparison of biological versus cultural technology.

	Biological Technology	Cultural Technology
Basic Process, Underlying Principle	• Living organisms are composed of cells. • Genes, traits are inherited through gene transmission. • Through evolution (variation, selection, replication), genetic changes in a population are inherited over several generations.	• Moore's Law • Law of Accelerating Returns • Creation of tools, external and internal to the body • Human ingenuity and creativity
Timescale	• Millions of years for primitive cells to develop into more complex organisms	• Decades or less for major paradigm shifts, progress is noticeably exponential
Selected Examples of Output	• Sentience • Numerus systems to support life (e.g., musculoskeletal, endocrine, nervous, digestive, circulatory, etc.) • 100-trillion-synapse brain • DNA • Language	• Digital technology • Computer languages • Wireless protocol • Algorithms • Prosthesis

3. On Being Biological and on Technologically Enhancing the Brain

As a product of biological evolution, the cerebral cortex is especially important as it shapes our interactions and interpretations of the world we live in [1]. Its circuits serve to shape our perception of the world, store our memories and plan our behavior. A cerebral cortex, with its typical layered organization, is found only among mammals, including humans, and non-avian reptiles such as lizards and turtles [15]. Mammals, reptiles and birds originate from a common ancestor that lived some 320 million years ago. For *Homo sapiens*, comparative anatomic studies among living primates and the primate fossil record show that brain size, normalized to bodyweight, rapidly increased during primate evolution [16]. The increase was accompanied by expansion of the neocortex, particularly its "association" regions [17]. Additionally, our cerebral cortex, a sheet of neurons, connections and circuits, comprises "ancient" regions such as the hippocampus and "new" areas such as the six-layered "neocortex", found only in mammals and most prominently in humans [15]. Rapid expansion of the cortex, especially the frontal cortex, occurred during the last half-million years. However, the development of the cortex was built on approximately five million years of hominid evolution; 100 million years of mammalian evolution; and about four billion years of molecular and cellular evolution. Compare the above timeframes with the timeframe for major developments in computing technology discussed next. Keep in mind that, among others, recent computing technologies are aimed at enhancing the functions of the cerebral cortex itself, and in some cases, this is done by implanting technology directly in the brain [5,18].

The recent history of computing technology is extremely short when considered against the backdrop of evolutionary time scales that eventually led to *Homo sapiens* and spans only a few centuries. But before the more recent history of computing is discussed, it should be noted that an important invention for counting, the abacus, was created by Chinese mathematicians approximately 5000 years ago. Some (such as philosopher and cognitive scientist Andy Clark) would consider this invention (and more recently digital calculators and computers) to be an extension of the neocortex. While interacting with technology external to the body extends cognition to that device, still, the neurocircuits controlling the device, remain within the brain. More recently, in 1801, Joseph Marie Jacquard invented a loom that used punched wooden cards to automatically weave fabric designs, and two centuries later early computers used a similar technology with punch cards. In 1822, the English mathematician Charles Babbage conceived of a steam-driven calculating machine that would be able to compute tables of numbers and a few decades later Herman Hollerith designed a punch card system which among others, was used to calculate the U.S. 1890 census. A half-century later, Alan Turing presented the notion of a universal machine, later called the Turing machine, theoretically capable of computing anything that could be computable. The central concept of the modern computer is based on his ideas. In 1943–1946 the Electronic Numerical Integrator and Calculator (ENIAC) was built which is considered the precursor of digital computers (it filled a 20-foot by 40-foot room consisting of 18,000 vacuum tubes) and was capable of calculating 5000 addition problems a second. In 1958, Jack Kilby and Robert Noyce unveiled the integrated circuit, or computer chip. And recently, based on efforts by the U.S. Department of Energy's Oak Ridge National Laboratory, the supercomputer, Summit, operates at a peak performance of 200 petaflops—which corresponds to 200 million billion calculations a second (a 400 million increase in calculations compared to ENIAC from 75 years earlier). Of course, computing machines and technologies integrated into the body will not remain static. Instead, they will continue to improve, allowing more technology to be used to repair, replace, or enhance the functions of the body with computational resources.

In contrast to the timeframe for advances in computing and the integration of technology in the body, the structure and functionality of the *Homo sapiens* brain has remained relatively the same for hundreds of thousands of years. With that timeframe in mind, within the last decade, several types of technology have been either developed, or are close to human trials, to enhance the capabilities of the brain. In the U.S., one of the major sources of funding for technology to enhance the brain is through the Defense Advanced Research Projects Agency (DARPA); some of the projects funded by DARPA are shown in Table 2. Additionally, in the European Union the Human Brain Project (HPB) is another major effort to learn how the brain operates and to build brain interface technology to enhance the brain's capabilities. One example of research funded by the HBP is Heidelberg University's program to develop neuromorphic computing [19]. The goal of this approach is to understand the dynamic processes of learning and development in the brain and to apply knowledge of brain neurocircuitry to generic cognitive computing. Based on neuromorphic computing models, the Heidelberg team has built a computer which is able to model/simulate four million neurons and one billion synapses on 20 silicon wafers. In contrast, simulations on conventional supercomputers typically run factors of 1000 slower than biology and cannot access the vastly different timescales involved in learning and development, ranging from milliseconds to years; however, neuromorphic chips are designed to address this by operating more like the human brain. In the long term, there is the prospect of using neuromorphic technology to integrate intelligent cognitive functions into the brain itself.

As mentioned earlier, medical necessity is a factor motivating the need to develop technology for the body. For example, to restore a damaged brain to its normal state of functioning, DARPA's Restoring Active Memory (RAM) program funds research to construct implants for veterans with traumatic brain injuries that lead to impaired memories. Under the program, researchers at the Computational Memory Lab, University of Pennsylvania, are searching for biological markers of memory formation and retrieval [20,21]. Test subjects consist of hospitalized epilepsy patients who have electrodes implanted deeply in their brain to allow doctors to study their seizures. The interest is to record the electrical activity in these patients' brains while they take memory tests in order to uncover the electric signals associated with memory operations. Once they have found the signals, researchers will amplify them using sophisticated neural stimulation devices; this approach, among others, could lead to technology implanted within the brain which could eventually increase the memory capacity of humans.

Other research which is part of the RAM program is through the Cognitive Neurophysiology Laboratory at the University of California, Los Angeles. The focus of this research is on the entorhinal cortex, which is the gateway to the hippocampus, the primary brain region associated with memory formation and storage [22]. Working with the Lawrence Livermore National Laboratory, in California, closed-loop hardware in the form of tiny implantable systems is being jointly developed. Additionally, Theodore Berger at the University of Southern California has been a pioneer in the development of a neuroprosthetic device to aid memory [23]. His artificial hippocampus is a type of cognitive prosthesis that is implanted into the brain in order to improve or replace the function(s) of damaged brain tissue. A cognitive prosthesis allows the native signals used normally by the area of the brain to be replaced (or supported). Thus, such a device must be able to fully replace the function of a small section of the nervous system—using that section's normal mode of operation. The prosthesis has to be able to receive information directly from the brain, analyze the information and give an appropriate output to the cerebral cortex. As these and the examples in Table 2 show, remarkable progress is being made in designing technology to be directly implanted in the brain. These developments, occurring over a period of just a decade, designed to enhance or repair the brain, represent a major departure from the timeframe associated with the forces of evolution which produced the current 100-trillion-synapse brain.

Table 2. Examples of brain enhancement programs (DARPA is gratefully acknowledged for use of the material).

DARPA Brain Initiative Program	Description	DARPA Brain Initiative Program	Description
Electrical Prescriptions (ElectRx)	This program aims to help the human body heal itself through neuromodulation of organ functions using ultraminiaturized devices, approximately the size of individual nerve fibers.	**Hand Proprioception and Touch Interfaces (HAPTIX)**	The HAPTIX program will create fully implantable, modular and reconfigurable neural-interface microsystems that communicate wirelessly with external modules, such as a prosthesis interface link, to deliver naturalistic sensations to amputees.
Neural Engineering System Design (NESD)	The NESD program's goal is to develop an implantable neural interface able to provide unprecedented signal resolution and data-transfer bandwidth between the brain and the digital world.	**Neuro Function, Activity, Structure and Technology (Neuro-FAST).**	The Neuro-FAST program seeks to enable unprecedented visualization and decoding of brain activity to better characterize and mitigate threats to the human brain, as well as facilitate development of brain-in-the-loop systems to accelerate and improve functional behaviors.
Next-Generation Nonsurgical Neurotechnology (N^3)	The N^3 program aims to develop a safe, portable neural interface system capable of reading from and writing to multiple points in the brain at once.	**Reliable Neural-Interface Technology (RE-NET)**	The RE-NET program seeks to develop the technologies needed to reliably extract information from the nervous system, and to do so at a scale and rate necessary to control complex machines, such as high-performance prosthetic limbs.
Restoring Active Memory (RAM)	The RAM program aims to develop and test a wireless, fully implantable neural-interface medical device for human clinical use. The device would facilitate the formation of new memories and retrieval of existing ones in individuals who have lost these capacities as a result of traumatic brain injury or neurological disease.	**Restoring Active Memory – Replay (RAM Replay)**	This program will investigate the role of neural "replay" in the formation and recall of memory, with the goal of helping individuals better remember specific episodic events and learned skills. The program aims to develop rigorous computational methods to help investigators determine not only which brain components matter in memory formation and recall, but also how much they matter.
Revolutionizing Prosthetics	The Revolutionizing Prosthetics program aims to continue increasing functionality of DARPA-developed arm systems to benefit Service members and others who have lost upper limbs.	**Systems-Based Neurotechnology for Emerging Therapies (SUBNETS)**	The SUBNETS program seeks to create implanted, closed-loop diagnostic and therapeutic systems for treating neuropsychological illnesses.
Targeted Neuroplasticity Training (TNT)	The TNT program seeks to advance the pace and effectiveness of cognitive skills training through the precise activation of peripheral nerves that can in turn promote and strengthen neuronal connections in the brain.	**Theodore Berger's Artificial Hippocampus**	The device works by mimicking the hippocampus' function of converting short term memory into long term ones by utilizing mathematical code that represents an entire memory.

4. Tool Use and Timeframe to Become Technology

As more technology is integrated into our bodies, and as we move away from the process of biological evolution as the primary force operating on the body, we need to consider exponential advances in computing technology that have occurred over the last few decades. One of the main predictors for the future direction of technology and particularly artificial intelligence is Ray Kurzweil, Google's Director of Engineering. Kurzweil has predicted that the technological Singularity, the time at which human general intelligence will be matched by artificial intelligence will be around 2045 [6]. Further, Kurzweil claims that we will multiply our effective intelligence a billion-fold by merging with the artificial intelligence we have created [6]. Kurzweil's timetable for the Singularity is consistent with other predictions, notably those of futurist Masayoshi Son, who argues that the age of super-intelligent machines will happen by 2047 [24]. In addition, a survey of AI experts (n = 352) attending two prominent AI conferences in 2015 responded that there was a 50% chance that artificial intelligence would exceed human abilities by around 2060 [25]. In my view, these predictions for artificial intelligence (if they materialize) combined with advances in technology implanted in the body are leading to a synergy between biological and cultural technology in which humans will be equipped with devices that will allow the brain to directly access the trillions of bits of information in the cloud and to control technology using thought through a positive feedback loop; these are major steps towards humans merging with technology.

In Figure 1, I am less specific about the date when the Singularity will occur compared to predictions by Kurzweil and Son, providing a range from 2050–2100 as a possibility. Further, in the Figure which also displays a timeframe for humans to merge with technology, I emphasize three time periods of importance for the evolution of human enhancement technology and eventual merger between humans and technology. These time periods suggest that the merging of humans with technology can be described as occurring in major stages (with numerous substages); for example, for most of the history of human technology, technology was external to the body, but as we move towards a being that is more technology that biology, we are entering a period of intermediate beings that are clearly biotechnical. However, returning to the first time period, it represents the preceding period of time that up to now is associated with human biological evolution and marked predominantly by humans using primitive (i.e., non-computing) tools for most of our history. Note that the tools designed by early humans were always external to the body and most frequently held by the hands. Such tools designed by our early ancestors allowed humans to manipulate the environment, but the ability to implant technology within the body in order to enhance the body with capabilities that are beyond what evolution provided humans is a very recent advancement in technology, one that could lead to a future merger between humans and technology.

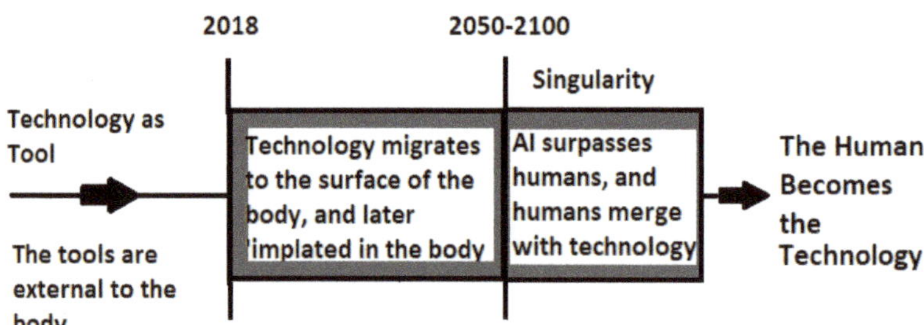

Figure 1. Timeframe for humans to merge with technology and for artificial intelligence to reach human levels of intelligence.

The second stage of human enhancement is now until the Singularity occurs [5]. In this short time period, technology will still be used primarily as a tool that is not considered as part of the human, thus humans will continue to be more biological than technological (but increasingly biotechnological beings will emerge). However, between now and the end of the century, we will see the development of more technology to replace biological parts, and to possibly create new features of human anatomy and brain functionality. In addition, during this time period people will increasingly access artificial intelligence as a tool to aid in problem solving and more frequently artificial intelligence will perform tasks independent of human supervision. But in this second stage of human enhancement, artificial intelligence will exist primarily on devices external to the human body. Later, artificial intelligence will be implanted in the body itself. Additionally, in the future, technology that is directly implanted in the brain will increase the computing and storage capacity of the brain, opening up new ways of viewing the world, and moving past the capabilities of the brain provided by biological evolution, essentially extending our neocortex into the cloud [18].

The third stage of technology development impacting human enhancement and our future to merge with technology is represented by the time period occurring after the technological Singularity is reached. At this inflection point, artificial intelligence will have surpassed humans in general intelligence and technology will have advanced to the point where the human is becoming equipped with technology that is superior to our biological parts produced by the forces of evolution; thus, the human will essentially become a form of nonbiological technology. Of the various implants that will be possible within the human body, I believe that neuroprosthetic devices will be determinative for allowing humans to merge with, and control technology, and ultimately to become technology. Given the focus of the discussion in this paper comparing the timescale for evolutionary forces which created humans versus the dramatically accelerated timescale under which technology is improving the functionality of humans in the 21st century, the figure covers no more than a century, which is a fraction of the timeframe of human evolution since the first *Homo sapiens* evolved a few hundred thousand years ago.

5. Moral Issues to Consider

As we think about a future in which human cognitive functions and human bodies may be significantly enhanced with technology, it is important to mention the moral and ethical issues that may result when humans equipped with enhancement technologies surpass others in abilities and when different classes of humans exist by nature of the technology they embrace [26]. Bob Yirka, discussing the ethical impact of rehabilitative technology on society comments that one area which is already being discussed is disabled athletes with high-tech prosthetics that seek to compete with able-bodied athletes [26]. Another issue where a technologically-enhanced human may raise ethical and legal issues is the workplace [5]. In an age of technologically enhanced people, should those with enhancements be given preferences in employment, or have different work standards than those who are not enhanced? Further, in terms of other legal and human rights, should those that are disabled but receive technological enhancements that make them "more abled" than those without enhancements, be considered a special class needing additional legal protections, or should able-bodied people in comparison to those enhanced receive such protections? Additionally, Gillet, poses the interesting question of how society should treat a "partially artificial being?" [27]. The use of technological enhancements could create different classes of people by nature of their abilities, and whether a "partial human" would still be considered a natural human, and receive all protections offered under laws, statutes, and constitutions remains to be seen. Finally, would only some people be allowed to merge with technology creating a class of humans with superior abilities, or would enhancement technology be available to all people, and even mandated by governments raising the possibility of a dystopian future?

Clearly, as technology improves, and becomes implanted within human bodies, and repairs or enhances the body, or creates new human abilities, moral and ethical issues will arise, and will need significant discussion and resolution that are beyond the scope of this paper.

6. Concluding Thoughts

To summarize, this paper proposes that the next step in human evolution is for humans to merge with the increasingly smart technology that tool-building *Homo sapiens* are in the process of creating. Such technology will enhance our visual and auditory systems, replace biological parts that may fail or become diseased, and dramatically increase our information-processing abilities. A major point is that the merger of humans with technology will allow the process of evolution to proceed under a dramatically faster timescale compared to the process of biological evolution. However, as a consequence of exponentially improving technology that may eventually direct its own evolution, if we do not merge with technology, that is, become the technology, we will be surpassed by a superior intelligence with an almost unlimited capacity to expand its intelligence [5–7]. This prediction is actually a continuation of thinking on the topic by robotics and artificial intelligence pioneer Hans Moravec who, almost 30 years ago, argued that we are approaching a significant event in the history of life in which the boundaries between biological and post-biological intelligence will begin to dissolve [7]. However, rather than warning humanity of dire consequences which could accompany the evolution of entities more intelligent than humans, Moravec postulated that it is relevant to speculate about a plausible post-biological future and the ways in which our minds might participate in its unfolding. Thus, the emergence of the first technology creating species a few hundred thousand years ago is creating a new evolutionary process leading to our eventual merger with technology; this process is a natural outgrowth of—and a continuation of—biological evolution.

As noted by Kurzweil [6] and Moravec [7], the emergence of a technology-creating species has led to the exponential pace of technology on a timescale orders of magnitude faster than the process of evolution through DNA-guided protein synthesis. The accelerating development of technology is a process of creating ever more powerful technology using the tools from the previous round of innovation. While the first technological steps by our early ancestors of a few hundred thousand years ago produced tools with sharp edges, the taming of fire, and the creation of the wheel occurred much faster, taking only tens of thousands of years [6]. However, for people living in this era, there was little noticeable technological change over a period of centuries, such is the experience of exponential growth where noticeable change does not occur until there is a rapid rise in the shape of the exponential function describing growth [6]; this is what we are experiencing now with computing technology and to a lesser extent with enhancement technologies. As Kurzweil noted, in the nineteenth century, more technological change occurred than in the nine centuries preceding it and in the first twenty years of the twentieth century, there was more technological advancement than in all of the nineteenth century combined [6]. In the 21st century, paradigm shifts in technology occur in only a few years and these paradigm shifts directed towards enhancing the body could lead to a future merger between humans and technology.

According to Ray Kurzweil [6], if we apply the concept of exponential growth, which is predicted by the law of accelerating returns, to the highest level of evolution, the first step, the creation of cells, introduced the paradigm of biology. The subsequent emergence of DNA provided a digital method to record the results of evolutionary experiments and to store them within our cells. Then, the evolution of a species who combined rational thought with an opposable appendage occurred, allowing a fundamental paradigm shift from biology to technology [6]. Consistent with the arguments presented in this paper, Kurzweil concludes that this century, the upcoming primary paradigm shift will be from biological thinking to a hybrid being combining biological and nonbiological thinking [6,28]. This hybrid will include "biologically inspired" processes resulting from the reverse engineering of biological brains; one current example is the use of neural nets built based on mimicking the brain's neural circuitry. If we examine the timing and sequence of these steps, we observe that the process has continuously accelerated. The evolution of life forms required billions of years for the first steps, the development of primitive cells; later on, progress accelerated and during the Cambrian explosion, major paradigm shifts took only tens of millions of years. Later on, humanoids eventually developed over a period of millions of years, and *Homo sapiens* over a period of only hundreds of thousands of years. We are now a tool-making species that may merge with and become the technology we are creating. This may happen by the end of this century or the next, such is the power of the law of accelerating returns for technology.

Funding: The writing of this paper was not funded through external sources.

Conflicts of Interest: The author declares no conflict of interest.

References

1. Wood, B. *Human Evolution*; Oxford University Press: New York, NY, USA, 2008.
2. *Number of Neurons in a Human Brain*. Available online: mmmhttps://hypertextbook.com/facts/2002/AniciaNdabahaliye2.shtml (accessed on 11 December 2018).
3. *The Human Memory*. Available online: jwww.human-memory.net/brain_neurons.html (accessed on 15 January 2019).
4. Liu, X.; Zeng, Y.; Zhang, T.; Xu, B. Parallel Brain Simulator: A Multi-scale and Parallel Brain-Inspired Neural Network Modeling and Simulation Platform. *Cogn. Comput.* **2016**, *8*, 967–981. [CrossRef]
5. Barfield, W. *Cyber Humans: Our Future with Machines*; Copernicus Press: Göttingen, Germany, 2015.
6. Kurzweil, R. *The Singularity is Near: When Humans Transcend Biology*; Penguin Books: New York, NY, USA, 2006.
7. Moravec, H. *Mind Children, The Future of Robot and Human Intelligence*; Harvard University Press: Boston, MA, USA, 1990.
8. Ambrosio, J. *Kurzweil: Brain will Extend to the Cloud, Computerworld*. 2012. Available online: https://www.computerworld.com/article/2491924/emerging-technology/kurzweil--brains-will-extend-to-the-cloud.html. (accessed on 29 December 2018).
9. Warwick, K. Cyborg Morals, Cyborg Values, Cyborg Ethics. *Ethics Inf. Technol.* **2003**, *5*, 131–137. [CrossRef]
10. Yokoyama, S. Molecular Evolution of Vertebrate Visual Pigments. *Prog. Retin. Eye Res.* **2000**, *19*, 385–419. [CrossRef]
11. Drewry, R.D. *What Man Devised He Might See*. Available online: http://www.woodlandseye.com/images/histoy_of_lenses_optical.pdf. (accessed on 12 January 2019).
12. Singer, E. *Implantable Telescope for the Eye*. MIT Technology Review. 2009. Available online: https://www.technologyreview.com/s/412835/implantable-telescope-for-the-eye/ (accessed on 8 December 2018).
13. Schaik, C.P.; O'Deaner, R.; Merrill, M.Y. The Conditions for Tool Use in Primates: Implications for the Evolution of Material Culture. *J. Hum. Evol.* **1999**, *36*, 719–741. [CrossRef] [PubMed]
14. Atchley, W.R.; Hall, B.K. A Model for Development and Evolution of Complex Morphological Structures. *Biol. Rev.* **1991**, *66*, 101–157. [CrossRef] [PubMed]
15. Falk, D.; Gibson, K.R. *Evolutionary Anatomy of the Primate Cerebral Cortex*; Cambridge University Press: New York, NY, USA, 2001.
16. Gingerich, P. Primate Evolution: Evidence from the Fossil Record, Comparative Morphology, and Molecular Biology. *Am. J. Phys. Anthropol.* **1984**, *27*, 57–72. [CrossRef]
17. Falk, D. 3.5 Million Years of Hominid Brain Evolution. *Semin. Neurosci.* **1991**, *3*, 409–416. [CrossRef]
18. Berger, T.W. Chap 12: Brain-implantable Electronics as a Neural Prosthesis for Hippocampal Memory Function. In *Toward Replacement Parts for the Brain*; Bradford Book: Cambridge, MA, USA, 2005.
19. Aamir, S.A.; Stradmann, Y.; Müller, P.; Pehle, C.; Hartel, A.; Grübl., A.; Schemmel, J.; Meier, K. An Accelerated LIF Neuronal Network Array for a Large-Scale Mixed-Signal Neuromorphic Architecture. *IEEE Trans. Circuits Syst. I* **2018**, *65*, 4299–4312. [CrossRef]
20. Ezzyat, Y.; Wanda, P.; Levy, D.; Kadel, A.; Aka, A.; Pedisich, I.; Sperling, M.R.; Sharan, A.D.; Lega, B.C.; Burks, A.; et al. Closed-loop Stimulation of Temporal Cortex Rescues Functional Networks and Improves Memory. *Nat. Commun.* **2018**, *9*, 365. [CrossRef] [PubMed]
21. Long, N.M.; Kahana, M.J. Modulation of Task Demands Suggests that Semantic Processing Interferes with the Formation of Episodic Associations. *J. Exp. Psychol. Learn. Mem. Cognit.* **2017**, *43*, 167–176. [CrossRef] [PubMed]
22. Mukamel, R.; Fried, I. Human Intercranial Recordings and Cognitive Neuroscience. *Annu. Rev. Psychol.* **2012**, *63*, 511–537. [CrossRef] [PubMed]

23. Song, D.; Robinson, B.S.; Hamson, R.E.; Marmarelis, V.Z.; Deadwyler, S.A.; Berger, T.W. Sparse Large-Scale Nonlinear Dynamical Modeling of Human Hippocampus for Memory Prostheses. *IEEE Trans. Neural Syst. Rehabil. Eng.* **2016**, *26*, 272–280. [CrossRef] [PubMed]

24. Galeon, D. Software CEO: The Singularity Will Happen by 2047. 2017. Available online: https://futurism.com/softbank-ceo-the-singularity-will-happen-by-2047// (accessed on 12 December 2018).

25. Revell, T. AI Will be Able to Beat us at Everything by 2060, Say Experts. *New Scientist.* 2017. Available online: https://www.newscientist.com/article/2133188-ai-will-be-able-to-beat-us-at-everything-by-2060-say-experts/. (accessed on 7 January 2019).

26. Yirka, B. *Is it Time to Access the Ethical Impact of Real Cyborgs on Modern Society?* 2017. Available online: https://techxplore.com/news/2017-06-ethical-impact-real-cyborgs-modern.html. (accessed on 29 December 2018).

27. Gillett, G. Cyborgs and Moral Identity. *J. Med. Ethics* **2006**, *32*, 79–83. [CrossRef] [PubMed]

28. Barfield, W.; Williams, A. Cyborgs and Enhancement Technology. *Philosophies* **2017**, *2*, 4. [CrossRef]

 philosophies

Article

Agency, Responsibility, Selves, and the Mechanical Mind

Fiorella Battaglia [1,2]

1   Faculty of Philosophy, Philosophy of Science and Religious Studies, Ludwig-Maximilians-University, 80539 Munich, Germany; Fiorella.Battaglia@lrz.uni-muenchen.de
2   Institute of Law, Politics, and Development, Sant'Anna School of Advanced Studies, 56127 Pisa, Italy; fiorella.battaglia@santannapisa.it

Abstract: Moral issues arise not only when neural technology directly influences and affects people's lives, but also when the impact of its interventions indirectly conceptualizes the mind in new, and unexpected ways. It is the case that theories of consciousness, theories of subjectivity, and third person perspective on the brain provide rival perspectives addressing the mind. Through a review of these three main approaches to the mind, and particularly as applied to an "extended mind", the paper identifies a major area of transformation in philosophy of action, which is understood in terms of additional epistemic devices—including a legal perspective of regulating the human–machine interaction and a personality theory of the symbiotic connection between human and machine. I argue this is a new area of concern within philosophy, which will be characterized in terms of self-objectification, which becomes "alienation" following Ernst Kapp's philosophy of technology. The paper argues that intervening in the brain can affect how we conceptualize the mind and modify its predicaments.

Keywords: brain–computer interface (BCI); human–robot interaction; mind; sense of agency; alienation

1. Introduction

The changes brought about by the use of neurotechnology informed by artificial intelligence (AI) can be dramatic, as Brain/Neural Computer Interfaces (BNCI) and robotics technologies can directly intervene in the brain and disrupt cognitive processes responsible for consciousness and identity raising questions about what it means to be a unique person. Our ordinary accounts of intentional behavior appeal to desires, habits, knowledge, reasons, and perceptions. These new interventions operate in a conceptual framework dominated by sub-personal categories. Such technologies are therefore a source for investigation in the field of moral theory. I propose that there are two possible paths of research in this area. On the one hand, questioning the impact of neurotechnology and AI includes philosophical, ethical, regulatory, legal, and social implications—an approach focusing on their applications. On the other hand, questioning neurotechnology and AI also includes concerns about how we make sense of ourselves and modify theoretical and practical concepts. By understanding and conceptualizing us in new, unexpected ways, we re-ontologize ourselves. The new philosophical framework cannot remain without practical implications as the new insights generated by this activity will trigger and motivate our actions. Therefore, as human enhancement technologies will make new claims, which impact competing approaches to the mind, they also represent a relevant area for moral theory. In this sense, neurotechnology motivates a relevant area of moral theory.

The philosophical framework behind this fascinating scenario was first named by James H. Moor in 1985 [1]; it is known as the anatomy of computer revolution, since he was using two stages for making explicit the perspective on the development of a cultural innovation initiated by the "introduction" and "permeation" of computers in social life. While the introduction stage does not alter social institutions and easily accommodate technology into the means-end relation, the permeation stage does have the potential to

Citation: Battaglia, F. Agency, Responsibility, Selves, and the Mechanical Mind. *Philosophies* **2021**, *6*, 7. https://doi.org/10.3390/philosophies6010007

Received: 14 December 2020
Accepted: 10 January 2021
Published: 19 January 2021

Publisher's Note: MDPI stays neutral with regard to jurisdictional claims in published maps and institutional affiliations.

disrupt social institutions and may bring unprecedented systematic thinking to bear on the very phenomenon of technology. His insights have undeniable interest and applications for theorizing about our merger with machines. It will be important for my purposes in this article to concentrate on the impact on moral theory. Therefore, I am not going to discuss in the paper the ethical issues related to the alterations following the interventions in the brain [2–5]. I will rather discuss some implications that these interventions might be able to trigger on scientific representing and understanding of the mind. The involved problems are not just ethical problems, but conceptual and philosophical ones.

While the first investigation will be concerned with challenges to human abilities, the latter investigation is concerned with the conceptual framework that these interventions are going to modify. My argument is focused on knowledge and aims to establish that interventions in the brain involve new understanding of the mind and on the basis of these new insights on the motivations for actions [6]. In other words, human enhancement technologies and our merger with machines call for the assessment of the adequacy of our representational devices for grasping what it means to be human from a moral, ethical, and legal standpoint. It rests on the idea that having access to the neural level of the brain might have an impact on both the knowledge about how it feels to have the experiences of that being and the knowledge about the fact that we evaluate each other's behavior and attitudes. It also challenges the idea that having physical knowledge about the mind might render redundant other sorts of knowledge such as the phenomenal and moral knowledge about mind and behavior. In a sense, inquiring features such as consciousness, intentionality, meaning, purpose, thought, and value; leads us to questioning whether they can be accommodated in a naturalistic image made of physical facts, whose access is granted by the physical sciences, this questioning renews a number of well-known philosophical disputes. Distilling this approach down even further, we might say that it aims to address the classical mind–body problem.

The mind–body problem has been addressed in a variety of ways. However, in the last decade, impressive advances have been made in the field of technological improvement of sensory, motor, and cognitive performances. This field, which has the potential to heal and enrich the mind, may also revolutionize the way in which the mind is understood and may have a bearing on how to model the mind. The reasons for avoiding a naturalistic approach to the mind are not reasons to avoid addressing the scientific treatment of it. A naturalistic approach to the mind, philosophically constrained by a compatibilist requirement (that free will is consistent with universal causal determinism) and supported by empirical evidence provided by the interplay with new technologies, generates a philosophically adequate and recognizable basis for a better understanding of the mind. This, in the most basic terms, is the thesis of this paper. Hence, the renewal of concerns against physicalist theories of consciousness and agency is currently being challenged by disruptive research and innovation, among which robotics and neurotechnology are prominent examples. At the metaphilosophical level, as technology is becoming an integral part of social practices, a relevant area of moral theory is consequently called into question. This is not a bad thing; on the contrary, it allows us to rethink how we know the characteristics of human mind. In this essay, I will present an analysis of the epistemic and ethical challenges posed by smart technologies about the way in which we understand the mind represented in its practical aspects as well [6,7]. Through a brief review of the three main approaches to the mind, first person perspective, second person perspective and third-person perspective, I will identify a main area of concern, deriving from accessing the mind and its features through neural mechanisms.

A survey of the literature in different fields provides a variety of suggestions: it is to be noted that the relevant impact is characterized in terms of enhancement technologies, new empirical evidence and transformation of vulnerability [8–13]. The nature of each of these categories, and the relationship between them, are both controversial. For example, some philosophers will question the improvement brought about by the enhancement technologies, while others will include thinking about human enhancement technologies as a very broad tent, including their impact on our epistemic sources concerning our sensory, motor, and cognitive abilities. Finally, others will think of enhancement as a shift in our vulnerable features. For the sake of space, this essay will focus on the last two points throughout. This opens up new epistemic improvements for a re-appraisal of notions such as agency. At the same time accessing neural mechanisms to modulate human behavior makes individuals vulnerable to "nudges", manipulation, and deception, which will be made practicable, not least by the adjustment of our representative features.

The final aim of this paper is to draw a critical and constructive analysis and to show that by merging with a machine we have as a result both a surplus of representational devices and a vulnerability warning. I will first describe this surplus of representational devices in terms of direct knowledge of our actions, which has been fed by empirical studies. I will then address this kind of concern in term of "alienation" following Ernst Kapp's philosophy of technology [14]. The claim is that "alienation" should inform the discussion on transformative effects of these technologies, that is, on those effects which cannot always be interpreted to clear cases of epistemic or ethical shortcomings. Thus, this argument is conducive to a different kind of epistemic and ethical discussion resulting from the facts obtained by the ongoing technological transformation of human motor, sensory, and cognitive traits. It is the objective of this paper to clearly distinguish between the various thematic strands (consciousness, agency, mind), which are also dealt with in quite different debates—philosophy of mind, theory of action, theory of rationality and the theories of personal identity, theory of cognitive science and to clearly show the connection between them, which is not drawn by the single theories.

2. The Limits of Treating the Mind as an Object

Let us start with the way in which we represent the mind to assess what kinds of transformations have been occurring by our merging with the machines in our representational devices. How do we conceptualize agency, responsibility, and consciousness? We normally can count on three competing approaches: First, the approach articulated in first person perspective and tailored to fulfill the requirements both of subjectivity and of consciousness as experience [15–18] second, the approach articulated in second person perspective [19–21] and tailored to serve the needs of the personality; finally, the approach articulated in third person perspective and tailored to fit the human brain facts [22].

The distinctive features of the phenomenological mind and of the personality can best be seen by contrasting them with one more approach, the physical approach. The quality of what "it feels like" to be in such a state cannot, in principle, be conveyed in a physical language tailored to fit the objects in the world. Such facts of experience show an irritating incompleteness of the objectifying description of the world. The same happens with the personality layer and its normative requirements. The problem with the objectification of the mind is that it exhibits two kinds of shortcomings. It cannot account for consciousness as experience [23] and cannot account for the causal role of mental events in the physical world [24]. Therefore, the process of objectification—while proving fruitful in interpreting, explaining and predicting a series of physical events—shows its limits especially in the case when we want to accommodate features of personality along with those of subjectivity. When ascribing someone's agency or moral responsibility what matters is whether the person in question is able to participate in normative discourse and practice. Whether she is able to give and take reasons for her actions, beliefs, and feelings. On the one hand, we rely on these two layers of complexity—subjectivity and personality—when representing normative relations between agents and accounting for meaningful experiences of the world. On the other hand, the physical approach makes it possible to treat the mind as an object. As a result, it is possible to make various predictions. The underlying idea is that rooting the study of the mind in the natural sciences will provide us with a comprehensive understanding of the main mental features. Moreover, motor, sensor, and cognitive abilities seem to be accommodated in this naturalistic version of the world. This is the case at least in psychophysical reductionism that says how the physical sciences could in principle provide an account of everything in the world.

The neural correlates of consciousness that cause the experience seem to be able to tell the whole story about the mind. A neural correlate of consciousness can be described as a minimal neural system that is directly related to states of consciousness. At least physicalist positions are keen to endorse this view. As fascinating as this kind of approach can be, as it promises to make the study of the mind scientific, it has costs, and I hope to show that they are not worth incurring. I will argue that there are philosophical costs to denying subjectivity and personality. I believe that physicalist approaches may allow for more philosophical insight, however. In this article, I will attempt to articulate what "getting at the mind" consists of, and what its merits and its faults are. A critical view says that psychophysical reductionism is a failure; the prospect that physical sciences could in principle provide a theory of everything is a failure in terms of explaining free actions and experience. In particular, the attempt to go from phenomenology to information processing in the brain is not going to happen in my view. Subjectivity and personality will be resistant to this kind of view. But interestingly, human enhancement technologies have introduced new insights in this discussion. By accessing the mind at the neural level, the human enhancement approach has introduced some major transformations in our knowledge.

There is a difference between being a mind and being an extended mind. It is an epistemic difference. The argument presented in this paper focuses on that difference. It is the distinct knowledge underling this specific self-objectification, which allows for detecting, influencing, and stimulating mental states. What presents itself in its relentlessness is the fact that the third approach may not be able to explain the other two dimensions elaborated in second- and first-person perspective and yet it does contribute to re-articulating a number of cognitive and affective features of the mind such as feeling, mood, perception, attention, understanding, memory, reasoning, and coordination of motor outputs and therefore high-level expressions of these properties such as selves and personality. In particular, the amount of knowledge available to be processed is remarkable if we consider the human–machine interaction. As a result, we are faced with a twofold novelty: some features of human behavior—prominently the "sense of agency", i.e., the kind of immediate knowledge we have of our actions– can be traced back to underlying mechanisms conceived as a causal chain of physical events at a sub-personal level and can be implemented and altered by accessing these very mechanisms at the neural level.

The debate about the "sense of agency" has been driven by empirical studies in the field of psychology and cognitive neuroscience [25–27]. What is more, robotics inspired by the idea of the extended mind has been another driver of this transformation. It says that the extended mind relies on sensorimotor synergies, and that only the understanding of their mechanisms will enable designing and building artificial systems that can generate a true symbiosis with the human. While we have learned much on both motor and sensory synergies in separation, not much attention has been paid before as to how the concept of synergies extends to the brain–body–environment interaction, and to the sensorimotor loop itself.

Pioneering work in the psychophysics and neurophysiology of the human hand and upper limb, and in the translation of this understanding in bionic concepts has been achieved [28]. The next level is to exploit the extended mind approach to a whole brain–body–environment interaction schema. In doing so, it will be possible both to explore the mechanisms by which the human and artificial system can remotely communicate and cognitively merge and to improve our comprehension of what it means to be human. This will in turn produce knowledge about our own actions. Whereas interacting with objects in the environment in which we live in is mostly immediate and unreflective, to engineer this very natural interaction (among our body, objects, and environment) is complicated. The success of engineered intuitive natural human–robot interaction (HRI) will critically depend on whether the operator will have no difficulty in recognizing the robot, its movements and sensations as part of their own body and their own actions: the exploitation of systems such as these requires the fusion of the human mind with robotic sensors and actuators. If this is going to happen, by interacting with the robot, then the operator will be released from the physical, cognitive and affective workload associated with traditional teleoperation. Crucially, such a system relies on a multilevel approach to ensure engineered actions to be as successful as the natural ones. Not only will our skilled capacities be improved, but our understanding of natural and intuitive human–robot interaction makes room for the improvement of our knowledge about our own actions and sense of agency. This opens up three levels of inquiry, which fully cover the interaction with the robot and hence provide both explanations on how intuitive natural human–robot interaction occurs and instructions about how to ergonomically and ethically design the robot platform. The levels, in which we choose to describe, explore, and discuss the system and its context, are the neural level, the ergonomic level, and the ethical level.

The benefits of the proposed exploration of the extended mind idea and its robotics applications are manifold. It is common to think of psychophysical reductionism as a failure or as a promise. This way of thinking certainly has its merits, if only to a limited extent. For, unfortunately, the nature of the mind–body problem is not well-understood philosophically, despite the fact that some important past debates have discussed a remarkable number of issues. There still remains the advantage, however, that the concept of the extended mind is empirically amenable and therefore less rich, obscure and slippery than that of mind, and hence easier to handle. So, a definition of mind grounded in the symbiosis with technological artifacts seems to be a good starting point. The mechanisms of the mind, which basically rely on causal explanations, cannot make sense of subjectivity and personality. Intervening in the brain will modify their physical extension and will result in an impact both on the physical and phenomenological level. This is very much in line with the version of the identity theory contended by Davidson [24]. It says that all mental events are identical with physical events. However, not all physical events have a mental extension. We cannot make sense of mental events by recurring to physical science. This is often referred to as the nothing-but reflex [24]. This construct characterizes the aptitude of sentences like these "Conceiving the "Ode to Joy" was nothing but a complex neural event".

By relating mental events to other mental events, we explain them. This occurs by appealing to desires, intentions, and perceptions of the human agent. Explanations such as these represent a conceptual framework dominated by reasons and are usually detached from sub-personal level instantiated and explored by physical science. And yet, there are some particular cases. In cases where we know particular identities and are able to correlate particular mental events with particular physical events, we can modify the mental events by intervening in the brain. For example, Miller et al. [29] were able to find the dynamics of sensorimotor cortices associated with the touch of a hand-held tool. When the identity is not questionable, then we may be able to obtain a real advance. Epistemic and practical challenges like these call for a reappraisal of our epistemic setting. There are not only feasibility questions involved such as "how can we facilitate the symbiosis between machines and us?" As more difficult questions begin to have positive answers, healing and protecting the mind, understanding the mind, enriching the mind, and modeling the mind can be correlated with the way in which we make sense of our experience. After two decades of research, the neural level of the human-like interaction with artifacts starts to be disentangled. First advances are registered concerning motor abilities, and much of the functioning of the motor system occurs without awareness. Nevertheless, we are aware of some aspects of the current state of the system and we can prepare and make movements in the imagination. These mental representations of the actual and possible states of the system are based on two sources: sensory signals from skin and muscles, and the stream of motor commands that have been issued to the system. Human-like interaction with the robots can lead to the augmentation of the motor system in the awareness of action as well as improvement in the control of action. At the same time this very enhancement adds another piece to the "mechanical mind" as I call this new epistemic device [30]. At some extent there is a convergence between "extended mind" and "mechanical mind". While the first is meant to connote the human–machine hybridization, the latter is supposed to mark the possible explanatory device arisen from this human machine hybridization. Blurring the line between persons and things will not be conducive to epistemic opacity. On the contrary, it will lead us to new insights.

Drawing on these new results, there are remarkable insights in the field of moral theory as well. It has been necessary to wait for the most recent developments of embodied cognition to give empirical evidence to the bodily involvement of the agent in the theory of action. The construct of the "sense of agency" provides new elements to overcome the vision of an agency characterized in the absence of bodily mechanisms of sensory processing and motor control. It provides a direct knowledge of our actions based neither on observation nor inference and can be implemented in a computational model of the agent. In addition, robotics fits well into this picture, because by investigating what happens at the neural level when we add an extension to the agent's body, it gives the possibility to modulate and correlate the knowledge related to the phenomenological experience with that related to the neural level. To combine a theoretical approach with a radical approach that relies on the construction of complete intelligent systems with a real apparatus of perception and movement that interacts with the environment is able to counteract misleading results. This approach presents strong evidence that the sense of agency originates in the neural processes responsible for the motor aspects of action. I contend that these recent results are compatible with a causal theory of action as suggested by our ordinary attitude that relates us to objects with a practical rather than simply theoretical stance. I believe that this contributes to a compatible perspective capable of integrating the knowledge that comes from different approaches to the same phenomenon. Giving scientific consideration to the physical events does not necessarily mean surrendering to naturalism; rather, it means not committing us in an alleged tension between two cultures—the scientific and the humanist.

3. Additional Epistemic Devices

In light of the fact that the traditional critical view has been revised and that accessing the neural correlates no longer means to subscribe to some form of reductionism, I will now argue that to integrate both the experiential dimension of consciousness and the space of reasons, in which we operate by appealing to reasons in order to explain actions, to make sense of our experience and to make room for normative claims, is fully compatible with the idea to treat our mind as an object. In short, the three rival accounts discussed above may not be competing against themselves anymore. The extra ingredients are the human enhancement technologies, which while allow our merging with the machine will produce an advancement in the understanding of the mind in terms of the same subject thought as necessarily united in its two components—the human and the machinic.

What is more, this kind of integration between human and machine will make room for new representational devices derived from our interaction with enhancement technologies. Its implied greater capabilities are conducive to more ethical sophistication in a conceptual framework. At this point in the discussion, I will now introduce some insights into human brain enhancements from a legal perspective. By looking to legal theory rather than moral theory, I may be able to better make sense of the transformative nature of human enhancements. As Brain/Neural Computer Interfaces and transparent robotics realize a seamless symbiosis with the human brain, they paradigmatically introduce new elements to be weighed in legal theory. A legal analysis will be able to elucidate some aspects of this new perspective, which I consider from a philosophical perspective in a second step of my analysis. Legal scholars seem to comply with the experiment to treat the mind as an object and also suggest that we need to re-shape the ontology accordingly. This is because, in the legal perspective, the main interest is a practical one. Mostly, it is a matter of regulating the human–machine interaction in the case that something goes wrong. This perspective, far from being departmental, turns out to be much more liberal, because it allows us to think of personhood outside metaphysical burdens. The legal perspective, while aims to address how to deal with regulatory issues, develops a conceptual and regulatory framework able to overcome the dualism between humans and things.

The nature of this interaction is also a fascinating philosophical matter, and human–machine interaction from a metaphysical perspective raises interesting questions of its own. Fundamentally, they call for a re-examination of our understanding of the mind and its properties. It is necessary to make a detour to value and integrate these new elements into philosophical analysis. If this is not performed, the philosophical analysis remains stuck with two equally harmful prongs, scientific naturalism and antireductionism. The two major recent developments in philosophy of mind—naturalism and antireductionism—are deeply opposed to each other in important ways, but there is a striking convergence on answering the question—even though differently—whether scientific facts are able to properly account for consciousness, intentionality, meaning, purpose, thought, and value. As I said, there is another detour available with a view to obtaining a re-examination of familiar, perhaps too familiar, ways of thinking.

For Jens Kersten, there are no metaphysical standards that loom over the way we conceive of the mind [31]. Jens Kersten succeeds in combining elements of the theory of the nature of technology with elements of personality theory. In short, he is committed to the idea that technology as an element of culture may offer the possibility to reflect and identify with one's own artifacts. According to Jens Kersten, the symbiotic connection between human and machine is better understood as the very expression of personality. In his view, over our merger with the machine lies the decisive difference between the instrumental and the symbiotic constellation. The instrumental constellation is characterized by the differentiation between the human being as person and the machine as a thing. In those cases, in which the machine is essential for the development of personality, this relationship can be charged according to legal personality provisions, without, however, overturning the fundamental distinction between person and thing. In the symbiotic constellation, this differentiation between person and thing is omitted: The machines become "Nobjects" [32], which are going to get rid of their legal status as things and, due to the claim for dignity and the right of personality of their bearers, are going to be protected as part of their human personality. In this way, the prostheses, the implants, and other devices are going to acquire in a parasitic way the same moral status of their human host. As already stated, applicable legal theory is mainly concerned with a practical solution of those cases in which something goes wrong. For this practical attitude, any metaphysical attitude regarding the moral status will not matter. Legal theory puts all metaphysical assumptions regarding the nature of the mind and its properties into brackets. I have been drawing on the legal analysis because it is able to express very clearly how our categorizing attitude is going to be transformed by this sort of technological intervention which makes it clear that it is not the differences between machine and human that matter. The symbiosis indeed highlights a space-in-between that allows human beings to be involved in an effective seamless relationship with the machine.

According to the more recent developments in robotics, the robot is no longer a mere extension of the human, as in traditional teleoperation, rather the human being, will fully merge with the machine, creating a unit or symbiont whose physical and cognitive capability is more than the sum of those of the human and machine symbionts. These new technologies have the potential to collapse this ontological threshold into a single dimension while offering intimate experiences that are triggered by the interaction directly. By looking to legal theory, we may be able to better address issues regarding the conceptual analysis of the mind. This is the case since the symbiosis does affect how we conceptualize our experience and reshapes the status of the objects closely linked to us. As a result, this will also motivate an alteration in the way we will consider the artifact. While particular events may be explained by physical fact, this cannot justify a different interpretation of subjectivity and other humanlike features. Spelling out how technologies find their parasitic way to accomplish a new status is an intriguing opportunity to perform a kind of thought-experiment. It leads to a sort of hybridization of the fully human character, blurring the distinctions between human and artifact. To be sure, major questions still stay open.

According to the analysis of the symbiosis, should we reverse the logical precedence of the experiential and normative aspects of the subjectivity and personality over the biological-machinic elements? Should we rely on a process of technical feasibility and usability in order to make sense of consciousness, personality, autonomy, and responsibility? Will self-objectification be the key feature that can make sense of the cultural dimension of human existence? Nonetheless, a consensus about the basic rudiments of the relationship between human and machine appears to be emerging. Hybridity is seen as a constitutive dynamics of techno-social civilization, which is linked with the dialectic of inclusion and exclusion of natural and artificial elements. Critics insist that we think of the notion of human in a different sense compared to its technological extensions, whereas their defenders describe new forms of human–machine interaction as indispensable forerunners to more inclusive and advanced forms of self-understanding. What is certain is that all these questions are released from metaphysical burdens. The seamless relationship between human and machine, by modifying personality and responsibility attribution, may contribute to a larger inquiry as an additional epistemic device. This feature of the self-objectification dynamics can be clarified by contrasting it with Kapp's view, as the classic example of alienation theories of technology.

4. Alienation

There is a dialectic involved here, insofar as even if human enhancement technologies have their positive impact in a liberal way of thinking over human nature, these same technologies carry within themselves other possibilities, which are often of quite contrary quality. Therefore, it is misleading to take the positive possibilities, which we have recorded through the legal clarification, for granted. Technology is a constitutive element of the practices and institutions of the human condition. Technology constitutes an area of ethical investigation, and at the same time presents challenges to moral theory because it transforms reality and demands that moral theory adapt to a changed situation such as the one that accompanies the gradual process of human merging with the machine. I will critically discuss the conceptualization of technology in terms of mere instrument. My intention is to contend the partiality of this way of understanding the technology. In my view, this initial conceptualization needs to be complemented by a second perspective, which is focused on technology as a process of mediation that tends to transform what is at stake.

When one is about to decide on the transformation whose source is technology, controversies may arise about the ways in which the relationships between technology and society can be conceptualized and interpreted. The interpretation of the social and ethical implications also depends on the characterization that we will give of the technology. The conceptualization of the relationship between technology and human is responsible for various questions about nature, the human nature, human action, autonomy, freedom and much more. It is not the intention of this contribution to explore all these dimensions. I will focus only on the conceptions of technology even if some of the results of my argument will have some impact on the other dimensions as well.

Reflecting on the way technology has been conceptualized can help to clarify its transformative effects. The process of our merging with machines is certainly one of the most remarkable. A first form of conceptualization has been developed within the philosophical anthropology. According to this philosophical movement, technology was conceived as compensation for the biological deficits of the human being. Cassirer observed that at the basis of technical action there is a teleological relationship that presupposes a rational actor or a multitude of actors acting cooperatively [33]. This first part of the characterization is the most conservative and does not question the current ontology. To a certain extent, it matches the "introduction stage" described by Moor [1]. However, it is partial and fails to formulate an adequate representation of the phenomenon that it wants to capture. There is, therefore, a second element, and that is culture. If we analyze technology from this point of view, then we will have to admit that it is part of creativity and freedom of spirit and therefore able to change the structure with which we are familiar. It is capable of having transformative effects such as transformation of the ontology. To a certain extent, Cassirer's cultural dimension matches the "permeation stage" described by Moor [1]. Drawing on this double characterization, further considerations have been derived. The first is that technology can be conceived as a tool that set us free from natural constraints and thus exonerates us, for example, from dull, dirty, and dangerous tasks.

However, technology can also be conceived as a risk of alienation of humans from themselves. In this interpretation, we place the profiles of dehumanization that have so much weight in current debates. In the philosophy of technology, there is the idea of conforming to the objective forms of human activity and therefore to attribute to them a function of mediation of self-understanding. By considering technology, it would be possible to know what becomes of us. Technology highlights the fact that the actions mediated by it always include a concrete intervention in the material environment and therefore always represents a form of a relationship with nature. Technology then is both a relationship with nature and, since the human being himself belongs to nature, it also includes a certain relationship with oneself and therefore also with the immaterial, symbolic and normative dimension. Human beings then not only find themselves in their artifacts, but above all they recognize themselves in them. This thesis allows Kapp to explain the human body in terms of an organism by using the objectification of the tools we create.

Kapp tries to explain the human body, its being organism through its self-made tools. As they are regarded as human unconscious alienation, human organs now appear to be extensions of artifacts, and these themselves can thus become models for their exploration and interpretation. Since artifacts and their functional context are meant to be subject to natural laws, it is clear that they provide a model for research on the human organism, provided that it too is subject to natural laws, and is therefore doomed to lose track of both the phenomenal character of consciousness and normative processes.

Kapp has focused on the interplay of reflections that exists between technology and the human. If the human is essentially technical, then technology is the ground of the human culture. In this way, then technology will not be limited to putting mechanical pieces side by side with human ones. In fact, the process of becoming technical will bring to the maximum development the use of biotechnological options by eroding the dominion of the natural that for a long time had remained beyond the possibilities of human influence. It will not be possible to hide that such a process of conquest of territories previously untouched by moral judgments has a potential impact on the theoretical and symbolic dimension. This is the typical shift that the philosophy of technology allows to implement with epistemic benefits. Thinking about human organs starting from artifacts, conceived as unconscious alienation of the human, organs now appear from this overturned perspective as extensions of artifacts. The objectification allowed by this epistemic deviation allows them to be considered models for their exploration and interpretation. There is, however, a caveat. Once the model for their understanding is that of artifacts, they will lose their inner face. Undoubtedly, artifacts and their functional context are constitutively subject to laws, it is clear that they can serve as a model for research on the human organism, provided, however, that it is also understood only in those configurations that are governed by natural laws. However, this epistemological move has an undoubted value. Following Ernst Kapp's, one might be convinced to appeal to his notion of alienation within his theory of technology as a self-knowledge device.

The central idea here is that, through technology, we may come across human features. Technology serves us as tool for knowing our constitution. It represents a reliable way to conceptualize us. My thesis is that while we can use such hybrid forms to explore the boundaries of our self-objectification, the direction of the human–machine interaction in the difficult ethical situations must however remain that of adapting the machine to the human. This mirroring dynamics exposes the transformations of the human vulnerability.

The alienation nexus includes vulnerability and direction. The normative indication is that the adaptation should be determined according to human and not to machinic standards. Overall, the risk of self-objectification, which becomes alienation, is very well identified by Kapp's theory. The recent attitude puts forward by engineers, computer scientist, and philosophers that contends that "buildings robot is philosophy making" [34,35] is somehow echoing Kapp's philosophy of technology. This means that the process of building robots forces us to reflect on what human capabilities are, and is therefore comparable to a process of self-knowledge. As the ancient Greeks relied on the Oracle of Delphi, so we can count on the construction of robots to obtain knowledge of ourselves. I will discuss this claim and in doing so I will borrow Ernst Kapp's terminology. In Kapp's account, the human being is to be found in her artifacts; in her technical culture, she recognizes herself there [14]. A traditional view says that there is no increment of self-understanding without mirroring of a mind-external world. So far there is no disagreement between Kapp [14] and Pfeiffer and Bongard [34] and Wallach and Allen [35]. With the exception that the Kapp's argument has two aspects, while the recent account restricts itself to the self-knowledge claim without further caveat.

On the one hand, Kapp argues for the epistemic potential of technology. On the other hand, he warns against the risks of sorting humans and things as items of similar shape and features. On Kapp's account these are the ethical boundaries of the self-objectivation.

This analysis may accommodate a number of worries connected to the image of the human being as a well-functioning machine. Therefore, we can address this tendency as the process of *mechanization of the human being*. For such reasons, the interventions in the brain are research topics and cannot serve as a normative tool. It would be wrong to use the insights from the process of self-objectivation to formulate ethical recommendations. Ethical recommendation concerning what does happen when humans share the same environment with highly automated systems can set the stage for a more integrated perspective on the human mind. As they identify the direction of the adjustment, they can also suggest that the project to understand the mind will root in self-reports and capacity to engage in normative practices and discourses. Experience and morality are to be moved out of the lab and to the street, office, kitchen, bar or wherever people happen to be when they feel, think, and act.

From a normative perspective, we can state that highly automated systems must be designed in a way that they must adapt more closely to the communication behavior of humans and not, conversely, demand increased adaptation performance from humans, who may not be willing to accomplish. If this adaptation goes the opposite path, we consider this—from an ethical perspective—as an undesirable development that may result in dehumanization processes. If the aim is to serve society, then innovation has to comply with human aims and desires. Otherwise, innovation without ethical dimension may result as a blind kind of innovation [36].

While the legal example shows how puzzling the hybrid constellation can be, the ethical analysis of human enhancement technologies is a source of hard moral questions. Intervening in the brain involves engineering issues and ethical problems, which can result in the conceptual and philosophical re-examination of our familiar insights.

5. Conclusions

The main objective of this paper was to discuss the ways in which human enhancement technologies such as Brain/Neural Computer Interfaces and robotics have a bearing on theoretical and methodological issues in philosophy. This is important, because the way we theorize and conceive of the mind's features during the process of building new technologies has an impact on normative discourse and practice. A deep understanding of how our mind perceives, thinks, and acts is not to be achieved just by means of one approach, which objectifies the mind. What we do need is an integrating perspective among the competing approaches. The feeling of being the owner of the action is just one successful area of investigation, which can benefit from different approaches and grow into a reliable account of our sense of agency, which can be enhanced in the presence of technical artifacts interacting with the body and the environment. In this sense, it is important to protect this symbiosis and prevent, control and monitor misuse, dual-use, and other non-intended applications, so that the benefits are not outweighed by the damages. The exploitation of the technological artifacts, their moral status, and their ethical governance are able to provide new understanding on aspects of the mind, such as the notions of self, agency, and responsibility.

Funding: This research was funded by the Institute of Law, Politics, and Development. Sant'Anna School of Advanced Studies, Pisa, Italy. Visiting and External Faculty Fellows 2020–2021 for the Strategic Project on "Governance for Inclusive Societies".

Conflicts of Interest: The author declares no conflict of interest.

References

1. Moor, J.H. WHAT IS COMPUTER ETHICS? *Metaphilosophy* **1985**, *16*, 266–275. [CrossRef]
2. European Group on Ethics in Science and New Technologies. *Ethical Aspects of ICT Implants in the Human Body*; EGE: Brussels, Belgium, 2005; p. 20.
3. Merkel, R.; Boer, G.; Fegert, J.; Galert, T.; Hartmann, D.; Nuttin, B.; Rosahl, S. *Intervening in the Brain. Changing Psyche and Society*; Springer: Berlin/Heidelberg, Germany, 2007.
4. Lucivero, F.; Tamburrini, G. Ethical monitoring of brain-machine interfaces. *AI Soc.* **2007**, *22*, 449–460. [CrossRef]
5. Steinert, S.; Friedrich, O. Wired Emotions: Ethical Issues of Affective Brain–Computer Interfaces. *Sci. Eng. Ethic.* **2020**, *26*, 351–367. [CrossRef] [PubMed]
6. Battaglia, F. The embodied self and the feeling of being alive. In *The Feeling of Being Alive*; Marienberg, S., Fingerhut, J., Eds.; Walter de Gruyter: Berlin, Germany; New York, NY, USA, 2012; pp. 201–221.
7. Battaglia, F.; Carnevale, A. Epistemological and Moral Problems with Human Enhancement. *J. Philos. Stud.* **2014**, *7*, 3–20. Available online: http://www.humanamente.eu/index.php/HM/article/view/111 (accessed on 18 January 2021).
8. Straub, J.; Sieben, A.; Sabisch-Fechtelpeter, K. Menschen besser machen. In *Menschen Machen. Die Hellen Und Die Dunklen SEITEN Humanwissenschftlicher Optimierungsprogramme*; Sieben, A., Sabisch-Fechtelpeter, K., Straub, J., Eds.; Transcript: Bielefeld, Germany, 2012; pp. 27–75.
9. Coeckelbergh, M. Drones, information technology, and distance: Mapping the moral epistemology of remote fighting. *Ethic- Inf. Technol.* **2013**, *15*, 87–98. [CrossRef]
10. Cockelbergh, M. *Human Being @Risk*; Springer: Berlin/Heidelberg, Germany, 2013.
11. Parens, E. (Ed.) *Enhancing Human Traits: Ethical and Social Implications*; Georgetown University Press: Washington, DC, USA, 1998.
12. Savulescu, J.; Bostrom, N. *Human Enhancement*; Informa UK Limited: London, UK, 2019; pp. 319–334.
13. Savulescu, J.; ter Meulen, R.; Kahane, G. (Eds.) *Enhancing Human Capacities*; Wiley-Blackwell: New York, NY, USA, 2011.
14. Kapp, E. Selections from Elements of a Philosophy of Technology. *Grey Room* **2018**, *72*, 16–35. [CrossRef]
15. Nagel, T. *Mind and Cosmos*; Oxford University Press: Oxford, UK, 2012.
16. Chalmers, D.J. *The Character of Consciousness*; Oxford University Press (OUP): Oxford, UK, 2010.
17. Jackson, F. Consciousness. In *The Oxford Handbook of Contemporary Philosophy*; Jackson, F., Smith, M., Eds.; Oxford University Press: New York, NY, USA, 2009; pp. 363–390.
18. Gallagher, S.; Zahavi, D. *The Phenomenological Mind*; Routledge: London, UK, 2020.
19. Darwall, S. *The Second-Person Standpoint*; JSTOR: New York, NY, USA, 2009.
20. Korsgaard, C.M.; O'Neill, O. *The Sources of Normativity*; Cambridge University Press (CUP): Cambridge, UK, 1996.
21. Scanlon, T.M. *What We Owe to Each Other*; Indiana University Press: Bloomington, IN, USA, 2000.
22. Craver, C.F. *Explaining the Brain*; Oxford University Press (OUP): Oxford, UK, 2007.
23. Jackson, F. What Mary Didn't Know. *J. Philos.* **1986**, *83*, 291. [CrossRef]
24. Davidson, D. Mental Events. In *Essays on Actions and Events*; Clarendon Press: Oxford, UK, 1980; pp. 207–224.
25. Haggard, P. Conscious intention and motor cognition. *Trends Cogn. Sci.* **2005**, *9*, 290–295. [CrossRef] [PubMed]
26. Gallagher, S. The Natural Philosophy of Agency. *Philos. Compass* **2007**, *2*, 347–357. [CrossRef]
27. Synofzik, M.; Vosgerau, G.; Newen, A. Beyond the comparator model: A multifactorial two-step account of agency. *Conscious Cogn.* **2008**, *17*, 219–239. [CrossRef] [PubMed]
28. Santello, M.; Bianchi, M.; Gabiccini, M.; Ricciardi, E.; Salvietti, G.; Prattichizzo, D.; Ernst, M.; Moscatelli, A.; Jörntell, H.; Kappers, A.M.; et al. Hand synergies: Integration of robotics and neuroscience for understanding the control of biological and artificial hands. *Phys. Life Rev.* **2016**, *17*, 1–23. [CrossRef] [PubMed]
29. Miller, L.E.; Fabio, C.; Ravenda, V.; Bahmad, S.; Koun, E.; Salemme, R.; Luauté, J.; Bolognini, N.; Hayward, V.; Farnè, A. Somatosensory Cortex Efficiently Processes Touch Located Beyond the Body. *Curr. Biol.* **2019**, *29*, 4276–4283.e5. [CrossRef] [PubMed]
30. Crane, T. *The Mechanical Mind*; Routledge: London, UK, 2017.

31. Kersten, J. Die maschinelle Person—Neue Regeln für den Maschinenpark? In *Roboter, Computer und Hybride. Was Ereignet Sich Zwischen Menschen und Maschinen*; TTN-Studien—Schriften aus dem Institut Technik—Theologie—Naturwissenschaften; Manzeschke, A., Karsch, F., Eds.; TTN-Studien: Baden-Baden, Germany, 2016; Volume 5, pp. 89–101.
32. Sloterdijk, P. Kränkung durch Maschinen. Zur Epochenbedeutung der neuesten Medizintechnologie. In *Nicht gerettet. Versuche nach Heidegger*; Sloterdijk, P., Ed.; Suhrkamp: Frankfurt, Germany; pp. 338–366.
33. Cassirer, E. Form and Technology. In *Ernst Cassirer on Form and Technology, Contemporary Readings*; Hoel, A., Folkvord, I., Eds.; Palgrave Macmillan: London, UK, 2012; pp. 15–53.
34. Pfeifer, R.; Bongard, J. *How the Body Shapes the Way We Think: A New View of Intelligence*; MIT Press: Cambridge, MA, USA, 2007.
35. Wallach, W.; Allen, C. *Moral Machines: Teaching Robots Right from Wrong*; Oxford University Press: Oxford, UK, 2009.
36. Fabris, A. *Ethics of Information and Communication Technologies*; Springer International Publishing: Heidelberg, Germany, 2018.

 *philosophies*

Review

Ethical Aspects of BCI Technology: What Is the State of the Art?

Allen Coin, Megan Mulder and Veljko Dubljević *

Department of Philosophy and Religious Studies, NC State University, Raleigh, NC 27695-8103, USA;
allen_coin@ncsu.edu (A.C.); mkmulder@ncsu.edu (M.M.)
* Correspondence: veljko_dubljevic@ncsu.edu

Received: 25 September 2020; Accepted: 21 October 2020; Published: 24 October 2020

Abstract: Brain–Computer Interface (BCI) technology is a promising research area in many domains. Brain activity can be interpreted through both invasive and non-invasive monitoring devices, allowing for novel, therapeutic solutions for individuals with disabilities and for other non-medical applications. However, a number of ethical issues have been identified from the use of BCI technology. In this paper, we review the academic discussion of the ethical implications of BCI technology in the last five years. We conclude that some emerging applications of BCI technology—including commercial ventures that seek to meld human intelligence with AI—present new and unique ethical concerns. Further, we seek to understand how academic literature on the topic of BCIs addresses these novel concerns. Similar to prior work, we use a limited sample to identify trends and areas of concern or debate among researchers and ethicists. From our analysis, we identify two key areas of BCI ethics that warrant further research: the physical and psychological effects of BCI technology. Additionally, questions of BCI policy have not yet become a frequent point of discussion in the relevant literature on BCI ethics, and we argue this should be addressed in future work. We provide guiding questions that will help ethicists and policy makers grapple with the most important issues associated with BCI technology.

Keywords: brain–computer interface (BCI); brain–machine interface (BMI); ethical; legal and social Issues (ELSI); neuroethics; narrative review

1. Introduction

Brain–Computer Interface (BCI) technology has been a promising area of research in recent decades, with advancements in the technology leading to a broadening of applications [1]. Researchers and clinicians are increasingly able to accurately interpret brain activity through both invasive (implanted) and non-invasive (outside the body) monitoring devices, allowing them to create better therapeutic solutions for patients suffering from disorders or diseases that inhibit their ability to interact with the world around them, e.g., patients suffering from the paralyzing locked-in syndrome who, with the use of a BCI device, are able to regain the ability to communicate. BCI technology is also being used for non-medical applications, such as gaming and human–technology interfaces. With this technology comes a number of ethical concerns that require consideration by all stakeholders involved (researchers, clinicians, patients and their families, etc.). Previous research into the ethics of BCI conducted by Burwell et al. [2] analyzed past publications that addressed ethical concerns associated with BCI technology. Burwell et al. identified common themes in the literature, including issues of responsibility for the consequences of BCI use; potential loss of autonomy, identity, and personhood as a result of BCI use; and security issues regarding the collection, analysis, and possible transmission of neural signals [2].

However, since Burwell et al. [2] conducted their original scoping review, there has been a rapid increase in the number of publications regarding BCI ethics. When conducting a literature search in

2020 using the same parameters that Burwell et al. used for their original search in 2016, we found that many additional and relevant articles [n = 34] had been published in the time since Burwell et al. conducted their study than had been published at any time before 2016, the last year included in the study [n = 42]. Additionally, there have been a number of advances in BCI technology in recent years, including commercial ventures that seek to utilize BCI in novel ways. One such example is the company Neuralink, founded by entrepreneur Elon Musk, which aims to achieve "a merger with artificial intelligence" [3]. There has been ample skepticism about Neuralink's goals and claims, with some referring to the company's public announcements and demonstrations as "neuroscience theater" [4]. Regardless of whether Neuralink's stated goals are feasible in the near-term future, the existence of commercial ventures like Neuralink in the BCI field certainly signals new areas of active development and may shed some light on where the technology could be heading.

One specific form of BCI development, Brain-to-Brain Interface (BBI), may lead to particularly novel social and ethical concerns. BBI technology combines BCI with Computer-to-Brain Interfaces (CBI) and, in newer work, multi-brain-to-brain interfaces—such as Jing et al.'s study [5]—real-time transfer of information between two subjects to each other has been demonstrated. Considering the rapid increase in publications about BCI ethics since 2016 and recent advances in the technology, a review of the state of the art of the ethical discussion of BCI is warranted. With these developments in mind, we review the academic discussion of the ethical implications of BCI in the last five years. Through this type of systematized qualitative analysis [6], we hope to provide a nuanced perspective on the complicated ethical implications of this technology and directions for its responsible development and use that will ensure it advances in an ethically sound manner.

In the following Background section, we will provide a detailed summary of Burwell and colleagues' findings before discussing our own research in the sections Materials and Methods, Results, Discussion, and Conclusion.

2. Background

In 2017, Burwell et al. published the first scoping review analyzing themes surrounding ethical issues in BCI. PubMed was used to find these articles using advanced searches combining various relevant terms (including Medical Subject Heading/MeSH terms) denoting "brain-computer interfaces" with keywords pertaining to ethics in general. From these articles, they identified narrower ethical concerns, such as "personhood," "stigma," and "autonomy." From an initial yield of around 100 documents, Burwell and colleagues selected 42 articles that met their inclusion criteria for this review. They included only papers that were written in English, presented discussions or empirical findings of the ethics of BCI, referred to human subjects, and considered BCI as technology that records data directly from the brain to a computer output. To provide direction for their study, Burwell et al. also consulted with four experts in the related research areas of clinical medicine, biomedical engineering, bioethics, and end-user perspectives.

After coding the selected articles, they found that most articles discussed more than one ethical issue, with the minority of the articles in question being empirical papers. The most frequently mentioned ethical issues included User Safety [57.1%, n = 24], Justice [47.6%, n = 20], Privacy and Security [45.2%, n = 19], and Balance of Risks and Benefits [45.2%, n = 19]. The authors focused primarily on the ethical issues of (i) User Safety, (ii) Humanity/Personhood, (iii) Stigma and Normality, (iv) Autonomy, (v) Responsibility, (vi) Research Ethics and Informed Consent, (vii) Privacy and Security, and (viii) Justice. Burwell and colleagues also briefly mentioned the less-cited concerns of military applications, enhancement and transhumanism, general societal impacts, and BCI technology regulation.

The issue of User Safety focused on potential direct physical harms to the user if the technology was to fail, an example being crossing the street with prosthetics when the BCI gives out. Ethicists that conducted research on this topic also discussed the unknown side effects of BCI, including the mental and physical toll of learning to use the technology.

The themes of Humanity and Personhood concerned whether the BCI would become part of the user's "body schema," and there was little uniformity in opinion on this issue among researchers. Some emphasized the fact that humans are already intricately linked to technology and that a BCI is no different, while one researcher went so far as to say we could evolve from *"homo sapiens"* to *"homo sapiens technologicus"* with technologies such as these [7]. BCI users interviewed in several qualitative empirical studies were discomforted by this possibility and distanced themselves from the idea of cyborgization.

The themes of Stigma and Normality addressed the BCI user's ability to influence or be influenced by the social stigma of disability. It is a possibility that individuals who feel stigmatized by a disability will consent to a BCI to counteract that. On the other hand, however, having a BCI could become stigmatized instead of the disability itself.

The issue of Responsibility was also salient in ethical discussions. If a negative action were to be carried out by someone using a BCI, would it be the user's fault or the fault of the technology? Many researchers claim that our legal system is not yet equipped to deal with this situation.

The issue of Autonomy refers to one's ability to act independently and of their own volition. If an action is only possible because of BCI technology, is that damaging to one's autonomy? On the other hand, autonomy may be increased on the basis that individuals living with BCI technologies may be able to do things on their own that once required constant supervision and assistance by a caregiver.

The nature of a BCI sending brain signals directly to a computer raises the possibility of hacking. This issue is connected to concerns regarding Privacy and Security. BCIs could also be used to extract information from a person such as their current mental state or truthfulness, or even be utilized to cause harm to the user or someone else.

Another major ethical issue identified was that of Research Ethics and Informed Consent. Many potential BCI users are people living in a locked-in state (almost complete paralysis), or with other conditions that limit their ability to give consent. Even if that is not the case, a severe disability may press an individual to consent to using BCI technology out of desperation without fully considering the risks.

The final main ethical concern discussed by Burwell et al. was Justice. Many researchers worry that the perspectives of those using BCIs are not fully considered in the research. There are also many questions pertaining to the fairness of providing only limited access to BCI technology. For instance, when experimental BCI studies are completed, do the participants get to keep the BCI? Should BCIs become widely available, there is also the concern that enhancement could potentially exacerbate existing inequalities between individuals and social strata, e.g., a scenario in which a BCI technology that provides cognitive or physical enhancement [8] is only available to the wealthy.

While it's not surprising to see many potential ethical issues and questions arising from use of a novel technology, what is surprising is the lack of suggestions to resolve them. One hope that Burwell and colleagues had for their scoping review was to facilitate informed recommendations for future research and development of neurotechnology. Our research, building upon the groundwork of Burwell and colleagues, sheds light on how the discourse on BCI ethics has evolved in the years since.

3. Materials and Methods

Building on prior work by Burwell et al., in April of 2020 we conducted a search using PubMed and PhilPapers in order to track academic discussion of this technology since 2016. The search terms were selected to mirror the search done by Burwell's original work on the subject. The following search queries were used.

PubMed: (("brain computer interface" OR "BCI" OR "brain machine interface" OR "Brain-computer Interfaces"[Mesh]) AND (("personhood" OR "Personhood"[Mesh]) OR "cyborg" OR "identity" OR ("autonomy" OR "Personal autonomy"[Mesh]) OR ("liability" OR "Liability, Legal"[Mesh]) OR "responsibility" OR ("stigma" OR "Social stigma"[Mesh]) OR ("consent"

OR "Informed Consent"[Mesh]) OR ("privacy" OR "Privacy"[Mesh]) OR ("justice" OR "Social Justice"[Mesh]))).

PhilPapers: ((brain-computer-interface|bci|brain-machine-interface)&(personhood|cyborg| identity|autonomy|legal|liability|responsibility|stigma|consent|privacy|justice)).

We sought to understand how the academic literature on the topic since 2016 addresses unique, new ethical concerns presented by emerging applications of BCI. While Burwell et al. identified 42 relevant articles published before 2016, our slightly modified search in 2020 using a similar methodology and exclusion/inclusion criteria yielded almost as many relevant articles discussing the ethics of BCI [n = 34] published since 2016. We used a limited, randomly selected sample [20.6%, n = 7] from the pool of 34 articles to identify trends and areas of concern or debate among researchers and ethicists, especially regarding topics like autonomy, privacy and security, and informed consent/research ethics. Additionally, while Burwell and colleagues only considered articles about BCI ethics specifying human subjects, we expanded our search and inclusion/exclusion criteria to include applications involving animals and other subjects, such as brain organoids [9].

Based on the abductive inference approach to qualitative research [10], we used the thematic framework developed by Burwell and colleagues to identify and map the overarching themes of ethical issues posed by BCIs (see Figure 1). The map identifies eight specific ethical concerns that define the conceptual space of the ethics of BCI as a field of research. Only one of the ethical concerns refers to physical factors specifically: User Safety. Two are explicitly about psychological factors: Humanity/Personhood and Autonomy; while the remaining five focus on social factors: Stigma and Normality, Responsibility and Regulation, Research Ethics and Informed Consent, Privacy and Security, and Justice.

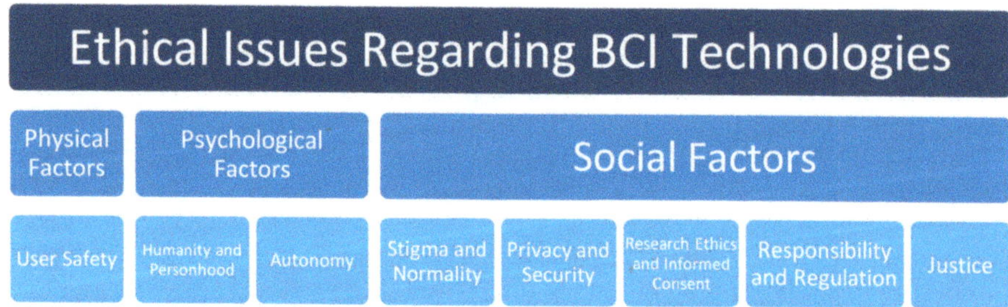

Figure 1. Overarching Themes in Brain–Computer Interface (BCI) Ethics.

4. Results

In the updated sample, we were able to locate mentions of all eight categories, with some ethical issues, such as Autonomy, being mentioned much more frequently [71.4%, n = 5] than others. While most articles discussed benefits in terms of the increases in autonomy and independence gained from using a BCI [11–14], the potential for autonomy to be compromised was also discussed. For example, Hildt [15] mentions the possibility of taking the information gained from BCI—or in this case, Brain-to-Brain Interface (BBI)—from the individual and using it without their consent or knowledge:

> "Participants in BBI networks depend heavily on other network members and the input they provide. The role of recipients is to rely on the inputs received, to find out who are the most reliable senders, and to make decisions based on the inputs and past experiences. In this, a lot of uncertainty and guessing will be involved, especially as it will often be unclear where the input or information originally came from. For recipients in brain networks, individual or autonomous decision-making seems very difficult if not almost impossible" [15] (p. 3).

Another frequently [57.1%, n = 4] discussed ethical issue was Humanity and Personhood, since BCIs could impact one's sense of self. In one specific study of BCI technology used in patients with epilepsy, there was a variety of resulting perspectives on sense of self, with some individuals saying that it made them feel more confident and independent, while others felt like they were not themselves anymore, with one patient expressing that the BCI was an " ... extension of herself and fused with part of her body ... " [11] (p. 90). Other articles more generally discussed the possibility of sense of self changing and the ways BCI technology could contribute to this. Sample and colleagues categorize three ways in which one's sense of self and identity could change: altering the users' interpersonal and communicative life, altering their connection to legal capacity, and by way of language associated with societal expectations of disability [14] (p. 2). Müller and Rotter argue that BCI technology constitutes a fusion of human and machine, stating that "the direct implantation of silicon into the brain constitutes an entirely new form of mechanization of the self ... [T]he new union of man and machine is bound to confront us with entirely new challenges as well" [13] (p. 4).

Research Ethics and Informed Consent was also a frequently [57.1%, n = 4] mentioned issue. The main consensus among the ethicists that discussed this was that it is very important to obtain informed consent and make sure that the subjects are aware of all possible implications of BCI technology before consenting to use it. Additionally, some ethicists warned against the possibility of exploiting potentially vulnerable BCI research subjects. As Klein and Higger note: "[t]he inability to communicate a desire to participate or decline participation in a research trial—when the capacity to form and maintain that desire is otherwise intact—undermines the practice of informed consent. Individuals cannot give an informed consent for research if their autonomous choices cannot be understood by others" [12] (p. 661).

User Safety was discussed as often as research ethics in the sample [57.1%, n = 4], with both psychological and physical harm being mentioned and explained as serious possibilities that need to be considered [13–15]. One article discussed the impacts of harm on the results of a BCI study, stressing the importance of stopping a clinical trial if the risks to the individual participants begin to outweigh the potential benefits to science [12].

Issues of User Safety led to discussions of Responsibility and Regulation when using BCI technology. While the term "regulation" was mentioned in several articles [57.1%, n = 4], only one went into significant detail about regulation in regard to BCIs specifically, discussing the issue of a "right to brain privacy," which can be understood in similar terms to existing privacy legislation, such as General Data Protection Regulation (GDPR) in the European Union or the Health Insurance Portability and Privacy Act (HIPAA) in the United States, to regulate the information gathered in BCI use [15] (p. 2). This was also the only time regulation was used in a legal sense for the technology, as opposed to regulating who it should be used for [11]. Responsibility was also mentioned multiple times [71.4%, n = 5], but again, only one article went into detail about who would be responsible for potentially dangerous or illegal BCI technology uses [16]. One discussed the "responsibility" of the research being divided among members of the research team [12], which was not Burwell et al.'s original meaning of the category Responsibility.

Privacy and Security concerns were mentioned somewhat less frequently [42.8%, n = 3] but still discussed in depth. Three articles [13,15,17] talked about the risks of extracting private information from people's brains and using it without their knowledge or consent, which is a significant concern for BCI technologies. Müller & Rotter connected this issue to User Safety, arguing that the increased fidelity of BCI data yields inherently more sensitive data, and that the "impact of an unintended manipulation of such brain data, or of the control policy applied to them, could be potentially harmful to the patient or his/her environment" [13] (p. 4).

Justice was also mentioned infrequently [28.6%, n = 2], with the main idea being inequality and injustice within the research. These discussions often related back to the aforementioned questions of when the trials would end and if the participants would get to subsequently keep the BCI technologies [12].

The final two social factors mentioned were Stigma and Normality [28.6%, n = 2] and Societal Implications [28.6%, n = 2]. Stigma was mainly discussed from the perspective of the device itself having a negative stigma around it, and the device itself being what is stigmatizing about the individual [12]. However, it was also mentioned that perhaps universalizing the technology instead of only targeting it toward a group that is considered "disabled" could reduce or eliminate stigma [14]. Societal Implications were discussed from several standpoints. One take was the possibility of BCI being used as a sort of social network instead of just for therapy for disabled individuals [15]. Another position was that—since society tends to universalize technology so that it is used by nearly everyone and hard to function without, e.g., cellphones—the use of BCI technology may become a " … precondition to the realization of personhood" [14]. This article also discussed the potential for BCI technology to reshape how society perceives disability.

Similar to Burwell et al.'s findings, Military Use and Enhancement were both mentioned rarely [14.3%, n = 1], and neither category was discussed any further than using it as an example for potential BCI use [15]. For the more prominent categories, the distribution can be seen in Table 1.

Table 1. The distribution of overarching themes of BCI ethics.

	Burwell et al.'s Distribution out of 42 Selected Papers (2016)	Our Distribution out of a Sample of 7 Selected Papers (2020)
User Safety	57.1%	57.1%
Humanity and Personhood	35.7%	57.1%
Autonomy	28.6%	71.4%
Stigma and Normality	26.2%	28.6%
Privacy and security	45.2%	42.8%
Research Ethics and Informed Consent	33.3%	57.1%
Responsibility and Regulation	31.0%	71.4%
Justice	47.6%	28.6%

5. Discussion

While there have been notable advancements in BCI and BBI technology and the body of literature of the ethical aspects of BCI technology has grown substantially since the original publication of Burwell and colleagues' research, our findings suggest that the original taxonomy developed by Burwell and colleagues remains a useful framework for understanding the body of literature specifically on the social factors of the ethics of BCI. We can use this taxonomy, with some slight modifications, which we outline below, to understand how the body of literature on the ethics of BCI is grappling with ethical issues arising from the applications of this rapidly advancing technology. Articles published since 2016 still mostly conform to the taxonomy and can be categorized using it in future iterations of the scoping review methodology.

There are, however, some areas within the growing body of literature on BCI ethics that have arisen since the original research was published that need to be incorporated into the taxonomy. We recommend the following modifications to the conceptual mapping outlined in Figure 1. First, expanding the discussion of the physical (e.g., harms to test animals) and psychological (e.g., radical psychological distress) effects of BCI technology. The publicly available information on commercial BCI endeavors (such as Neuralink) frequently mentions experiments with increasingly complex and even sentient animals, such as Neuralink's demonstration of their technology on live pigs [4]. The lack of ethical scrutiny of these studies is an essential cause for concern [18]. Thus, ethical discussions should be expanded to include public awareness of private industry research into BCI using animals. Secondly, while the risks of physical harm from BCI are fairly well-understood and covered in the literature, further research is needed to understand emerging psychological factors in BCI ethics, examining how human–AI intelligence symbiosis, brain-to-brain networking, and other novel applications of the technology [2] may affect psychological wellbeing in humans. For instance, in the interview study by

Gilbert and colleagues, one patient mentioned that "she was unable to manage the information load returned by the device," which led to radical psychological distress [11] (p. 91).

Going forward, it is imperative to expand on the connection between ethics and policy in discussions of BCI technology and conduct more empirical studies that will help separate non-urgent policy concerns, which are based on not-yet attained effects of BCI, from the more urgent concerns based on the current state of science in regards to BCI technology. In this, we echo Voarino and colleagues [19] in stating that we must advance the discussion from merely mapping ethical issues, into an informed debate that explains which ethical concerns are high priority, which issues are moderately important, and what constitutes a low priority discussion of possible future developments.

That said, it is important to make sure that the ethics literature keeps pace with engineering advances and that policy does not lag behind. In that vein, following Dubljević [20], we propose that the key ethical question for future work on BCI ethics is:

> What would be the most legitimate public policies for regulating the development and use of various BCI neurotechnologies by healthy adults in a reasonably just, though not perfect, democratic society?

Additionally, we need to distinguish between ethical questions regarding BCI technology that ethicists and social scientists can answer for policy makers and those that cannot be resolved even with extensive research funding [21]. Therefore, following Dubljević and colleagues [22], we posit that these four additional questions need to be answered to ensure that discussions of BCI technology are realistic:

1. What are the criteria for assessing the relevance of BCI cases to be discussed?
2. What are the relevant policy options for targeted regulation (e.g., of research, manufacture, use)?
3. What are the relevant external considerations for policy options (e.g., international treaties)?
4. What are the foreseeable future challenges that public policy might have to contend with?

By providing answers to such questions (and alternate or additional guiding questions proposed by others), ethicists can systematically analyze and rank issues in BCI technology based on an as-yet to be determined measure of importance to society. While we have not completed such analyses yet, we do provide a blueprint above, based on conceptual mapping and newly emerging evidence, of how this can be done.

6. Conclusions

This article builds on, and updates, previous research conducted by Burwell and colleagues to review relevant literature published since 2016 on the ethics of BCI. Although that article is now somewhat outdated in terms of specific references to and details from the relevant literature, the thematic framework and the map we created—with the eight specific categories that it provides—and the nuanced discussion of overarching social factors have withstood the test of time and remain a valuable tool to scope BCI ethics as an area of research. A growing body of literature focuses on each of the eight categories, contributing to further clarification of existing problems. BCI ethics is still in its early stages, and more work needs to be done to provide solutions for how these social and ethical issues should be addressed.

Despite seeing the significance of these eight categories continue into more recent research, it is worth noting that we found that the distribution of the eight categories was different in recent years, compared with the distribution previously identified by Burwell and colleagues in the literature published before 2016. For instance, among our sample of articles, we found that Autonomy was mentioned most frequently [71.4%, n = 5] along with Responsibility and Regulation [71.4%, n = 5], with Research Ethics, User Safety and Humanity and Personhood each discussed in 4 out of 7 [57.1%] of the articles in the sample. However, despite Responsibility and Regulation being mentioned in five out of the seven papers, it was only discussed at length in one. None of these categories were among

Burwell and colleagues' top four most frequently mentioned (see Table 1). It seems that while the eight issues mapped are still ethically significant with regards to BCI research, the emphasis among them may be shifting toward concerns of psychological impact.

On that note, psychological effects (e.g., radical psychological distress) need to be carefully scrutinized in future research on BCI ethics. Additionally, one aspect that was not explicitly captured in the original thematic framework or the map we reconstructed from it is physical harm to animals used in BCI experimentation [18]. Finally, more detailed proposals for BCI policy have not yet become a frequent point of discussion in the relevant literature on BCI ethics, and this should be addressed in future work. We have provided guiding questions that will help ethicists and policy makers grapple with the most important issues first.

Author Contributions: A.C. contributed with Conceptualization, Data curation, Formal analysis, Methodology, Project administration, Writing—original draft, Writing—review & editing. M.M. contributed with Data curation, Writing—original draft, Writing—review & editing. V.D. contributed with Conceptualization, Formal analysis, Methodology, Project administration, Supervision, Writing—original draft, Writing—review and editing. All authors have read and agreed to the published version of the manuscript.

Funding: This research received no external funding.

Acknowledgments: The authors would like to thank for their valuable discussion and feedback the members of the NeuroComputational Ethics Research Group at NC State University—in alphabetical order, Elizabeth Eskander, Anirudh Nair, Sean Noble, and Abigail Presley. Additionally, the authors would like to thank Joshua Myers (NC State University) for his assistance with the early stages of this paper.

Conflicts of Interest: The authors declare no conflict of interest.

References

1. Ramadan, R.A.; Vasilakos, A.V. Brain computer interface: Control signals review. *Neurocomputing* **2017**, *223*, 26–44. [CrossRef]
2. Burwell, S.; Sample, M.; Racine, E. Ethical aspects of brain computer interfaces: A scoping review. *BMC Med. Ethics* **2017**, *18*, 1–11. [CrossRef] [PubMed]
3. Marsh, S. Neurotechnology, Elon Musk and the Goal of Human Enhancement. In *The Guardian*. Available online: https://www.theguardian.com/technology/2018/jan/01/elon-musk-neurotechnology-human-enhancement-brain-computer-interfaces (accessed on 1 January 2018).
4. Regalado, A. Elon Musk's Neuralink Is Neuroscience Theater. In *MIT Technology Review*. Available online: https://www.technologyreview.com/2020/08/30/1007786/elon-musks-neuralink-demo-update-neuroscience-theater/ (accessed on 30 August 2020).
5. Jiang, L.; Stocco, A.; Losey, D.M. BrainNet: A multi-person brain-to-brain interface for direct collaboration Between Brains. *Sci. Rep.* **2019**, *9*, 6115. [CrossRef] [PubMed]
6. Saigle, V.; Dubljević, V.; Racine, E. The impact of a landmark neuroscience study on free will: A qualitative analysis of articles using Libet et al.'s methods. *AJOB Neurosci.* **2018**, *9*, 29–41. [CrossRef]
7. Zehr, E.P. The potential transformation of our species by neural enhancement. *J. Mot. Behav.* **2015**, *47*, 73–78. [CrossRef] [PubMed]
8. Coin, A.; Dubljević, V. The authenticity of machine-augmented human intelligence: Therapy, enhancement, and the extended mind. *Neuroethics* **2020**. [CrossRef]
9. Hyun, I.; Scharf-Deering, J.C.; Lunshof, J.E. Ethical issues related to brain organoid research. *Brain Res.* **2020**, *1732*, 146653. [CrossRef] [PubMed]
10. Timmermans, S.; Tavory, I. Theory construction in qualitative research: From grounded theory to abductive analysis. *Sociol. Theory* **2012**, *30*, 167–186. [CrossRef]
11. Gilbert, F.; Cook, M.; O'Brien, T.; Illes, J. Embodiment and estrangement: Results from a first-in-human "intelligent BCI" trial. *Sci. Eng. Ethics* **2019**, *25*, 83–96. [CrossRef] [PubMed]
12. Klein, E.; Higger, M. Ethical considerations in ending exploratory brain–computer interface research studies in locked-in syndrome. *Camb. Q. Healthc. Ethics* **2018**, *27*, 660–674. [CrossRef] [PubMed]
13. Müller, O.; Rotter, S. Neurotechnology: Current developments and ethical issues. *Front. Syst. Neurosci.* **2017**, *11*, 93. [CrossRef] [PubMed]

14. Sample, M.; Aunos, M.; Blain-Moraes, S.; Bublitz, C.; Chandler, J.A.; Falk, T.H.; McFarland, D. Brain–Computer interfaces and personhood: Interdisciplinary deliberations on neural technology. *J. Neural Eng.* **2019**, *16*, 063001. [CrossRef] [PubMed]

15. Hildt, E. Multi-person brain-to-brain interfaces: Ethical issues. *Front. Neurosci.* **2019**, *13*, 1177. [CrossRef] [PubMed]

16. Nasuto, S.J.; Hayashi, Y. Anticipation: Beyond synthetic biology and cognitive robotics. *BioSystems* **2016**, *148*, 22–31. [CrossRef] [PubMed]

17. Agarwal, A.; Dowsley, R.; McKinney, N.D.; Wu, D.; Lin, C.T.; De Cock, M.; Nascimento, A.C. Protecting privacy of users in brain-computer interface applications. *Trans. Neural Syst. Rehabil. Eng.* **2019**, *27*, 1546–1555. [CrossRef] [PubMed]

18. *Neuroethics and Nonhuman Animals*; Johnson, S.L.; Fenton, A.; Shriver, A. (Eds.) Springer: Berlin/Heidelberg, Germany, 2020.

19. Voarino, N.; Dubljević, V.; Racine, E. tDCS for memory enhancement: A critical analysis of the speculative aspects of ethical issues. *Front. Hum. Neurosci.* **2017**. [CrossRef] [PubMed]

20. Dubljević, V. *Neuroethics, Justice and Autonomy: Public Reason in the Cognitive Enhancement Debate*; Springer: Berlin/Heidelberg, Germany, 2019.

21. Parens, E.; Johnston, J. Does it make sense to speak of neuroethics? *EMBO Rep.* **2007**, *8*, S61–S64. [CrossRef] [PubMed]

22. Dubljević, V.; Trettenbach, K.; Ranisch, R. The socio-political roles of Neuroethics and the case of Klotho. *AJOB Neurosci.* **2021**, in press.

Publisher's Note: MDPI stays neutral with regard to jurisdictional claims in published maps and institutional affiliations.

 *philosophies*

Article

The Ethics of Genetic Cognitive Enhancement: Gene Editing or Embryo Selection?

Marcelo de Araujo [1,2]

1   Faculty of Law, Federal University of Rio de Janeiro, Rio de Janeiro 20211-340, Brazil;
    marceloaraujo@direito.ufrj.br or marcelo.araujo@uerj.br
2   Department of Philosophy, State University of Rio de Janeiro, Rio de Janeiro 20550-900, Brazil

Received: 1 July 2020; Accepted: 31 August 2020; Published: 3 September 2020

Abstract: Recent research with human embryos, in different parts of the world, has sparked a new debate on the ethics of genetic human enhancement. This debate, however, has mainly focused on gene-editing technologies, especially CRISPR (Clustered Regularly Interspaced Short Palindromic Repeats). Less attention has been given to the prospect of pursuing genetic human enhancement by means of IVF (In Vitro Fertilisation) in conjunction with in vitro gametogenesis, genome-wide association studies, and embryo selection. This article examines the different ethical implications of the quest for cognitive enhancement by means of gene-editing on the one hand, and embryo selection on the other. The article focuses on the ethics of cognitive enhancement by means of embryo selection, as this technology is more likely to become commercially available before cognitive enhancement by means of gene-editing. This article argues that the philosophical debate on the ethics of enhancement should take into consideration public attitudes to research on human genomics and human enhancement technologies. The article discusses, then, some of the recent findings of the SIENNA Project, which in 2019 conducted a survey on public attitudes to human genomics and human enhancement technologies in 11 countries (France, Germany, Greece, the Netherlands, Poland, Spain, Sweden, Brazil, South Africa, South Korea, and United States).

Keywords: gene editing; embryo selection; CRISPR; cognitive enhancement; assisted reproductive technologies (ART); public opinion; in vitro gametogenesis (IVG); genome-wide association studies (GWAS)

---

1. Introduction

The conceptualization of the human body as a complex machine is not new among philosophers and scientists. Descartes, Hobbes, and La Mettrie, for example, are commonly associated with the "mechanical philosophy" of the 17th and 18th centuries [1–3]. But it was not until the first half of the 20th century that the mechanical conception of the human body took a different turn: If the human body is a complex machine, maybe the machine could be re-built and improved. This idea emerged in the aftermath of the World War I, not as a matter of philosophical inquiry, but as a matter of necessity and as a problem for policymakers.

As thousands of soldiers returned from the front with missing limbs, engineers and physicians had to work together in order to devise prosthetics limbs that might enable veterans to get back to work. A booklet published by the Red Cross in 1918, for instance, was entitled *Reconstructing the Crippled Soldier* [4]. The booklet contains several pictures of men using prosthetic limbs that functioned as tools such as pliers and hammers. The governments, especially in Germany, were eager to convince the new working force that the "scientific prostheses" [5]—as they were known—would make the disabled men even better and more productive than before [6–8]. Not everyone was convinced, though. In a short essay called "The Prosthetic Economy," published in 1920, the Austrian artist Raoul Hausmann sarcastically suggested that since the "prosthetic limb never gets tired," a "25-hour workday" should be adopted. He also went on to say: "The prosthetic man is therefore a better man, raised to a superior class thanks to the world war" [9]. The philosophical discussion on the ethics of human enhancement, as a topic for academic research, is rather recent, but the ethical concerns over the attempt to re-engineer the human body for the purpose of enhancement is not. It harks back to a broad societal debate that took place during the interwar years [6–8,10].

Now, 100 years later, policymakers, again, have to deal with technologies that have the potential to improve the human body. The conceptualization of the human body as a machine remains in place, but this time the machine is seen in a different light. As the human brain (wrongly or not) is increasingly compared to a supercomputer, and the whole human genome has been sequenced and represented as a six-billion string of letters (A, T, C, and G), the mechanical conception defended in the past has given rise to a new image of the human body, to a vision of human beings not so much as composed of mechanical parts, but as a machine that can be understood at a molecular level, as a source-code that can be read and re-written by means of gene-editing tools. In this article, I intend to focus on the more recent development: I will focus on genetic engineering, rather than on prosthetic limbs, as a means to enhance the human body.

The philosophical debate on the ethics of human enhancement involves a distinction between two basic ideas, which have not always been carefully kept apart. One concerns the *means* of enhancement, the other concerns the *human capacity* to be enhanced. These are, for short, the *"enhancement by which means question"* and the *"enhancement of what question."* One can defend a pro-enhancement attitude as regards, for instance, the use of drugs such as modafinil to boost one's cognitive functions, but argue against the use of drugs for the purpose of enhancing one's physical capacities in professional sports. One can also have a pro-enhancement stance as regards the use of drugs to improve cognitive and physical capacities, but argue that human moral capacities should not be modulated by means of drugs. Or one can alternatively advance a broader pro-enhancement attitude as regards the use of drugs for improving one's cognitive, physical, and moral capacities, but still strongly oppose the use of genetic engineering to promote the enhancement of any human capacities. It is possible to think of some other combinations between the *means* of enhancement, such as prosthetic limbs, deep brain stimulation, or surgeries, and *human capacities* such as creativity, the propensity to form long-lasting affective bonds with other persons, or aesthetic appearance. Thus, compelling arguments for the use of one specific method of enhancement, targeted at one or more specific human capacities, may not be equally compelling as regards other means of enhancement, targeted at other human capacities. The attempt to advance an all-encompassing approach to the ethics of human enhancement, on the other hand, can obscure important issues or conflate problems that should be dealt with separately. In this article, I intend to examine the use of genetic engineering as a means of pursuing cognitive enhancement. "Genetic engineering" itself is too broad a concept, for it can refer to technologies that, as I intend to show later, give rise to different ethical questions. I will focus here on two such technologies, namely gene editing (henceforth GE) and embryo selection (henceforth ES).

In addition to the *"enhancement by which means question"* and the *"enhancement of what question,"* two further questions must be addressed in the human enhancement debate. One is the *"enhancement for whom question."* This is especially important in the case of genetic engineering as a means of cognitive enhancement, for the persons who are going to be affected by the enhancement procedure do not yet exist. This means that they cannot give their consent to the use of techniques that will affect not only their personal identities as bearers of certain human capacities, but will ultimately also determine whether or not they are going to come into existence. This, as I intend to show later, raises ethical questions that do not emerge when we deploy other methods of human enhancement such as, for instance, drugs or prosthetic limbs.

Finally, a perennial question in the ethics of human enhancement debate concerns our very understanding of what human enhancement is about: *What is human enhancement* and how does it differ from *treatment*? In this article, I will not try to draw a clear-cut line between enhancement and treatment, which could be applied to all cases involving the quest for human enhancement. Neither shall I attempt to advance a definition of human enhancement that covers every case that has been described in the philosophical literature as instances of human enhancement [11].[1] For reasons I am going to spell out later, I suggest we should rather speak of "family resemblance," as Wittgenstein used the expression in the *Philosophical Investigations*, instead of proposing a very comprehensive definition of human enhancement [12].

I will speak here of "genetic cognitive enhancement" (GCE) to refer to GE and ES as procedures that, in the future, might enable prospective parents to increase the odds of having a child with increased intelligence. To the extent that one has increased intelligence, one can on average perform better in a wide range of cognitive domains such as, for instance, information processing speed, working memory, executive function, episodic memory, sustained attention, selective attention, etc. [13] (pp. 11–12), [14] (p. 279), [15] (pp. 53–54). One can of course have better cognitive performance without having increased intelligence in the sense I will focus on in this article. One can, for example, use some artificial intelligence device in order to improve information processing speed. One can even suggest that the device itself is an extension of the human mind, whether or not it is internal to the human body [16], (for Reference [17] see pp. 29–44). The device, however, cannot be genetically inherited. Although different kinds of technologies can be deployed to improve cognitive performance, I will focus here only on GCE.

The question, then, is whether pursuing GCE, whether through GE or ES, is a morally acceptable goal. Different moral theories and philosophical traditions, and even theological ideas, have been deployed in the philosophical debate on the ethics of human enhancement. On the other hand, far less attention has been given to public attitudes related to the development of human enhancement technologies. Some recent surveys on the public perception of human enhancement technologies and human genomics either focus too narrowly on a single country (United States) or on a single region (Europe), or they focus instead too narrowly on public perception of one specific technology (CRISPR, Clustered Regularly Interspaced Short Palindromic Repeats), and not on the public perception of human enhancement technologies at large [18–21]. In this article, I intend to focus on two recent surveys conducted by the SIENNA Project, of which the author of this article is a partner.[2] The SIENNA Project has advanced probably the most comprehensive account of public attitudes towards human genomics and human enhancement to date. The SIENNA Project has also delivered a corresponding account of public perception of artificial intelligence and robotics, which I will not take into consideration in this article. The surveys were conducted in 11 countries, spanning across different economic, cultural, and geographical landscapes (France, Germany, Greece, the Netherlands, Poland, Spain, Sweden, Brazil, South Africa, South Korea, and United States). The researchers employed the CATI method (computer assisted telephone interviewing) to interview one thousand respondents in each country [22,23].

2. Therapeutic and Non-Therapeutic Uses of Assisted Reproductive Technologies (ART)

Assisted reproductive technologies (henceforth ART) have been developing at a rapid pace over the last decades. These technologies include, for instance:

- Contraceptive pills
- IVF (in vitro fertilization)
- PGD (preimplantation genetic diagnosis)
- Egg-freezing (oocyte cryopreservation)

---

[1] For a recent review of the literature on human enhancement, see, e.g., Jensen and Nagel, 2018 [11].

[2] The SIENNA Project (stakeholder-informed ethics for new technologies with high socio-economic and human rights impact) has received funding under the European Union's H2020 research and innovation programme under grant agreement No 741716. Kantar Public was commissioned by the University of Twente on behalf of the SIENNA project to prepare the surveys. Web: https://www.sienna-project.eu/.

- CRISPR (clustered regularly interspaced short palindromic repeats)
- In vitro gametogenesis (IVG)
- Time-lapse analysis of embryos
- Genome-wide association studies (GWAS)
- Artificial intelligence (AI)

Some of these technologies (CRISPR, GWAS, and AI) do not qualify as ART in their own right, but in combination with other technologies they constitute some of the most recent developments in the field of ART. Contraceptive pills (or other science-based contraceptive methods) qualify as ART to the extent that they provide a negative control over human reproduction. I call it "negative" because, in this case, one's primary intention is to avoid a pregnancy. One needs the technology, then, to avert the occurrence of a process that would otherwise naturally occur—or might naturally occur. Some technologies, on the other hand, provide a "positive" control over human reproduction. One needs the technology, then, to stimulate the occurrence of a process that, for some reason or other, would not otherwise naturally occur.

The positive control of human reproduction can be pursued for both therapeutic and non-therapeutic goals. The best known technology for the positive control of human reproduction is IVF. Human embryos generated by means of IVF do not have, though, equal chances of adhering to the uterus, or equal chances of developing into a healthy child. For this reason, the number of embryos generated in each IVF cycle is usually greater than the number of embryos eventually transferred into the uterus. Further development in the field of ART enabled the creation of the PGD, which is usually undertaken by collecting a few cells from a two- to five-day old embryo (blastocyt) for genetic analysis [24] (p. 34). PGD enable clinicians to screen embryos for health issues such as, for instance, Huntington disease, Marfan Syndrome, cystic fibrosis, and other known genetic disorders, before they can be transferred into the uterus. A new technique called time-lapse analysis has been recently employed to carry out a non-invasive assessment of patterns of development in embryos images in order to help clinicians to decide which embryos have the best prospect for transfer into the uterus [25–27]. And more recently still, time-lapse analysis has also been performed in combination with the use of artificial intelligence in order to further improve the assessment of embryo viability [28]. This procedure enhances the clinicians' ability to select an embryo that has more chances of adhering to the womb and developing into a healthy child.

Although most ART have been originally developed for medical reasons, it is clear by now that ART can be used for non-therapeutic purposes as well. Yet, in in some cases, a clear-cut distinction between "therapeutic" and "non-therapeutic" uses cannot be easily drawn. This is one of the reasons I will not propose a clear-cut criterion to distinguish enhancement from treatment in this paper. Contraceptive pills, for example, are a pharmaceutical product, but we do not ordinarily think of contraceptive pills as a sort of "medicine," for they are not usually meant to treat any disease. The IVF and PGD, on the other hand, can indeed be used to address human infertility, which the World Health Organization (WHO) understands to be "a disease of the reproductive system defined by the failure to achieve a clinical pregnancy after 12 months or more of regular unprotected sexual intercourse" [29] (p. 1522). But when IVF and PGD are deployed to detect genetic disorders in human embryos, it is not entirely clear whether these technologies are being used for therapeutic purpose or as a device for selective contraception. If some genetic disorder is detected, clinicians and prospective parents have to decide whether or not to transfer the embryo into the uterus [30]. Treating the embryo, though, is not an option, at least not on legally and ethically acceptable grounds, as I am going to explain later. Sometimes, on the other hand, a distinction between "therapeutic" and "non-therapeutic" is more apparent: "Social egg-freezing" and GCE are two such cases. In the remainder of this section, I would like to describe briefly the former, before examining the latter more carefully in the next sections.

A woman may decide to resort to egg-freezing (or oocyte cryopreservation) in order to have some of her egg cells retrieved and frozen during the early years of her adult life. This enables her, for instance, to focus on her professional career while she is young, or while she waits until she has met the partner with whom she wants to have a child. Later, after she has secured the material means to give her child the best start in life, or met a partner she considers suitable to her, one or more egg cells, which had been previously frozen and stored in a fertility clinic, may be fertilized. For this, the woman does not even have to have a male partner. She can simply order a vial of semen from a private sperm bank and have it delivered to a fertility clinic. By the time the egg cells are fertilized, the woman may be less fertile than she was earlier, when the egg cells were retrieved from her ovaries. But this does not mean that she has become ill later, as fertility naturally decreases with age. The procedure, thus, is not used to treat an illness. It could be described, rather, as a technology for the enhancement of human reproduction, even though fertility is not a human capacity usually taken into consideration in the human enhancement debate. In the current literature on oocyte cryopreservation, it has become usual to distinguish "social egg freezing" from "medical egg freezing." The latter is deployed for medical reasons, when a woman has to undergo, for instance, cancer treatment, which may compromise her fertility after the treatment. Social egg freezing, on the other hand, is undertaken for non-therapeutic reasons [31–35].

Now, further development of new technologies such as IVG and CRISPR has the potential to give rise to other non-therapeutic applications of ART, including human enhancement. IVG has never been used for the generation of human gametes intended to start a pregnancy. CRISPR, on the other hand, has been reportedly used for the purpose of human reproduction only once, in 2018 in China, under circumstances that the Chinese government considered illegal, and the international community recognized as ethically unacceptable. The question, then, is whether further development of IVG and CRISPR might give rise to circumstances in which their use as ART, whether for therapeutic or non-therapeutic goals, might be considered ethically acceptable.

3. IVG and CRISPR as a Means of GCE

Embryonic cells, in their early stages of development, can turn into any other types of cells. These cells are also known as stem cells. During the embryonic development, stem cells become specialized cells, such as nerve cells, heart cells, neurons, bone cells, and several other types of cells that make up the human body. Specialized cells are more commonly called somatic cells. Normally, somatic cells can only reproduce and become another somatic cell of the same type, but new technologies now allow researchers to turn somatic cells back into stem cells. Somatic cells that have been induced to behave like stem cells are called "induced pluripotent stem cells" (iPSCs). Because iPSCs can turn into virtually any other type of cell, they can also become reproductive cells or gametes. The generation of reproductive cells from iPSCs is known as in vitro gametogenesis (IVG), or sometimes also as "in vitro generated gametes," "artificial gametes," or "synthetic gametes" [36–43], (for Reference [44] see p. 4). IVG was carried out for the first time in Japan, in 2016, on mice. Researchers managed to generate several eggs cells from a small sample of mouse skin cells. The eggs cells were then fertilized in vitro, and the resulting embryos were used to start a pregnancy. The mice produced by the Japanese researchers were healthy and fertile [45]. Now, further research on IVG might perhaps enable clinicians to use this technology for the generation of human gametes in the future. The range of therapeutic and non-therapeutic applications would be enormous. But then, too, the number and scope of ethical concerns.

A man, for instance, could request the use of IVG to turn a tiny sample of his skin cells into egg cells and, thus, become a mother. In like manner, a woman could use IVG for the generation of sperm cells and, thus, become a father. Two men, or two women, could also have a child genetically related to them. Prospective parents who currently rely on IVF to start a pregnancy would also benefit from IVG. Instead of undergoing a series of hormone injections and a surgical procedure to obtain the eggs cells, the woman would only have to provide a small sample of her skin cells. This would make the entire IVF procedure far less invasive—and possibly less expensive—for the women [46] (p. 127).

IVG could also disrupt the market for human gametes. Some women who currently undergo the egg retrieval procedure do not intend to become mothers themselves. Their intention is, rather, to donate or sell some of their egg cells, which in turn can be used for research or for human reproduction. This practice, however, has raised some ethical concern relative to informed consent and exploitation of women [47], (for Reference [48] see p. 44). Men, too, can donate sperm cells for altruistic or pecuniary reasons [49–52], but prohibition of anonymous donation over the last years has led to a shortage of sperm cells in many sperm banks around the world. Recently, some bioethicists have suggested that the institutionalization of presumed consent for post-mortem sperm cells donation might be introduced as a solution for the problem relative to the current shortage of male gametes in sperm banks [53]. IVG of human cells, on the other hand, could reduce drastically the demand for donation of male and female reproductive cells, making access to human gametes much easier and cheaper for prospective parents and researchers.

Many sperm banks offer some information on the educational background or academic achievements of their individual donors. Prospective parents may use the information, then, as a rough predictor of the cognitive capacities of their offspring. But IVG, in conjunction with IVF and other technologies such as GWAS and AI, could provide prospective parents with a more reliable and effective method for assessing their chances of having a child with increased intelligence. The problem, though, is that the quest for GCE by means of IVG, in conjunction with IVG and GWAS, would require the generation of far more embryos than the amount of embryos that are currently generated in fertility clinics during every IVF cycle. The idea here is that the more embryos from the same parents one has, the more chances one also has of spotting one particular embryo with more odds of developing into a child with increased intelligence [39,40,46,54] (pp. 191–202), [55] (pp. 88, 93, 106–110).

Currently, PGD enables clinicians to screen embryos for some genetic diseases before implantation into the womb. Unviable embryos are discarded or, depending on local legislation, used for scientific research. But as genetic sequencing technologies becomes faster and cheaper, clinicians might be able to screen vast amounts of embryos from the same couple for both medical and non-medical issues. And as human intelligence itself is highly inheritable [56–58], clinicians could screen the embryos for this specific trait as well. There is not, though, one human gene for intelligence. It is believed that hundreds or thousands of different genes, scattered at different loci of the human genome, are involved in the expression of most psychological, behavioral, and physical human traits. It does not mean, of course, that genes alone determine all features one individual will ultimately have, for the expression of human traits very often depends on diverse degrees of interaction between genome and environment.

GWAS enable researchers to determine which regions of the human genome are associated with the expression of specific psychological, behavioral, or physical traits. This association can be measured and translated in terms of "polygenic scores," which is a probabilistic—thus not deterministic—estimation of the occurrence of a certain trait as the result of the influence of a large group of genes, [46] (p. 77). GWAS, thus, could help prospective parents and clinicians to select one embryo, amongst hundreds or thousands of embryos, with increased odds of developing into a child with higher intelligence. On the other hand, as Henry Greely points out, the vast amount of genetic data resulting from the assessment of polygenic scores could also lead prospective parents to make decisions that, in the end, might compromise the well-being of their child. How to decide, for instance, between $Embryo_1$ and $Embryo_2$ when $Embryo_1$ has $x\%$ of having $trait_1$ ($IQ > 100$), $y\%$ of developing $trait_2$ (minor propensity to obesity), and $z\%$ of developing $trait_3$ (propensity to depression), while $Embryo_2$ presents different polygenic scores for $trait_1$, $trait_2$, and $trait_3$? The decision becomes increasingly more difficult when the number of embryos to choose from is >100 and the number of eligible traits (both medically and non-medically relevant traits) is also very high [46] (pp. 193–196). In face of this "information overload"—as Sonia Suter aptly puts the problem—prospective parents would probably have to rely on the expertise of genetic counselors and computer algorithms to assess the polygenic scores their future offspring and adjust their preference for a specific trait like increased intelligence against medically relevant traits [46] (pp. 193–196), [54] (pp. 273–276). Once a decision has been made in favor of a specific embryo, the remaining embryos would have to be discarded.

Currently, depending on local legislation, unclaimed viable embryos can be "adopted." In 2017, for example, a woman in the United States gave birth to a baby that developed from an embryo that had been frozen for 24 years [59]. The mother was just over one year old when the embryo was generated. In adopting the unclaimed embryo, the woman became the gestational mother, and not the biological mother of the baby. Frozen embryos may remain unclaimed for several reasons: The couple may have divorced after the generation of embryos and, thus, they may have changed their minds as to become father or mother; or the father or mother (or both) may have died after the generation of the embryo; or the couple may have simply decided, after a successful gestation, not to use the remaining embryos for the generation of yet another child. Yet, even if many women would volunteer to adopt surplus embryos, a significant amount of embryos would still have to be discarded.

GCE by means of GE, on the other hand, would not require the generation of vast amounts of human embryos. Instead of selecting one among hundreds of embryos, clinicians could generate only one embryo by means of IVF—with or without the use of IVG. Then, the embryo could be edited with CRISPR to increase the chances that it will later develop into a child with increased intelligence. Thus, no embryo would have to be discarded. It must be emphasized, though, that at its current stage of development CRISPR is not considered a reliable kind of ART—neither for therapeutic nor non-therapeutic goals. Moreover, as mentioned earlier, thousands of genes are involved in the expression of human intelligence. Even if CRISPR becomes safe and reliable for therapeutic goals within the next years, it is still unclear whether it could be deployed for the purpose of GCE without the risk of off-target mutations and mosaicism [60]. Maybe other forms of human enhancement by means of GE will become acceptable before it can be used for cognitive enhancement, especially in areas where a clear-cut line between enhancement and treatment cannot be drawn.

Consider, for instance, the first actual use of CRISPR as a form of ART, which occurred in November 2018 in China. He Jiankui, a Chinese researcher, edited the genome of two twin girls at an early stage of embryonic development. He Jiankui's goal was to enable HIV positive men to generate children free of the HIV virus. The usual measure, in these cases, consists in treating the semen with a process known as sperm-wash before proceeding to IVF. This alone should have been enough to enable a man to generate a child uncontaminated by the HIV virus. However, He Jiankui's goal was not simply to guarantee the birth of an HIV-free child. His intention was to generate a child *immune* to the HIV virus. It is clear, then, that the procedure, in this case, was deployed for non-therapeutic reasons. It aimed at the genetic enhancement of two human beings. The whole procedure might be called "immune system genetic enhancement." Incidentally, some authors have speculated that the procedure might have also indirectly provided the twin girls with some degree of GCE [61,62].

The international scientific community vehemently repudiated He Jiankui's procedure, as it represented a clear violation of the ethical guidelines proposed by many international bodies over the last years [63–65]. However, the fact that the procedure was deployed as a means of human enhancement, rather than as a method of treatment for a genetic disorder, was not the main focus of criticism. Many articles, published shortly after He Jiankui announced the birth of the twin girls, did not even take notice of this aspect of his experiment. The main reason for criticism, as I have mentioned above, is the fact that CRISPR cannot yet be considered a safe and reliable kind of ART. It represents a threat not only to the well-being of children whose genome have been edited, but also to the well-being of their descendants.

Now, let us assume, for the sake of argument, that IVG and CRISPR will become available as reliable and safe kinds of ART within the next decades, and that further research on human genomics will provide solid knowledge on the genetics of human intelligence. Would there be good moral reasons, then, to favor one procedure or the other—ES or GE—for the purpose of GCE? Or should we entirely abandon the prospect of pursuing the ES and GE as kinds of ART, whether for therapeutic or non-therapeutic goals? In the next section, I will address these questions by comparing some of the ethical implication each of these technologies entails. Then I will focus on two reports recently published by the SIENNA Project on the public perception of scientific research on human enhancement technologies and human genomics.

4. Public Attitudes towards GCE

How should we assess the ethical implications resulting from the use of technologies such as IVG and CRISPR for the purpose of GCE? Philosophical investigation into the principles of morality often relies on or starts with what *we* understand by concepts such as duty, virtue, or morality. Consider, for instance, Kant's statement in the Preface to the *Groundwork of the Metaphysic of Morals*:

> *"In this work, I have adopted the method that is, I believe, most fitting if one wants to take one's route analytically from common cognition [gemeinen Erkenntnisse] to the determination of its supreme principle and in turn synthetically from the examination of this principle and its sources back to common cognition, in which we find it used"* [66] (p. 13).

We have to start somewhere, and our common understanding (or "common cognition") of some elementary moral ideas is a good starting point—indeed, perhaps the only one we have. But Kant also assumes that we all share a "common idea of duty and of moral laws" [66] (p. 7). The principles of morality, on this analysis, do not have to be agreed upon through a social contract, or through societal debate or similar decisional procedure. The principles have to be, rather, "brought to light" (*aufgeklärt*), for they already dwell human beings' "natural sound understanding" (*natürlichen gesunden Verstande*) [66] (p. 23). More often than not, though, *our* common understanding of moral concepts is *common* only to the extent that it is shared by some specific group of individuals, living in this or that century, in this or that country, as members of this or that social group. Our current understanding of moral duties and moral laws was certainly not shared, for instance, by Aristotle and other thinkers in the tradition of virtue ethics. And upon examination, we also realize that some ideas at the core of Kant's moral theory, such as the assumption that "ought implies can," are not shared by most people [67,68]. It might be argued, on the other hand, that Kant was not interested in a psychological account of our common understanding of moral concepts, but in a normative account. If someone, or even a large group of people, happen to lack "natural sound understanding," or does not recognize what the proper understanding of moral laws amounts to, that person or group of persons should be enlightened. On this account, public attitudes should not be taken into consideration in our analysis of the moral implications resulting from GCE, for moral reasoning itself cannot be enlightened by public opinion.

Philosophical investigation should certainly not pander to public opinion, as public opinion itself can be scrutinized, clarified, and corrected in the light of philosophical analysis, but neither should philosophical investigation underestimate the plurality of points of view at play in the moral debate on issues that matter for the public in general. This is especially important when philosophical expertise is deployed for the elaboration of public policies involving controversial issues, or for the regulation of technologies that may challenge our common understanding about what is ethically acceptable or morally deplorable. As Neil Levy aptly puts the problem:

> *"Moral expertise is not the exclusive preserve of moral philosophers; instead, it is a domain in which a multitude of thoughtful people outside the academy make important contributions. The arena in which moral debate occurs is, accordingly, not limited to the peer-reviewed journals. Instead moral debate also, and almost certainly more importantly, takes place in newspapers and on television; in novels and films: everywhere moral conflicts are dramatized and explored"* [17] (pp. 309–310).

Recall that the first broad societal debate on the ethical implications from the quest for human enhancement by means of new technologies, as I mentioned at the outset of this paper, was mainly driven by stakeholders such as former soldiers, policy-makers, physicians, engineers, and artists rather than by philosophers. The methodological challenge for the assessment of ES and GE as means of GCE consists, then, in our capacity to take into consideration a variety of points of views. From the point of view of a person who was generated by means of new kinds of ART, for instance, it may be important to know which procedure was used before his or her birth: ES or GE? If the embryo has been manipulated with CRISPR, the person has every reason to be grateful or, as the case may be, resentful for any changes that have been made in his or her genome. Suppose that the attempt to promote some cognitive enhancement by means of CRISPR has not been successful and that, for this reason, the person suffers from some minor cognitive disorder such as for instance dyslexia—a condition that makes learning to read more difficult. Let us call that person $Person_1$. $Person_1$ could hold her parents or the fertility clinic responsible for her suffering, for the learning disability compromises her chances of success in achieving goals she would like to pursue in life. After all, if the GE procedure had not been used, $Person_1$ would not have dyslexia. Now, let us suppose there is another person ($Person_2$) who also suffers from dyslexia and has the same degree of learning difficulty as $Person_1$. However, $Person_2$ was not submitted to GE at an embryonic stage of development. $Person_2$ was generated by means of a combination of IVG and IVF, along with hundreds of sibling embryos, and then selected for implantation into the uterus. All sibling embryos were discarded afterwards. $Person_2$ may feel as frustrated as $Person_1$, for the learning disability also compromises her chances of success in achieving goals she would like to pursue. But as some philosophers have noticed, unlike $Person_1$, $Person_2$ cannot simply say that some kind of wrongdoing was inflicted on her, or that she has the right to blame her parents or the fertility clinic for being a dyslectic person. For $Person_2$, the only alternative would be not to have been selected in the first place and, thus, never to have been born. Darek Parfit and Thomas Schwartz are among the first philosophers who drew attention to this kind of paradox [69–71]. Schwartz called it the "fallacy of beneficiary-conflation" [71] (p. 7), but it is better known as the "non-identity-problem," as Parfit called it early in the 1980s [70] (p. 359). It is a paradox because a person may apparently present good reasons for wishing that a certain course of actions had never occurred in the past, but without that particular course of actions, which she would prefer had never occurred, she herself would not exist. One might argue, however, that from a legal point of view and as a matter of social justice, we should treat $Person_1$ and $Person_2$ the same way: If $Person_1$ has a right to reparation due to the disorder caused by the GE procedure, then $Person_2$ should also have the same right, as both cases, at least in the tradition of common law, can be characterized as a tort of wrongful life [72,73].[3] A recent survey conducted online among 763 participants also suggests that most people believe that the "non-identity-problem" is not really relevant for an assessment of situations like this, involving the claims of $Person_1$ and $Person_2$ [74].

Alternatively, one might argue that in view of the ethical issues that underlie ES and GE, both procedures should be banned for the purposes of human reproduction. The only non-problematic form of human reproduction—one might suggest—is natural reproduction, that is sexual intercourse of a man and a woman, unaided by ART. However, banning the two procedures does not solve the problem, it just gives rise to a further ethical concern. Consider $Person_3$, afflicted by dyslexia just as $Person_1$ and $Person_2$. $Person_3$, unlike $Person_1$ and $Person_2$, was generated naturally—her conception and birth did not involve the use of ART. Does $Person_3$ have the right to blame her parents or clinicians for being a person who has dyslexia? The answer to this question will depend on the stage of technological development at the time that $Person_3$ was conceived, and on the availability of the relevant technologies to her parents. For $Person_3$, it could be frustrating to learn that her cognitive impairment could have been avoided by means of GE. From the perspective of $Person_3$, it would have been better if her parents had indeed resorted to GE—even if it had involved some reasonable degree of risk. ES, on the other hand, would not have solved *her* problem, as she would not exist if some other embryo, free of dyslexia, had been chosen in her place. The choice for natural reproduction, therefore,

---

[3] See e.g., Frasca [72] (p. 185): "The difference between wrongful birth and wrongful life is that in wrongful birth the parents bring the legal action and seek damages; whereas, in wrongful life it is the child who seeks damages and the child or the child's representative who brings the lawsuit." For a recent account of the legal issues surrounding the use of gene editing technologies within the context of American law, see Macintosh [73].

is not morally neutral. In circumstances where prospective parents have access to the relevant medical technologies, which allow them to have a child without a condition that, later, may compromise her well-being, the couple is morally accountable both for using the technology and for failing to use it. As Anthony Kenny puts the problem: "As technology increases our knowledge of evils and our power to remove them, it increases our responsibility for not removing them" [75] (p. 124).

For the time being, though, neither ES (in conjunction with IVG and GWAS) nor GE may be used as a kind of ART—whether for therapeutic or non-therapeutic goals. Leaving aside the question as to which technology will (from a purely technical point of view) mature first, there is some reason to believe that GCE by means of ES might become commercially available prior to GCE by means of GE. This would be a reason, then, to focus the ethical debate on ES rather than on GE.

One reason for the assumption that GCE by means of ES may become commercially available prior to GCE by means of GE is that some fertility clinics, especially in the United States, are now gradually advertising a wider range of medical conditions they can screen for [76–79]. Given the thin line between therapeutic and non-therapeutic uses of ART, it may be just a matter of time before intelligence becomes available for screening along with medical conditions such as diabetes, propensity to depression, or intellectual disability, even if the procedure is not overtly advertised as GCE, but rather as therapeutic measure to reduce the odds of having a child with some mental conditions that might compromise his or her well-being later in life. Some people might even argue that ES (whether or not in conjunction with IVG and GWAS) does not qualify as a human enhancement technology at all. Thus, from their perspective, they would be circumventing at least some of the ethical concerns involved in the quest for GCE. Their argument might also be explored by private companies and fertility clinics that cannot advertise their services as genetic enhancement because local legislation explicitly prohibits the pursuit of human enhancement at a genetic level. Whether they argue from genuine philosophical reasoning or from commercial opportunism is not the point. What matters is that they can argue that ES (unlike GE) does not do anything to the embryo. In the normal distribution curve for IQ scores, some people will naturally have IQ > 100 while others will have IQ < 100. GCE by means of ES consists in an attempt to optimize the chances of selecting an embryo that, among other things (e.g., having improved information processing speed, improved working memory, improved executive function, etc.) will develop into a child who will be in the former group. Thus, most definitions of "human enhancement," discussed in the philosophical literature, will not straightforwardly include ES as a means of human enhancement, for the "enhanced" person, in this case, has not been enhanced at all—i.e., the person herself has not been subjected to an enhancing process. The person, rather, developed from an embryo that had better odds of growing into a person with increased intelligence, quite independently of any modification in the embryo itself. If GCE requires a modification in the embryo, then ES clearly does not qualify as GCE. Yet, there is some reason to believe that GCE does not require embryo modification.

Some questions of social justice will emerge regardless of the procedure one chooses to increase the odds of having a child with increased intelligence. This is the reason why I have stated at the outset of this article that I would eschew the attempt to propose an all-encompassing definition of *human enhancement* and speak, instead, of "family resemblance." There are some relevant resemblances between ES and GE that enable us to consider both procedures as human enhancement technologies: (i) Both procedures enable prospective parents to improve the odds of having a child with increased intelligence, even if the child (or the embryo) is not modified when one opts for ES; and (ii) both procedures give rise to the same issues of social justice that require proper regulation from the relevant policy-makers. To see this, let us recall Person$_1$ and Person$_2$.

Person$_1$ was born by means of GE and Person$_2$ was born by means of ES. Let us suppose this time, though, that both procedures (GE and ES) were quite successful, as Person$_1$ and Person$_2$ have, among other things, an IQ of 145. Person$_1$ and Person$_2$ will be able to pursue goals that most people will not be able to pursue, at any rate not as effortlessly as Person$_1$ and Person$_2$. This in turn gives rise to questions of social justice: Is it fair that Person$_1$ and Person$_2$ have these advantages over most people, especially considering that neither Person$_1$ nor Person$_2$ did anything to deserve an IQ of 145?

Some influential theories of justice suggest it would not be fair. As John Rawls put the problem in *A Theory of Justice*:

> *"The unequal inheritance of wealth is no more inherently unjust than the unequal inheritance of intelligence. It is true that the former is presumably more easily subject to social control; but the essential thing is that as far as possible inequalities founded on either should satisfy the difference principle"* [80] (p. 245).

Rawls is not suggesting here, of course, that intelligence itself should be redistributed in line with the difference principle, but that the benefits that people obtain, thanks to their increased intelligence, should also benefit those who have been less fortunate in the natural lottery. After all, they have not deserved their talents—or lack of talents, as the case may be. It is a matter of pure chance that the natural lottery has allotted some people with more intelligence, or that some people were born into wealthier families.[4] But Rawls also suggests, in a another passage, that maybe intelligence, too, could be directly subject to social control: "In the original position, then, the parties want to insure for their descendants the best genetic endowment (assuming their own to be fixed)" [80] (p. 92). The caveat in parentheses at the end of the sentence may be an indication that Rawls himself did not believe that the genetic endowment of our descendants cannot be changed. Maybe it can (Rawls must have thought) and if it can, it should be subject to social control in line with principles of justice, just in the same way inheritance of wealth is. Rawls, understandably, does not develop this topic further, dismissing it as a "speculative and difficult matter" [80] (p. 92). But the prospect of using ES and GE as means of bequeathing our descendants a better genetic endowment makes this topic now less speculative, though certainly not less difficult, than it was in 1971, when *A Theory of Justice* was published [81–84].[5]

If ES and GE become commercially available within the next decades, prospective parents may suffer under the financial and emotional pressure to enhance their children, even if they would otherwise prefer not to enhance them. They may think, for instance, that it would be better to invest their savings in the education rather than in the genetic endowment of their future children. But if, on the other hand, most parents opt for the GCE, they may believe that a non-enhanced child will lag behind in the competition for goods such as university admittance or better positions in the job market. Now, the social pressure here is neither less nor more intense if prospective parents take into consideration that ES (unlike GE) will not modify the genome of their future offspring. Some prospective parents might perhaps favor GE over ES, if they believe, for instance, that abortion is morally wrong, for ES (unlike GE) will require the discarding of human embryos and this, too, may be perceived as equally wrong. But that will be a different kind of pressure: A pressure to favor GE over ES while under the social pressure to enhance their future offspring. From the perspective of policymakers, ES counts as GCE, for ES gives rise to the same questions of social justice that GE does. But policymakers cannot take into consideration here only philosophical arguments. This might lead them to believe, for instance, that the non-identity problem requires that in cases involving wrongful life only the demands of people who were born through GE are valid. This would clearly conflict with the perception that most people, to whom policymakers are accountable, have of the problem. One of the surveys recently conducted by the SIENNA Project revealed, for example, that, on average, 79% of people across 11 countries believe that prospective parents will feel "pressured to have genetic

---

4   See Rawls [80] (p. 87): "Those who have been favored by nature, whoever they are, may gain from their good fortune only on terms that improve the situation of those who have lost out. The naturally advantaged are not to gain merely because they are more gifted, but only to cover the costs of training and education and for using their endowments in ways that help the less fortunate as well. No one deserves his greater natural capacity nor merits a more favorable starting place in society. But, of course, this is no reason to ignore, much less to eliminate these distinctions. Instead, the basic structure can be arranged so that these contingencies work for the good of the least fortunate. Thus we are led to the difference principle if we wish to set up the social system so that no one gains or loses from his arbitrary place in the distribution of natural assets or his initial position in society without giving or receiving compensating advantages in return."

5   Some authors have been trying to address these difficulties over the last few years. See, e.g., [81–84].

testing done on their unborn baby," if genetic testing becomes increasingly common (see Scheme 1). This, too, raises questions of social justice, as many prospective parents may suffer emotionally or financially under the pressure to have their unborn children genetically tested. Some genetic testing may, for instance, reveal an underlying condition that affects not only the unborn baby, but also the living siblings (or the mother, or the father), who might perhaps prefer not to be informed about that particular condition [85]. It is a matter of societal debate to establish, then, at a national level, how policymakers are expected to address this question. No general philosophical moral principle can be deployed here without failing to do justice to public attitudes towards this problem.

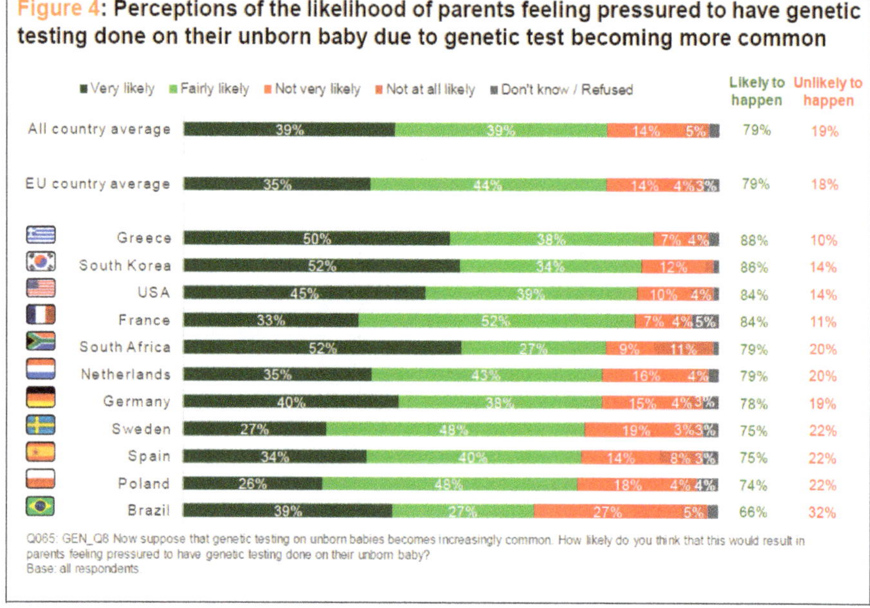

Scheme 1. "Perceptions of the likelihood of parents feeling pressured to have genetic testing done on their unborn baby due to genetic test becoming more common" [23] (p. 35).

One might suggest now that the quest for GCE has implications for social justice at both national and international levels. At a national level, for reasons I have suggested above, and which are not new in the ethics of human enhancement debate: The rich will bequeath to their children not only their wealth, but also better genetic endowment. Their children, thus, will be better off both financially and cognitively. Lack of appropriate regulation at national level, therefore, might aggravate existing social inequalities. At an international level, countries that are in a position to develop research on human enhancement technologies and human genomics are on average richer than countries that do not have the capability to advance new knowledge and new technologies in these areas. Human enhancement, therefore, might aggravate social inequality at a global level as well.

Yet, some of recent findings of the research conducted by the SIENNA Project seem to suggest a different picture. Scientifically and technologically developed countries are not especially supportive of research on human enhancement technologies. Quite the opposite, as one of the reports puts it: "Overall, perceptions of human enhancement technology were most positive in South Africa, Greece and Brazil and least positive in Germany, the USA and France" [22] (p. 6). As far as support for cognitive enhancement is concerned, a similar pattern emerged: Brazil, South Africa, and Poland were most supportive, while the Netherlands, Germany, and France were least supportive of cognitive enhancement technologies (see Scheme 2). As for research to better understand the genetics of human intelligence, Brazil, South Africa, and South Korea were most supportive, whereas the Netherlands, Germany, and France, again, were the least supportive countries in this area (see Scheme 3). As for research on human embryos in order to better understand "how to treat or cure health conditions," a different pattern emerged: Spain, the Netherlands, and Sweden were most supportive, whereas Greece, the United States, and Germany were least supportive (see Scheme 4).

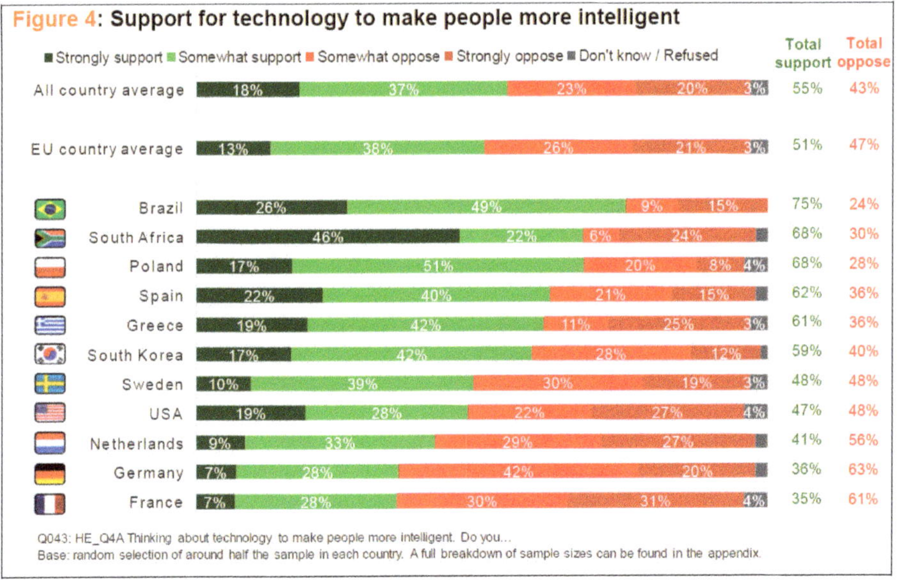

Scheme 2. "Support for technology to make people more intelligent" [22] (p. 29).

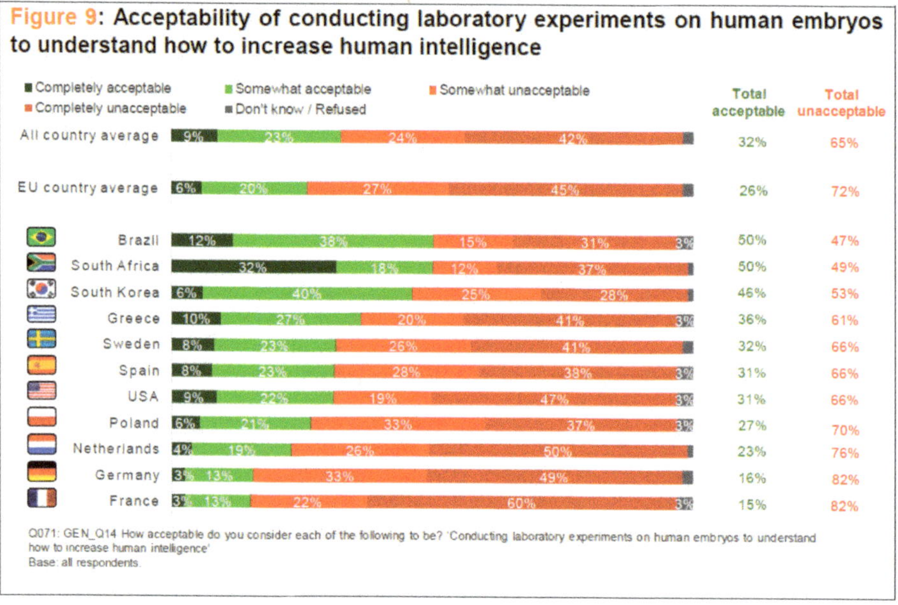

Scheme 3. "Acceptability of conducting laboratory experiments on human embryos to understand how to increase human intelligence" [23] (p. 44).

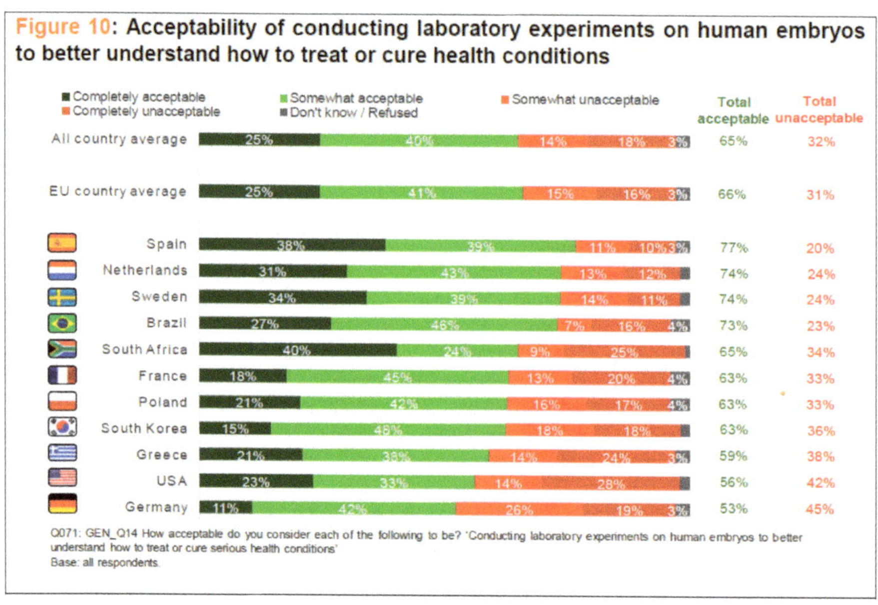

Scheme 4. "Acceptability of conducting laboratory experiments on human embryos to better understand how to treat or cure health conditions" [23] (p. 45).

One possible implication of these findings is that the unregulated use of GCE technologies might perhaps lead to a greater increase of social inequality at a national level rather than at an international level. Countries such as the United States might perhaps even have more demand for GCE abroad, from countries that are more supportive of human enhancement technologies, than internally. But this tendency will also depend, among other factors, on national laws for the importation of genetic technologies and regulation for the range of procedures that fertility clinics are allowed to perform [86].[6] This is a topic, though, I cannot delve into in this article, even recognizing that it is worth pursuing for a broader understanding of the ethics of GCE at an international level. My intention here was rather to call attention to the worldwide plurality of public attitudes towards human genomics and human enhancement, as recently surveyed by SIENNA Project, and suggest that these attitudes, too, should be taken into consideration in the debate on the ethics of human enhancement.

5. Conclusions

Even prior to the emergence of CRISPR, around 2012, philosophers and bioethicists had been discussing the ethical implications of the quest for GCE. However, most contributions in this area have focused on gene-editing and germline modification. Far less attention has been given to GCE by means of ES. Recent developments in human genomics, including the further development of GWAS and the prospect of using IVG for the generation of human gametes in the future, open up new possibilities for parents who would like to have a child with increased intelligence. In assessing the ethics of GCE, philosophers should be cautious not to rely uncritically on *our* moral intuitions, for human attitudes to human enhancement technologies and research on the genetics of human intelligence vary greatly across different economic, cultural, and social landscapes, as some recent findings published by the SIENNA Project reveal. And even when our moral intuitions are supported by the best possible philosophical arguments, philosophical arguments cannot be the only source of reason in a broader societal debate on how to regulate new technologies such as CRISPR and IVG.

Funding: This research was funded by the Alexander-von-Humboldt-Foundation (Germany, 2018), CNPq (The National Council for Scientific and Technological Development, Brazil, 2019–2023), and FAPERJ (Research Support Foundation of the State of Rio de Janeiro, Brazil, 2019–2022). The author is a member of the SIENNA Project (Stakeholder-Informed Ethics for New technologies with high socio-ecoNomic and human rights impAct) funded by the European Union's H2020 research and innovation program under grant agreement No 741716). This article and its contents reflect only the views of the author and does not intend to reflect those of the European Commission. The European Commission is not responsible for any use that may be made of the information it contains.

Acknowledgments: An early draft of this paper was presented at the Philosophy Department of the University of Graz (June 2019), at the Philosophy Department of the University of Konstanz (July 2019), at the Institute of Biomedical Ethics and History of Medicine in Zurich (September 2019), and in a workshop at University of Bochum (*New Perspectives on the Ethics of Human Enhancement*, February 2020). The author thanks the participants for their critical comments. The author also thanks Peter Stemmer (University of Konstanz) and Philipp Stehr (University of Bochum) for reading and commenting on an earlier version of the present article.

Conflicts of Interest: The author declares no conflict of interest. The funders had no role in the design of the study; in the collection, analyses, or interpretation of data; in the writing of the manuscript, or in the decision to publish the results.

References

1. Rossi, P. *Rossi, Paolo. I Filosofi e le Macchine (1400–1700)*; Feltrinelli: Milan, Italy, 1962; ISBN 978-88-07-88947-9.
2. Rabinbach, A. *The Human Motor: Energy, Fatigue, and the Origins of Modernity*; University of California Press: Berkeley, CA, USA, 1992; ISBN 978-0-520-07827-7.
3. Roux, S. From the mechanical philosophy to early modern mechanisms. In *The Routledge Handbook of Mechanisms and Mechanical Philosophy*; Glennan, S., Illari, P., Eds.; Routledge: London, UK; New York, NY, USA, 2018; pp. 26–45.

---

[6] For a quick overview on the regulation of the use of gene-editing technologies on human cells in different parts of the world, see Ledford [86].

4. McMurtrie, D. *Reconstructing the Crippled Soldier*; Red Cross Institute for Crippled and Disabled Men: New York, NY, USA, 1918. Available online: http://resource.nlm.nih.gov/101560506 (accessed on 1 September 2020).

5. Amar, J. The re-education of war-cripples. The scientific prosthesis. In *The Physiology of Industrial Organisation and the Re-Employment of the Disabled*; The Library Press Limited: London, UK, 1918; pp. 257–310.

6. Neumann, B. Being prosthetic in the First World War and Weimar Germany. *Body Soc.* **2010**, *16*, 93–126. [CrossRef]

7. Harrasser, K. *Körper 2.0: Über die Technische Erweiterbarkeit des Menschen*; Transcript: Bielefeld, Germany, 2013; ISBN 978-3-8376-2351-2.

8. Spreen, D. Prothese. In *Der Körper in der Enhancement-Gesellschaft*; Transcript: Bielefeld, Germany, 2015.

9. Hausmann, R. Prothesenwirtschaft: Gedanken eines Kapp-Offiziers. *Die Aktion* **1920**, *47*, 669.

10. Araujo, M.d. World War I to the Age of the Cyborg: The Surprising History of Prosthetic Limbs. Available online: http://theconversation.com/world-war-i-to-the-age-of-the-cyborg-the-surprising-history-of-prosthetic-limbs-64451 (accessed on 13 August 2020).

11. Jensen, S.R.; Nagel, S.; Brey, P. *D3.1: State-of-the-Art Review [WP3—Human Enhancement]*. SIENNA Project (Stakeholder-informed ethics for new technologies with high socio-economic and human rights impact); European Union's H2020 Research and Innovation Programme under Grant Agreement No 741716: Luxembourg, 2018. Available online: https://cordis.europa.eu/project/id/741716/results (accessed on 1 September 2020).

12. Wittgenstein, L. *Philosophical Investigations*; Basil Blackwell: Oxford, UK, 1958; ISBN 978-0-631-11900-5.

13. Harvey, P.D.; Keefe, R.S.E. Methods for delivering and evaluating the efficacy of cognitive enhancement. In *Cognitive Enhancement*; Kantak, K.M., Wettstein, J.G., Eds.; Handbook of Experimental Pharmacology; Springer: Berlin, Germany, 2015; pp. 5–25. ISBN 978-3-319-16521-9.

14. Maslen, H. Toward an ethical framework for regulating the market for cognitive enhancement devices. In *Cognitive Enhancement: Ethical and Policy Implications in International Perspectives*; Jotterand, F., Dubljević, V., Eds.; Oxford University Press: Oxford, UK; New York, NY, USA, 2016; pp. 275–292, ISBN 978-0-19-939681-8.

15. Flynn, J. *What is Intelligence? Beyond the Flynn Effect*; Cambridge University Press: Cambridge, UK, 2007; ISBN 978-0-521-88007-7.

16. Clark, A.; Chalmers, D. The extended mind. *Analysis* **1998**, *58*, 7–19. [CrossRef]

17. Levy, N. *Neuroethics*; Cambridge University Press: Cambridge, UK, 2007; ISBN 978-0-521-68726-3.

18. Gaskell, G.; Bard, I.; Allansdottir, A.; da Cunha, R.V.; Eduard, P.; Hampel, J.; Hildt, E.; Hofmaier, C.; Kronberger, N.; Laursen, S.; et al. Public views on gene editing and its uses. *Nat. Biotechnol.* **2017**, *35*, 1021–1023. [CrossRef] [PubMed]

19. Blendon, R.J.; Gorski, M.T.; Benson, J.M. The public and the gene-editing revolution. *N. Engl. J. Med.* **2016**, *374*, 1406–1411. [CrossRef] [PubMed]

20. Funk, C.; Kennedy, B.; Sciupac, E.U.S. *Public Wary of Biomedical Technologies to 'Enhance' Human Abilities*; Pew Research Center: Washington, DC, USA, 2016; Available online: https://www.pewresearch.org/science/2016/07/26/u-s-public-wary-of-biomedical-technologies-to-enhance-human-abilities/ (accessed on 1 September 2020).

21. Howell, E.L.; Yang, S.; Beets, B.; Brossard, D.; Scheufele, D.A.; Xenos, M.A. What do we (not) know about global views of human gene editing? Insights and blind spots in the CRISPR era. *CRISPR J.* **2020**, *3*, 148–155. [CrossRef]

22. Prudhomme, M. *D3.5: Public Views of Human Enhancement Technologies in 11 EU and Non-EU Countries*; SIENNA Project (Stakeholder-informed ethics for new technologies with high socio-economic and human rights impact); European Union's H2020 Research and Innovation Programme under Grant Agreement No 741716; European Commission: Luxembourg, 2019; Available online: https://ec.europa.eu/research/participants/documents/downloadPublic?documentIds=080166e5c70baba5&appId=PPGMS (accessed on 1 September 2020).

23. Hanson, T. *D2.5: Public Views on Genetics, Genomics and Gene Editing in 11 EU and Non-EU Countries*; European Union's H2020 Research and Innovation Programme under Grant Agreement No 741716; European Commission: Luxembourg, 2019; Available online: https://ec.europa.eu/research/participants/documents/downloadPublic?documentIds=080166e5c70bb819&appId=PPGMS (accessed on 1 September 2020).

24. Fishel, S.; Dowell, K.; Thornton, K. Reproductive possibilities for infertile couples: Present and future. In *Infertility in the Modern World: Present and Future Prospects*; Bentley, G.R., Mascie-Taylor, C., Eds.; Cambridge University Press: Cambridge, UK, 2004; pp. 17–45.

25. van de Wiel, L. Cellular origins: A visual analysis of time-lapsed embryo imaging. In *Assisted Reproduction across Borders: Feminist Perspectives on Normalizations, Disruptions and Transmissions*; Lie, M., Lykke, N., Eds.; Routledge: New York, NY, USA; Abingdon, UK, 2017; pp. 288–301.

26. Diamond, M.P.; Suraj, V.; Behnke, E.J. Using the Eeva Test^{TM} adjunctively to traditional day 3 morphology is informative for consistent embryo assessment within a panel of embryologists with diverse experience. *J. Assist. Reprod. Genet.* **2014**, *32*, 61–68. [CrossRef] [PubMed]

27. Kieslinger, D.C.; Gheselle, S.D.; Lambalk, C.B. Embryo selection using time-lapse analysis (Early Embryo Viability Assessment) in conjunction with standard morphology: A prospective two-center pilot study. *Hum. Reprod.* **2016**, *31*, 2450–2457. [CrossRef]

28. Zaninovic, N.; Rocha, C.J.; Zhan, Q. Application of artificial intelligence technology to increase the efficacy of embryo selection and prediction of live birth using human blastocysts cultured in a time-lapse incubator. *Fertil. Steril.* **2018**, *110*, e372–e373. [CrossRef]

29. Zegers-Hochschild, F.; Adamson, G.D.; de Mouzon, J.; Ishihara, O.; Mansour, R.; Nygren, K.; Sullivan, E.; Vanderpoel, S. International Committee for Monitoring Assisted Reproductive Technology (ICMART) and the World Health Organization (WHO) revised glossary of ART terminology. *Fertil. Steril.* **2009**, *92*, 1520–1524. [CrossRef]

30. Daar, J.; Benward, J.; Collins, L.; Davis, J.; Francis, L.; Gates, E.; Ginsburg, E.; Klipstein, S.; Koenig, B.; La Barbera, A.; et al. Transferring embryos with genetic anomalies detected in preimplantation testing: An Ethics Committee Opinion. *Fertil. Steril.* **2017**, *107*, 1130–1135. [CrossRef]

31. Baldwin, K.; Culley, L.; Hudson, N.; Mitchell, H. Running out of time: Exploring women's motivations for social egg freezing. *J. Psychosom. Obstet. Gynecol.* **2019**, *40*, 166–173. [CrossRef] [PubMed]

32. Bhatia, R.; Campo-Engelstein, L. The Biomedicalization of Social Egg Freezing: A Comparative Analysis of European and American Professional Ethics Opinions and US News and Popular Media. *Sci. Technol. Hum. Values* **2018**, *43*, 864–887. [CrossRef]

33. Inhorn, M.C. The egg freezing revolution? Gender, technology, and fertility preservation in the twenty-first century. In *Emerging Trends in the Social and Behavioral Sciences: An Interdisciplinary, Searchable, and Linkable Resource*; Scott, R.A., Kosslyn, S.M., Eds.; Wiley: New York, NY, USA, 2015; pp. 1–14, ISBN 978-1-118-90077-2.

34. Goold, I.; Savulescu, J. In favour of freezing eggs for non-medical reasons. *Bioethics* **2009**, *23*, 47–58. [CrossRef] [PubMed]

35. Savulescu, J.; Goold, I. Freezing eggs for lifestyle reasons. *Am. J. Bioeth.* **2008**, *8*, 32–35. [CrossRef] [PubMed]

36. Cohen, I.G.; Daley, G.Q.; Adashi, E.Y. Disruptive reproductive technologies. *Sci. Transl. Med.* **2017**, *9*, eaag2959. [CrossRef]

37. Chen, D.; Gell, J.J.; Tao, Y.; Sosa, E.; Clark, A.T. Modeling human infertility with pluripotent stem cells. *Stem Cell Res.* **2017**, *21*, 187–192. [CrossRef]

38. Fang, F.; Li, Z.; Zhao, Q.; Li, H.; Xiong, C. Human induced pluripotent stem cells and male infertility: An overview of current progress and perspectives. *Hum. Reprod.* **2018**, *33*, 188–195. [CrossRef]

39. Hendriks, S.; Dancet, E.A.F.; van Pelt, A.M.M.; Hamer, G.; Repping, S. Artificial gametes: A systematic review of biological progress towards clinical application. *Hum. Reprod. Update* **2015**, *21*, 285–296. [CrossRef]

40. Hendriks, S.; Dondorp, W.; de Wert, G.; Hamer, G.; Repping, S.; Dancet, E.A.F. Potential consequences of clinical application of artificial gametes: A systematic review of stakeholder views. *Hum. Reprod. Update* **2015**, *21*, 297–309. [CrossRef]

41. Smajdor, A.C.; Cutas, D. Artificial gametes and the ethics of unwitting parenthood. *J. Med. Ethics* **2014**, *40*, 748–751. [CrossRef]

42. Newson, A.J.; Smajdor, A.C. Artificial gametes: New paths to parenthood? *J. Med. Ethics* **2005**, *31*, 184–186. [CrossRef] [PubMed]

43. Palacios-González, C.; Harris, J.; Testa, G. Multiplex parenting: IVG and the generations to come. *J. Med. Ethics* **2014**, *40*, 752–758. [CrossRef] [PubMed]

44. Nuffield Council on Bioethics. *Background Paper: Artificial Gametes*; Nuffield Council on Bioethics: London, UK, 2015. Available online: https://www.nuffieldbioethics.org/publications/artificial-gametes (accessed on 1 September 2020).

45. Cyranoski, D. Mouse eggs made from skin cells in a dish. *Nature* **2016**, *538*, 301. [CrossRef] [PubMed]
46. Greely, H. *The End of Sex and the Future of Human Reproduction*; Harvard University Press: Cambridge, MA, USA, 2016; ISBN 978-0-674-98401-1.
47. Niemiec, E.; Howard, H.C. Germline genome editing research: What are gamete donors (not) informed about in consent forms? *CRISPR J.* **2020**, *3*, 52–63. [CrossRef]
48. Ishii, T.; Beriain, I.D.M. Safety of germline genome editing for genetically related "future" children as perceived by parents. *CRISPR J.* **2019**, *2*, 370–375. [CrossRef]
49. Mohr, S. Beyond motivation: On what it means to be a sperm donor in Denmark. *Anthropol. Med.* **2014**, *21*, 162–173. [CrossRef]
50. Woestenburg, N.O.M.; Winter, H.B.; Janssens, P.M.W. What motivates men to offer sperm donation via the internet? *Psychol. Health Med.* **2016**, *21*, 424–430. [CrossRef]
51. Thijssen, A.; Provoost, V.; Vandormael, E.; Dhont, N.; Pennings, G.; Ombelet, W. Motivations and attitudes of candidate sperm donors in Belgium. *Fertil. Steril.* **2017**, *108*, 539–547. [CrossRef]
52. Freeman, T.; Jadva, V.; Tranfield, E.; Golombok, S. Online sperm donation: A survey of the demographic characteristics, motivations, preferences and experiences of sperm donors on a connection website. *Hum. Reprod.* **2016**, *31*, 2082–2089. [CrossRef]
53. Hodson, N.; Parker, J. The ethical case for non-directed postmortem sperm donation. *J. Med. Ethics* **2020**, *46*, 489–492. [CrossRef] [PubMed]
54. Suter, S. The tyranny of choice: Reproductive selection in the future. *J. Law Biosci.* **2018**, *5*, 262–300. [CrossRef] [PubMed]
55. Suter, S.M. In vitro gametogenesis: Just another way to have a baby? *J. Law Biosci.* **2015**, *3*, 87–119. [CrossRef] [PubMed]
56. Sniekers, S.; Stringer, S.; Watanabe, K.; Jansen, P.R.; Coleman, J.R.I.; Krapohl, E.; Taskesen, E.; Hammerschlag, A.R.; Okbay, A.; Zabaneh, D.; et al. Genome-wide association meta-analysis of 78,308 individuals identifies new loci and genes influencing human intelligence. *Nat. Genet.* **2017**, *49*, 1107–1112. [CrossRef]
57. Asbury, K.; Plomin, R. *G is for Genes: The Impact of Genetics on Education and Achievement*; Wiley: Hoboken, NJ, USA, 2014.
58. Flynn, J. *Intelligence and Human Progress. The Story of What was Hidden in Our Genes*; Elsevier: Amsterdam, The Netherlands, 2013.
59. Gabbatt, A. Woman Gives Birth to Baby That Grew from Embryo Frozen 24 Years Ago. Available online: https://www.theguardian.com/us-news/2017/dec/20/woman-gives-birth-to-baby-who-spent-24-years-as-a-frozen-embryo (accessed on 15 August 2020).
60. Ledford, H. CRISPR gene editing in human embryos wreaks chromosomal mayhem. Three studies showing large DNA deletions and reshuffling heighten safety concerns about heritable genome editing. *Nature* **2020**, *583*, 17–18. [CrossRef]
61. Regalado, A. China's CRISPR Twins might Have Had Their Brains Inadvertently Enhanced. 2019. Available online: https://www.technologyreview.com/2019/02/21/137309/the-crispr-twins-had-their-brains-altered/ (accessed on 1 September 2020).
62. Cohen, J. Did CRISPR help—or harm—the first-ever gene-edited babies? *Science.* 21 February 2019. Available online: https://www.sciencemag.org/news/2019/08/did-crispr-help-or-harm-first-ever-gene-edited-babies (accessed on 1 September 2020).
63. National Academies of Sciences, Engineering, and Medicine. *Human Genome Editing. Science, Ethics, and Governance*; The National Academies Press: Washington, DC, USA, 2017; ISBN 978-0-309-45288-5.
64. Deutsche Akademie der Naturforscher Leopoldina. *Ethische und Rechtliche Beurteilung des Genome Editing in der Forschung an Humanen Zellen*; Nationale Akademie der Wissenschaften: Halle, Germany, 2017. Available online: https://www.leopoldina.org/uploads/tx_leopublication/2017_Diskussionspapier_GenomeEditing.pdf (accessed on 1 September 2020).
65. Nuffield Council on Bioethics. *Genome Editing: An Ethical Review*; Nuffield Council on Bioethics: London, UK, 2016. Available online: https://www.nuffieldbioethics.org/wp-content/uploads/Genome-editing-an-ethical-review.pdf (accessed on 1 September 2020).
66. Kant, I. *Groundwork of the Metaphysic of Morals. A German*; English ed.; Timmermann, J., Ed.; Gregor, M., Translator; Cambridge University Press: Cambridge, UK, 2011; ISBN 978-1-107-61590-8.

67. Chituc, V.; Henne, P.; Sinnott-Armstrong, W.; De Brigard, F. Blame, not ability, impacts moral "ought" judgments for impossible actions: Toward an empirical refutation of "ought" implies "can". *Cognition* **2016**, *150*, 20–25. [CrossRef]
68. Chituc, V.; Henne, P. The Data against Kant. *The New York Times*. 19 February 2016. Available online: https://www.nytimes.com/2016/02/21/opinion/sunday/the-data-against-kant.html (accessed on 1 September 2020).
69. Parfit, D. On doing the best for our children. In *Ethics and Population*; Bayles, M.D., Ed.; Schenkman Publishing Company Inc.: Cambridge, UK, 1976; pp. 100–115.
70. Parfit, D. The non-identity problem. In *Reasons and Persons*; Oxford University Press: Oxford, UK, 1984; pp. 351–390.
71. Schwartz, T. Obligations to posterity. In *Obligations to Future Generations*; Siroka, R.I., Barry, B., Eds.; Emple University Press: Philadelphia, PA, USA, 1978; pp. 3–13.
72. Frasca, R.R. Negligent beginnings: Damages in wrongful conception, wrongful birth and wrongful life. *J. Forensic Econ.* **2006**, *12*, 85–205. [CrossRef]
73. Macintosh, K.L. *Enhanced Beings: Human Germline Modification and the Law*; Cambridge University Press: Cambridge, UK, 2018; ISBN 978-1-108-47120-6.
74. Doolabh, K.; Caviola, L.; Savulescu, J.; Selgelid, M.J.; Wilkinson, D. Is the non-identity problem relevant to public health and policy? An online survey. *BMC Med. Ethics* **2019**, *20*, 46. [CrossRef]
75. Kenny, A. *What I Believe*; A&C Black: London, UK, 2006; ISBN 978-0-8264-8971-5.
76. The Economist Modern Genetics will Improve Health and Usher in "Designer" Children. It may also Provoke an Ethical Storm. *The Economist*. 7 November 2019. Available online: https://www.economist.com/science-and-technology/2019/11/07/modern-genetics-will-improve-health-and-usher-in-designer-children (accessed on 1 September 2020).
77. Regalado, A. The World's First Gattaca Baby Tests Are Finally Here. The DNA Test Claims to Let Prospective Parents Weed Out IVF Embryos with a High Risk of Disease or Low Intelligence. *MIT Technol. Rev.* **2019**. Available online: https://www.technologyreview.com/2019/11/08/132018/polygenic-score-ivf-embryo-dna-tests-genomic-prediction-gattaca/ (accessed on 1 September 2020).
78. Wilson, C. Choose your child's intelligence. *New Sci.* **2018**, *240*, 6–7. [CrossRef]
79. Devlin, H. IVF Couples could be Able to Choose the 'Smartest' Embryo. US Scientist Says It will Be Possible to Rank Embryos by 'Potential IQ' within 10 Years. *The Guardian*. 24 May 2019. Available online: https://www.theguardian.com/society/2019/may/24/ivf-couples-could-be-able-to-choose-the-smartest-embryo (accessed on 1 September 2020).
80. Rawls, J. *A Theory of Justice. (Revised Edition, 1999)*; Harvard University Press: Cambridge, UK, 1999; ISBN 0-674-00078-1.
81. Resnik, D.B. Genetic engineering and social justice: A Rawlsian approach. *Soc. Theory Pract.* **1997**, *23*, 427–448. [CrossRef]
82. Sneddon, A. Rawlsian decision making and genetic engineering. *Camb. Q. Healthc. Ethics* **2006**, *15*, 35. [CrossRef]
83. Allhoff, F. Germ-line genetic enhancement and Rawlsian primary goods. *J. Evol. Technol.* **2008**, *18*, 10–26.
84. Papaioannou, T. New life sciences innovation and distributive justice: Rawlsian goods versus Senian capabilities. *Life Sci. Soc. Policy* **2013**, *9*, 1–13. [CrossRef]
85. Fulda, K.G.; Lykens, K. Ethical issues in predictive genetic testing: A public health perspective. *J. Med. Ethics* **2006**, *32*, 143–147. [CrossRef] [PubMed]
86. Ledford, H. Where in the world could the first CRISPR baby be born? A look at the legal landscape suggests where human genome editing might be used in research or reproduction. *Nat. News* **2015**, *526*, 310–311. [CrossRef] [PubMed]

 *philosophies*

Article

Marketing the Prosthesis: Supercrip and Superhuman Narratives in Contemporary Cultural Representations

Chia Wei Fahn

Department of Foreign Languages and Literature, National Sun Yet-sen University, Kaohsiung 80424, Taiwan; celinefahn@gmail.com

Received: 24 February 2020; Accepted: 7 July 2020; Published: 10 July 2020

Abstract: This paper examines prosthetic technology in the context of posthumanism and disability studies. The following research discusses the posthuman subject in contemporary times, focusing on prosthetic applications to deliberate how the disabled body is empowered through prosthetic enhancement and cultural representations. The disability market both intersects and transcends race, religion, and gender; the promise of technology bettering the human condition is its ultimate product. Bionic technology, in particular, is a burgeoning field; our engineering skills already show promise of a future where physical impediment will be almost obsolete. I aim to cross-examine empowering marketing images and phrases embedded in cinema and media that emphasize how disability becomes super-ability with prosthetic enhancement. Though the benefits of biotechnology are most empowering to the disabled population, further scrutiny raises a number of paradoxical questions exposed by the market's advance. With all these tools at our disposal, why is it that the disabled have yet to reap the rewards? How are disabled bodies, biotechnology, and posthuman possibilities commodified and commercialized? Most importantly, what impact will this have on our society? This paper exemplifies empowering and inclusive messages emphasized in disabled representation, as well as raising bioethical concerns that fuel the ongoing debate of the technological haves and have-nots. Furthermore, this paper challenges the ideals of normative bodies while depicting the disabled as an open, embodied site where technology, corporeality, and sociology interact. To conclude, I believe that an interdisciplinary approach that balances the debate between scientific advance, capital gain, and social equality is essential to embracing diverse forms of embodiment.

Keywords: bionics; prosthesis; biotechnology; disability; marketing; cultural studies; Disney; supercrip

---

1. Introduction

Studies in both natural sciences and the humanities have been devoted for much of the last two decades to exploring how biotechnology revolutionizes human interaction with our material surroundings. From bifocal glasses to high performance prosthetic limbs, hearing aids, and pacemakers, the conjuncture between the human body and biotechnology is expanded by various forms of prosthetic extensions. Human bodies are now seen by some commentators as "machines to be fine-tuned and perfected through add-ons," as the body is continuously augmented and its capabilities enhanced [1] (p. 248). A humanist view of the body as a coherent, self-contained, and autonomous self is no longer applicable; biotechnology has led to a most significant shift in body politics that views the body as a network or assemblage that is constantly evolving with technology. The relationship between human bodies and biotechnology is symbiotic, working "in between the points of contact linking bodies and technologies to configure them differently" [2] (pp. 3–4). This reconfiguration of the body and technology is most prominently seen in bionic engineering, with the infinite possibilities of technological progress materialized through prosthetic limbs, optics, and organs [3]. Traditionally perceived as a visual reminder of our vulnerability that signifies "physical and metaphysical lack," prosthetics have become a corporeal extension that empowers the disabled body, allowing the human body to transcend boundaries and limitations [4].

The interrogation of the permeability of corporeal boundaries sparked an ongoing body of work that examines the "continuous deconstruction and reconstruction" of embodiment [5]. Prosthetic technology destabilizes the human boundaries, with Donna Haraway's famous question: "Why should our bodies end at the skin?" urging discourses that extend posthuman concepts of the body to rise in response [6]. Applied prosthetics demonstrate "the ultimate expression of human control—helping us shape and define the way we would like to be," contributing to new forms of embodiment in the posthuman subject [7]. This posthuman subject embraces the body as unstable and "refers to the destabilization and unsettling of boundaries between human and machine, nature and culture, and mind and body that digital and biotechnologies are seen to engendering" [8]. In other words, the posthuman traces a different discursive framework that affirms forms of alternative embodiment and opens the body as an extended, distributed, interconnected, and relational entity. Posthumanist thought views human subjectivity as an assemblage that evolves in conjunction with machines, animals, and the environment.

Our bodies have become fluid and ambiguous, disrupting binaries such as human versus non-human, and able-bodied versus disability. With prosthetic technology constantly pushing the boundaries of physical limitation, the disabled body is now regarded as an intricately diverse site of interdisciplinary discourse. Scholars in the field focus their efforts on subverting our body concept, social constructions, and disputes by exploring the relationship between disabled bodies and applied prosthetics [9]. Their efforts place disabled bodies at the nucleus of progress in assistive technology that no longer casts the body in a "dualistic frame, but bears a privileged bond with multiple others and merges with one's technologically mediated planetary environment" [10] (p. 92). Social and material debate in disability studies has emerged with the "new connections among rediscovered agencies and thereby dethrones the ideal of a human cognitive, physical, emotional normal" [11] (p. 310). Therefore, a posthuman approach to disability recognizes a new ethics of the body that is embodied in "the conceptual edges between 'the human' and 'the posthuman,' the organic and the mechanic, the evolutionary and the postevolutionary, and flesh and its accompanying technologies" [2] (p. 3). I believe the disabled body represents a versatile embodiment of these conceptual edges and is most reflective of social change. By a juxtaposition of textual analysis and case examples in both popular culture and practical use, this paper examines the rising trend of empowering disabled bodies that focuses on the bionic industry.

Groundbreaking research in bionic limbs and neurotransmission poses a direct challenge to the limitations of humanity and human performance. Bionic technology is a burgeoning market; our engineering technologies already show promise of a future where physical impediment will be almost obsolete. The following research reflects on the posthuman subject in contemporary times, focusing on prosthetic applications to deliberate how the disabled body is empowered through prosthetic enhancement and cultural representations. I aim to cross-examine marketing images and phrases embedded in the fields of sports, cinema, and media that portray the prosthesis as empowering to emphasize how disability becomes super-ability with prosthetic enhancement. Bionic prosthetics are heavily promoted in contemporary film and media, a trend I will thoroughly review by exemplifying innovative work in the field as well as addressing future possibilities that include the freedom to alter our bodies in all forms and functions.

Though the benefits of biotechnology are most empowering to the disabled population, further scrutiny raises a number of paradoxical questions exposed by the market's advance. With all these tools at our disposal, why is it that the disabled have yet to reap the rewards? How are disabled bodies, biotechnology, and posthuman possibilities commodified and commercialized? Most importantly, what impact will this have on our society? Living in the times of the posthuman, I advocate various ways of becoming that embrace a sustainable ethics of transformations that emphasize the future of embodiment in real bodies. In other words, the future of posthuman embodiment should not only be in pursuit of "glamorous cyborgs," but also in creating a sustainable future that provides equal access to advanced technology and all its promised opportunities. This should especially apply to the "anonymous masses of the underpaid, digital proletariat who fuel the technology-driven global economy without ever accessing it themselves" [10] (p. 90). However, under the current market mechanism, I am more concerned that the uneven distribution of technological resources and restrictive access to biotechnology create a new social caste, an issue I aim to address in the following analysis of enhancement technologies.

2. Disabled Bodies and the Prosthesis—Transcending Physical Boundaries and Embodying Posthuman Possibilities

Posthuman embodiment focuses on the body as a constant state of becoming that includes shifting social paradigms and political subjectivities [1,10,12]. The disabled body embodies new interactions with our material environment as well as with each other; disability studies is therefore "perfectly at ease with the posthuman" because "disability has always contravened the traditional classical humanist conception of what it means to be human" and responds directly to the complex diversities in our contemporary society [13] (p. 342). An intersecting dialogue, however, has only emerged in the last decade, with Rosi Braidotti's *The Posthuman* (2013). Braidotti asserted that disability studies "combine the critique of normative bodily models with the advocacy of new, creative models of embodiment," affirming the disabled body as an open, embodied site where technology, corporeality, and sociology interact [10] (p. 146). With bionic technology, the human body is transformed into "a module onto which various technological additions can be attached … technology is not separate but part of the body" [14] (p. 241). Braidotti's viewpoint resonates with Davis' concept of dismodernism; both challenge the body's boundaries to explore the openness of disabled bodies and forge interdisciplinary connections. In this regard, disability studies becomes an exemplary network of various intersecting discourses.

The work of Margrit Shildrick furthers the idea of a fluid, posthuman body that is embodied by the disabled; she believes that the disabled body, the prosthetic body, and normative bodies are all a "profound interconnectivity of all embodied social relations" that is "simply one mode among multiple ways of becoming" [15] (p. 8). Shildrick's perspective expounds upon Katherine N. Hayles' posthuman analysis; both addressed the relationship between human limitations and the utilization of supplementary devices in facilitating day-to-day living conveniences. Hayles advocated for a "posthuman view … of the body as the original prosthesis we all learn to manipulate, so that extending or replacing the body with other prostheses becomes a continuation of a process that began before we were born" [5] (p. 3). Consequently, technological growth and innovative design are now seen as having a unique influence regarding disability and posthumanism; disabled bodies are a "dynamic hybrid" that is focused "not on borders but on conduits and pathways, not on containment but on leakages, not on stasis but on movements of bodies, information and particles" that transcend corporeal boundaries and join the biological to the technological in posthuman embodiment [12] (p. 10).

According to Braidotti, the human condition is expanded by "the four horsemen of the posthuman apocalypse: nanotechnology, biotechnology, information technology and cognitive science" [10] (p. 59), causing "the boundaries between 'Man' and his others [to] go tumbling down, in a cascade effect that opens up unexpected perspective" [10] (pp. 66–67). The disabled subject affirms connections between the body and technologies that is the quintessential posthuman condition [16]. Nayar built upon this concept of transcending physical boundaries to study posthuman embodiment in disabled bodies, stating that "the differently abled … exist on the other side of the border. It is this issue of boundary-marking and personhood that brings disability studies and bioethics into the ambit of critical posthumanism" [12] (p. 101). Disabled bodies interrogate existing normative standards of the human subject, contributing significantly to posthumanist thought by calling attention to a deconstruction of the deviance and normalcy that regulate the body. Disability disarms and disrupts the normative assumptions about what it means to be human; it extends and expands into perspectives that look affirmatively towards alternative modes of embodiment.

A posthuman subjectivity is based on creativity and the opening up of possibilities. With prosthetic technology we will have the tools to redesign and reinvent the body; prosthetics are an invitation to reconsider notions of normality, function, and ability. The push to realize posthuman possibilities and alternative embodiment has led to progressive development in bionic augmentation and neurotransmission technology. Today, we have a selection of prosthetics that range from plastic surgery injections to full-bodied exoskeleton suits that enable the paralyzed to stand, walk, and even run again—the development of prosthetic tools that not only seek to mimic human function, but also enhance the body's capacity. Disabled bodies present a means to enhance humanity and transcend physical limitations, and with that recognition came a shift in the portrayal of disability as well as an emerging disability-related market. The following discussion explores how disability is increasingly portrayed in empowering cultural representations as posthuman, examining the "embodiment and embeddedness" of the human being that integrate the biological with the technological [17] (pp. xv–xvi). Following a transitory dialogue from disability to the posthuman, I aim to bring an interdisciplinary discourse on posthuman embodiment into consideration with the biotechnologies that evoke fundamental changes in the body and our society, with a special focus on prosthetic limbs and the bionic industry.

3. Disability (Broadly Defined) and Commercial Trends: Subverting Norms and Creating a Limitless Market

In 2010, Snyder and Mitchell identified contemporary times as a new era of social inclusion that "open[s] rhetorical claims of a new era of inclusion for people with disabilities" [18] (p. 116). Their work calls attention to the global reality of disabled populations, with a special focus on the role of disability and disabled bodies in the consumer market. Disability was seen in the past as a visual reminder of human imperfection; however, with recent developments in biotechnology, the disabled body is transformed into a canvas for augmentation as technology turns the disabled body into a marketable commodity by "selling one's capacities (and, now, incapacities) in a market economy" [18] (p. 121). Over the last two decades, disability-related markets have been identified as the largest and fastest-growing subgroup in the United States, with *Forbes* magazine estimating the scope of this "overlooked and growing market" to be limitless [19]. Marketing groups such as Fifth Quadrant Analytics, Sutton Marketing, and the LinkedIn platform now offer consulting services to companies that wish to capitalize upon a targeted disabled demographic that is seen as the world's "next big consumer segment" [19].

Technological designs not only target people with disabilities as a core consumer base, but appeal to "healthy" people who share the inevitability of disability [20] (p. 8). Disability is an inherent part of the human condition; the understanding that "everyone will be disabled if they live long enough" has been "an incredibly generative disability studies insight," from contact lens to pacemakers, products that address the broad definition of disability have led related businesses to form the world's fastest growing market [21] (p. 200). The human condition is "in fact disabled, only completed by technology and by interventions," establishing how the boundaries between ability and disability are open and subject to change as a result of productive needs and the utilization of extended tools [15] (p. 241). Davis sees a new ethics of the body in this commercial trend and focuses on the care of the body as "a requirement for existence in a consumer society"; our bathrooms alone hold an arsenal of tools catering to individual needs [22]. The knowledge required to design these products originates from an understanding of our physical lack and impairments, rendering forms of biological data into the ultimate commodity of the twenty-first century.

Progress in medical and reproductive fields is built upon the study of disabled and abject bodies and has bettered the lives of those who live with disability and disease. Disabled people provide physical data for the "micro-management of information and bodies" that lead to biotechnological breakthroughs [18] (p. 117). Disability is commercialized "in the popular sphere of product advertisements," disabled bodies are ubiquitously featured in commercials for "pharmaceuticals, prosthetically engineered bodies and minds, mutating organisms that may be better adapted for a future world yet to come" [18] (p. 117). The niche markets of disability create opportunities that "catalyze radical new approaches—approaches that might subsequently have more widespread applications" resulting in innovative designs that are marketable to both disabled and non-disabled groups [23]. Disabled bodies become both a saleable commodity and a target demographic as technological development expands the new market for body and body matters. This phenomenon benefits the disabled both physically and financially; biotechnological businesses are creating job positions that tap into disabled people's unique insights and embodied experience. Tusler focused on instrumentalizing disabled people for corporations to serve the technological industry in *How to Create Disability Access to Technology*, arguing that the "challenge of being successful has led many people with disabilities to be skilled problem solvers and consumers—making them valuable assets" [24]. In 2014, LinkedIn published an article titled "Ten Ways to Target the Disability Market & Keep a Competitive Edge," urging companies to develop an inclusive and comprehensive diversity strategy that incorporates product design, marketing, and, most of all, insures "a marketing strategy and mechanism that builds a company's brand as an inclusive workplace committed to equality and to hiring, retaining and promoting people with disabilities" [25]. This foray into disabled employment and the communicative understanding of disabled needs that focuses on providing product choices has opened employment opportunities catering to the disabled population and subverted able-bodied norms previously dictating the job market.

Inclusive efforts in product and strategy design have furthered empowerment through the increased visibility and representation of disabled persons, while allowing the disabled to contribute their unique perspective to professional knowledge in product design. The disabled perspective and embodied experience are highly instrumental in the development of assistive technology, leading to exponential industrial growth in the area of bionic prostheses. Subversive narratives of physical empowerment and accessibility are most prevalent in marketing prosthetic and bionic limbs. The advent of social media and transparent communications has also resulted in considerable changes in brand image presentation, with tales of inclusive products heavily promoting demonstrations of empowering people with disabilities [26]. Representations of prosthetic technology have progressed in a drastic shift from supplementing the disabled body and creating a normalized image to representing the physical embodiment of enhanced power through prosthetic augmentation.

The disability-related market both intersects and transcends race, religion, and gender; the promise of technology bettering the human condition is its ultimate product. An analysis of media advertisements featuring disability finds that messages of empowerment and inclusion, as opposed to stigmatizing portrayals, are a common theme [27]. Market recognition, universal product design, and inclusive representation all contribute to furthering social and individual empowerment in the disabled community. With increasing advances in prosthetic limbs and bionic technology, portrayals of people who use such devices accentuate enhanced physical competence that surpasses even the able-bodied, a narrative that rewrites social attitudes toward the prostheses and also disrupts the disabled stereotype. At the same time, an increase in disabled representation serves as a positive advertisement for brands and merchandise alike, promoting corporal images in an era that markets social inclusion. In turn, these portrayals also subvert the traditional stigma of normalized bodies and display how the disabling aspects of a differently-abled body can also be empowering. This transition from disabled to super-abled changes, albeit slowly, how we regard the disabled body and acclimatizes the public to alternative forms of posthuman embodiment in contemporary practices. In the following sections I will elaborate on the shift in social narrative, with a special focus on limb disability and the cultural representations of prosthetic enhancement.

4. From Disabled to Super-Abled: Commercializing the Supercrip Narrative in Film and Media

Supercrip is an encompassing term traditionally representing inspirational, disabled people. In the past, the supercrip was extensively criticized for its depiction of disabled people who overcome spiritual and physical challenges to perform everyday tasks, revealing the low expectations from normalized society [28–34]. An affirming support of supercrip narratives has risen in recent years, however, to support the portrayal of disabled people who are "brave for defying expectations"; disabled people are urged to recognize that "our actions are purposeful, our art exciting, or our words meaningful, we do inspire" [35] (p. 198). Building upon this support is Goodley and Runswick-Cole's argument that the disabled body has always demanded to be recognized not as an embodiment of lack, but of possibility [20]. This demonstrates a shift in perspective that forms the nexus of this paper; subversive representations of disability promote empowerment not in the ability to overcome, but in the different abilities presented through an alternative embodiment that transcends the confines of physical normalcy.

The 2015 film *Mad Max: Fury Road* provides an inspiring example of physical empowerment with the disabled character "Furiosa." In a movie filled with male characters and battle scenes, Furiosa stands out as an emblem of indisputable power. In a much-discussed scene, Furiosa punches the titular Max in the face with her amputated stub, exuding an on-screen strength that subverts both norm and form. Furiosa's disability is not depicted as a challenge to be overcome, nor is her physical lack portrayed as a vulnerability. Furiosa is defined by her abilities, not her infirmities; the character radiates power that transcends the physical category [36]. Furiosa's scene is powerful because of the matter-of-fact treatment of her body. She embraces her form and utilizes it fully; Furiosa's body just *is*. This supercrip portrayal subverts the norm of frail and fragile disabled imagery and draws the viewer's attention to ability, not disability. The result is undoubtedly empowering, yet also raises concerns over a new form of ableist expectation that subjugates the body, an issue I will discuss in the following sections.

In recognition of empowering portrayals of the disabled body, Sami Schalk called for a reevaluation of the supercrip stereotype to build a "future scholarship to interrogate supercrip representations in a variety of cultural arenas," arguing that "to dismiss outright all representations of supercrips as 'bad' is to disregard potentially entire genres of popular culture productions, ones which tend to have very large audiences" [29] (p. 84). Schalk focused on the mechanisms of supercrip narratives, proposing that "supercrip narratives produce . . . representations of purportedly extraordinary disabled people of three distinct, yet related types" [29] (p. 79). A "regular" supercrip narrative exemplifies disabled persons who are able to execute normalized tasks; the "glorified" supercrip narrative portrays accomplishments that challenge the non-disabled; and the "superpowered" supercrip narrative emphasizes transhuman power [29]. Advances in prosthetics and assistive technology further transformed the supercrip narrative from overcoming physical trials to sensationalizing the technological superhuman [37]. The rapid change in the lived-in reality of disability is reflected in supercrip representations embodying messages of power and interdependency, as the supercrip "amazes and inspires the viewer by performing feats that the nondisabled viewer cannot imagine doing" [38] (p. 71). These representations embody possible futures in posthuman development and demonstrate paradigm shifts in the representations of disability.

Science and technology offer a transcendent way of life that far surpasses rehabilitation or even the normalized body [39–42]. Technological cures "enable . . . an enhanced functionality" that allows the body to become "stronger, faster, or more responsive" [43]. Prosthetic augmentations create a techno-posthuman world "with science as the rescuer of the human from its mortal self . . . the human becomes an assemblage of parts, conceived of in terms of a machine that can be . . . repaired, and redesigned" [1] (p. 260). With augmented technology rapidly becoming accessible to the disabled population, the rebranding of prosthetics as a means of physical empowerment is most prominently seen in film and popular media, exploring "deviant and disabled bodies . . . whose embodiments are situated along the entire spectrum of ability" [42] (p. 2). A vast body of works that subvert the abject to heroic either popularizes heroes whose marked disability becomes their superpower, or showcases individuals who gain superhuman capabilities through prosthetic augmentation. From the paraplegic Professor X and the blind Daredevil, to the bionically-sustained Cyborg, disability is portrayed as the capability to "conjure a world in which freaks, crips, gimps ... cultivate a unique power and perspective outside ... conformity" [44]. Prosthetically-enhanced supercrips, in particular, became inspiring images designed to appeal to the general public, images that also represent alternative, posthuman embodiments. These portrayals fuel the disabled consumer market, advertising physical empowerment and unlimited possibilities that can be gained through the use of advanced prosthetics.

Empowering prosthetics are prominently featured in Marvel Studio's *Iron Man*, with titular figure Tony Stark transforming from a disabled and disillusioned playboy to world-saving superhero in an exoskeleton suit. At the beginning of the film, Stark is fatally wounded by an explosion that sends shrapnel flying into his chest cavity, entering his bloodstream to gradually make its way toward his heart. To prevent certain death, Stark inserts a miniature reactor into his chest that generates a magnetic protective field with the added benefits of powering a full-body bionic suit. Though the *Iron Man* comics debuted in 1968, the flawed and disabled hero is now an iconic figure in the contemporary marketing of bionic enhancement. The *Iron Man* franchise does not label Tony Stark as a triumphant hero rising above physical impairment; Stark's disability neither defines nor confines him. Instead, the piece of shrapnel in Stark's chest is motivation to continuously navigate new technology that improves his state of living. Physical empowerment through prosthetic technology is a consistent theme throughout the series, as Stark continues to improve upon his initial creation. With each installment of the *Iron Man* trilogy, viewers see an upgrade in Stark's armor, with Stark becoming increasingly powerful through the integration of mechanical engineering and biotechnological science. Stark is a prime example of the superpowered, supercrip narrative utilized in contemporary film that aims to change society's perception of physical disability through empowering representations of high-functioning prosthetics.

Contemporary technology has yet to match Stark's AI-navigated, G-force-defying Iron Man suit, though the concept of a bionic bodysuit has been manufactured and proven highly functional. A product test video released in 2014 by California-based company Ekso Bionics portrays paraplegic woman Amanda Boxtel walking for the first time in 22 years while encased in a 3D printed exoskeleton suit. A bionic exoskeleton suit of similar design was featured two years later in *Captain America: Civil War*, allowing Stark's friend to walk again after suffering severe spinal injuries. The exoskeleton suit allows the paralyzed to regain mobility, inspiring various designs and working prototypes from bionic engineering companies such as Rex Biotics, ReWalk Robotics, and SuitX. The ReWalk exoskeleton, for example, uses inertial sensors to detect subtle changes in the user's center of gravity to provide a natural gait and stair-climbing abilities [45]. Exoskeleton suits were prominently featured in a series of promotion videos for the Cybathlon 2016 competitions held in Kloten, Switzerland, as an integral part of the unique championship for people with disabilities to compete in completing everyday tasks with the aid of state-of-the-art assistive technologies [34] (pp. 14–15). In China, preliminary tests on a wheelchair and exo-suit hybrid were conducted and results published in 2019, with engineers working to elevate motion assistance and motor ability for wheelchair users. Users can move from sitting to standing positions by changing the mechanical configuration on motorized wheelchairs, and an exoskeleton function provides movement in knee and hip joints that enables the user to walk. Apart from an increase in mobility, this also puts forth a solution for complications related to long-term sitting, such as pressure sores, muscle atrophy, and bladder infections [46]. Although the user's body remains encased in an assistive prosthesis, both suit and wheelchair represent empowerment instead of confinement.

The bionic bodysuit demonstrates a future where physical disabilities are transcended through technology. This injection of science-fiction-inspired technology into the source material not only serves as a demonstration of technological capacity, but also blurs the line between literary imagination and scientific reality, as science inspired by fiction intrinsically weaves into the fabric of contemporary reality. Later installments in the film franchise introduced an antihero named the Winter Soldier, an amputee with an ultra-strong cybernetic arm that lends him extreme strength and dexterity in battle. Marvel's Winter Soldier brought the franchise back from futuristic possibilities in augmented physicality to a more conventional form of prosthesis, leaving fans enthralled by the return to an identifiable narrative as the Winter Soldier struggles with PTSD while performing extraordinary feats in hand-to-hand combat. The following year, Marvel's mother company Disney Inc. announced a new line of prosthetic limbs co-developed by the UK company Open Bionics that were based on the same designs as worn by the main characters in their *Star Wars* and *Iron Man* franchises. Open Bionics is part of the Disney Accelerator program, an incubating branch of the Disney Corporation that invests in new technology and media companies specializing in products that span virtual reality, artificial intelligence, and robotics. Disney makes equity investments in each of the 11 participating companies through undisclosed amounts of funding, with the companies receiving guidance from Disney marketing executives, investors, and established leaders in the industry. The start-up tech company specializes in creating prosthetic cybernetic arms that operate using sensors attached to the skin in order to detect muscle movements. These muscle movements control the hand and open and close the fingers, allowing users to enjoy mobility currently unavailable in traditional prosthetics. This new line of bionic limbs either features concept art inspired by fan favorite characters, or in the case of the *Winter Soldier* and *Iron Man*, boasts identical designs to the original.

Open Bionics' company website expounds upon its goal of building and developing the next generation of bionic limbs and turning disabilities into superpowers. Named the "Hero Arm," the superhero-inspired, Disney-branded bionic arms aim to promote the same sense of empowerment as their film counterparts, making users "feel like superheroes" [47]. With the Hero Arm, a user's "limb difference is your very own superpower" [47]. Limbitless Solutions, a 3D printing company, also designs bionic arms inspired by Iron Man's super-powered gauntlets. The company's promotional video of Robert Downey Jr. delivering a 3D-printed prosthetic arm to a disabled child in 2015 is an example of how science fiction is "used in contemporary social media to promote new technology" [48]. These collaborations between film franchises and the bionic industry mark a first in brand marketing as well as indicate a clear recognition of the expansive opportunities in the disability market. High-functioning prosthetics have now been branded with a "superhero" image that lends power to those who use them. The superpowered, supercrip narrative not only offers "counter discursive forays" into the public perception of disability, but also empowers the disabled through physical advantages that have been widely publicized in popular media [49]. The following section examines these augmentative advantages in media representation.

5. From Supercrip to Superhuman—Disability as an "Unfair" Advantage

In addition to film representations of prosthetic empowerment, popular media have also become a platform for the superpowered supercrip narrative and product demonstration. Websites such as YouTube, Facebook, and Instagram provide an open and indiscriminative medium for the disabled to share their stories and for innovators to display the latest breakthroughs in technology. Through YouTube videos, the prosthetically enhanced are able to construct and popularize narratives that emphasize their capabilities instead of disability. The disabled gain power through public viewership that subverts the oppressive, able-bodied gaze on the disabled body, and seize control of rhetoric and perception. Rosemarie Garland-Thomson elaborated upon the "visual relation between a spectator and a spectacle," stating that staring at the disabled body is an act that "registers the perception of difference and gives meaning to impairment by marking it as aberrant" [38] (p. 56). The gaze able-bodied people direct upon the disabled body is usually furtive: we are taught that it is rude to stare so "the disabled body is at once the to-be-looked-at and not-to-be-looked-at, further dramatizing the staring encounter by making viewers furtive and the viewed defensive" [38] (p. 57). The spectator's stare "creates disability as a state of absolute difference rather than simply one more variation in human form," while at the same time "constitutes disability identity by manifesting the power relations between the subject positions of disabled and able-bodied" [38] (p. 57). Media platforms turn this fascination with physical disability into a subversive opportunity to reverse the power structures between the spectacle and the speculator "that elicit responses or persuade viewers to think or act in certain ways" [38] (p. 58). The disabled want to be seen, "inviting the gaze of ableist culture" to focus on embodiments of "all that is good with a hyper-ableist philosophy: blurring man-machine, re-enabling disability, blurring the lines between disability and ability" [50] (p. 145). Through the media, disabled people are able to control, construct, and mold social narratives to fit their agendas. The supercrip narrative, once stigmatizing, has become a marketing vehicle of education and product advertising, as well as empowerment for the prosthetically enhanced, with differently constructed bodies striving to break the constraints of a binary dis/ability discourse [33] (p. 80).

Double amputee Hugh Herr is widely known as a staunch advocate of the empowering proprieties of prosthetic technology and a physical representation of the empowered supercrip. Herr regularly appears in YouTube videos, marketing campaigns, or TED Talk presentations to expound upon the possibilities of prosthetic applications, heavily emphasizing a narrative that focuses on the empowering freedom of engineered prosthetics and performing physical feats that are elusive to even "normal" bodies. Herr was severely injured during a mountain climbing incident in his twenties and lost both legs to frostbite. After surveying the "normalizing" prosthetic limbs that were currently available on the market, Herr decided that he was "not handicapped; the technology is" [51]. He began to experiment with unconventional designs that fit a purpose, rather than adhering to normative restrictions. Within weeks of the amputation, Herr resumed climbing and at the same time, he began modifying prosthetics to fit his needs. Working with traditional materials at the time, he cut off a heel to reduce weight, increased his legs' stiffness when it was necessary, added garden rake-like spikes for ice climbing, made feet narrow enough to insert into in small cracks, and even altered his height [52]. Within a year of the amputation, Hugh Herr was able to climb heights beyond the reach of his previously "normal" body.

Herr's progress has been well documented since the time of his accident, transitioning from the glorified supercrip hero of the 1980s into an embodiment of superhuman prowess. In a 2014 TED Talk speech, Herr stated,

> A human being can never be 'broken.' Technology is broken. Technology is inadequate. This simple but powerful idea was a call to arms, to advance technology for the elimination of my own disability, and ultimately the disability of others. I began by developing specialized limbs that allowed me to return to the world of rock and ice climbing. I quickly realized that the artificial part of my body is malleable; able to take on any form, any function—a blank slate for which to create, perhaps, structures that could extend beyond biological capability. [53]

In the years that followed, Herr has made continuous breakthroughs, starting with a computer-controlled artificial knee in 2003. In 2004, Herr created the Biomechatronics group at MIT that combines the fields of biology, mechanics, and electronics to restore function to those in need. In 2007, the Biomechatronics team produced a powered ankle–foot prosthesis, which allows an amputee to walk with speed and effort comparable to those with biological legs, called the emPower; a lightweight apparatus that houses 12 sensors, three computers, tensioning springs, and muscle-tendon actuators. The emPower success led Herr's team to further their innovative designs in extreme bionics and build prosthetic legs that not only allow the disabled to walk, but to run, climb, and even dance with ease [53].

The spokesperson for the center of Extreme Bionics as well as product prototype, Herr quips, "I'm kind of what they're selling" [54]. Each presentation Herr gives as a frequent public speaker is a visual display of the agility and capability of his bionic legs. Herr's prosthetics, once makeshift alterations of garden tools he repurposed from his father's garage, are now sleek, balanced, and intuitively responsive. Herr wears pants that are tailored just below his knees, exposing the technology of the bionic legs to the public eye with the company logo proudly displayed on each artificial ankle. With each agile step taken, Hugh Herr embodies the superpowered supercrip narrative as a techno-marvel that attracts the attention of both disabled and normative-bodied consumers. During talks and interviews, Herr constantly directs audience attention to a narrative that emphasizes superhuman capabilities achieved through technology, explaining that with innovative technology his amputation was an advantage, not a disability. Though specialized prosthetics gave him the ability to reach new heights in rock and ice climbing, Herr explained that the double amputation also became a physical advantage, as "his body got colder and achier as he climbed but his legs did not. He was able to move faster and higher than before, in part because the amputations had left him 14 pounds lighter" [55]. In an interview, Herr said: "My climbing colleagues first labeled me as 'courageous,' which is always very demeaning . . . The second I became competitive; I became a threat. I had a few people threaten to amputate their own limbs to achieve the same advantage" [56]. The term "advantage" would continue to appear in conjunction with Herr's prosthetics.

The advantages of prosthetic augmentation as compared to the "natural" body are also frequently debated in sports, with double amputee Oscar Pistorius a prime example. The first amputee to compete in an Olympic track event, Pistorius' qualification to run in the Olympics came into question with the IAAF (International Association of Athletics Federations) in 2008. The IAAF published an amendment of its competition rules in 2007 to enforce a ban on the use of "any technical device that incorporates springs, wheels or any other element that provides a user with an advantage over another athlete not using such a device"; this became a defining moment in prosthetic development [57]. Following the ban, IAAF invited Pistorius to a joint research project that revealed his advantage while running with prosthetics to be overwhelmingly high, leading to the IAAF ruling Pistorius' prostheses ineligible for use in competitions conducted under its rules. Additional studies showed that once a runner on blades accelerates to top speed, one potential advantage lies in the ability to move the prostheses faster and with less effort [58,59]. Similarly, the IAAF barred German Paralympic long jumper Markus Rehm from competing in the 2016 Rio Olympics after Rehm failed to prove his prosthetic leg did not give him "unfair" advantages. Rehm and Pistorius are prominently disabled under the current social definition, yet have been accused of unfair advantages over their peers. This dispute can be understood as a fundamental inquiry into disability and capability. If all sports are meant to be a celebration and display of bodies that transcend physical limitations, why is a celebration of prosthetic athleticism rejected? Traditional sports feature the "human" capacity, with athletes performing "natural" skills; however, with health supplements and analytical training, modern day sports can hardly be seen as exclusive to the boundaries of human physicality [60]. Soon, augmented bodies will offer a performance capacity with which "no 'natural' body could possibly compete"; therefore, the body's boundaries need to expand to include the variations of bodies integrated with technologies we define as posthuman [61].

With articles, interviews, and public speeches meticulously building upon the empowering aspects of technological enhancement; prosthetics, once representing immobility and exclusion, carry infinite possibilities with the application of enhancive technologies. Disability is transformed into capability and disadvantage into advantage [62]. The posthuman body is temporal, malleable, and should be understood in part through advances in technology that "destabilize any 'absolutes' in body construction" [61] (p. 128). The assumptions we rely upon when constructing the rhetoric of ableism rely on a notion of "correct" and "normal" human bodies to which we all must conform in order to be declared fully human [63] (p. 183). In turn, humanity is the end product of a normalizing process in corporeality and embodiment; the body is not simply a natural and enclosed subject but part of a social institution; our understanding of embodiments and corporealities is constantly deconstructed and rebuilt according to social needs, structure, and narrative. Bodies are no longer simply categorized as abled, disabled, or super-abled, but are part of constantly changing sociocultural frames and techno-capacity; what is "natural" is change and adaptation to the alternatives. The posthuman body reveals constant, performative constructions where bodies can be altered and even redesigned. The examples of prosthetic enhancement of real bodies are testimony to the fact that prosthetic technology will continuously push the boundaries of physical limitation, leading to fundamental shifts in normative social paradigms that transform the prosthesis from a symbol of stigma into a highly desirable commodity for generations to come.

6. Transitioning Boundaries—Building a Posthuman Body of Your own Design

With this in mind, the body as a social institution is unstable, its boundaries negotiable and in constant transition. The transition from an emphasis on the natural body to the posthuman is, as Vaccari stated, an "erosion of the distinction between organism and machine, nature and art, and the biological and engineering sciences" [64] (p. 138). Vaccari's argument focuses on Descartes' philosophical departure from Aristotle's principles of biology. Aristotle's biology principles kept living organisms ontologically bound and whole, differentiating artifacts and organisms through metaphysics that separate the corporeal and material [64–66]. Vaccari cited Descartes' Cartesian mechanism as a rhetoric of posthuman thought, stating that Descartes "blurred the unity and boundaries of living bodies" by uniting bodies and artifacts under the concept of matter [64] (p. 140). He argued that "the only natural norm of the body is the mutual interrelation of the organs; we can easily envision the replacement of certain elements of this arrangement without affecting the 'nature' of the whole" [64] (p. 155). Conventional body boundaries must be replaced by fluidity, in other words, contemporary society is in a position where a hybridity between the body and prostheses is embraced, pursued, and constantly reinvented. Our bodies are posthuman; the cyborg is an evolutionary state as well as the next step of human materiality [67].

In 2011, Herr embellished upon the fluidity of the posthuman body to predict that the prosthesis will become an extension of our physical form:

> As we march into this 21st century, the changes we'll see in prosthetic designs [will be that] the artificial prosthetic will become more intimate with the biological human body. There will be a mergence, if you will. The prosthesis will be attached to the body mechanically by a titanium shaft that goes right into the residual bone, wherein you can't take the artificial limb off. Another intimate connection will be electrical. The nervous system of the human will be able to communicate directly with the synthetic nervous system of the artificial limb. [68]

Neurologically controlled, feeling prosthetics are much more than a supplementary addition to the body; they are evolving into an actual extension of our being. It is prudent to assume that soon our bodies may have little to do with the original state of our birth. Apart from the development in external prosthetics, scientists have also discovered ways to bring the prosthesis into our body by means of transplantation. Organ transplants are an established procedure; however, the need for organ donors far exceeds the supply. An exciting development in the field has been 3D printing technology. Scientists are now able to "print" artificial organs using biomaterials that promise to one day transform the market and become an answer to the donor shortage and organ rejection. Disabled or aging bodies will have the option to "change parts" in order to maintain and even increase physical functionality. As Herr states, "My biological body will degrade in time due to normal, age-related degeneration. But the artificial part of my body improves in time because I can upgrade" [68]. This is no doubt a great step in shaping the human condition at will; we are able to replace our organic bodies with the customizable and upgradable. By means of science and technology we can look forward to a world where physical normalcy and normative boundaries are displaced by a new, creative world of our imaginative design.

In addition to high performance bionic limbs and prosthetic organs, the image of prosthetics is rapidly evolving past the human form to present a variety of designs and functions. Herr is fond of reminding audiences how he "viewed the missing biological part of my body as an opportunity, a blank palette for which to create" [69]. Under the posthuman design, the prosthesis no longer represents a lack in the disabled body, but the freedom and infinite possibilities in altering and transforming the human form [58,70,71]. The infinite possibilities of the prosthesis are best embodied by Aimee Mullins, whose public career began as a record-breaking athlete at the Paralympic Games in 1996. Aimee Mullins' legs captured the world's attention not because of their functional innovation, but because of their unconventional design. Mullins opened British fashion designer Alexander McQueen's London show on a pair of hand-carved wooden prosthetic legs made from solid ash. With intricately carved flowers and grape vines, the prosthetics were designed to look like Victorian knee-length boots with a Louis heel, pointed toe, and slim ankle. She wears her "Barbie legs" to formal events and is able to alternate her height between 5 feet, 8 inches to 6 feet, 1 inch. In Matthew Barney's *Cremaster 3*, Mullins changes between legs shaped like a jellyfish to painted cheetah limbs, legs cast out of dirt, and transparent polyethylene legs complete with heels; each change of prosthetics represents a shift in her role and identity in the film. Mullin's legs have come to be associated with discussions about the unstable materiality of human bodies, "her prosthesis … extends her bodily boundaries and open and problematizes her sense of identity" [71] (p. 33). The fact that the significance of prosthetic devices "has shifted from trying to mimic the human form in order to camouflage, disguise and replace, exhibited symbolic entities which look like machines and evoke a sci-fi futuristic multiplicity of human and post-human bodily imaginaries" ushers in an age of transgressive design and function that reconceptualizes body, material, and embodiment [71] (p. 45).

Mullins believes that the new age of prosthetics "can stand as a symbol that the wearer has the power to create whatever it is that they want to create in that space" [72]. With the infinite possibilities of prosthetic creation, the disabled can become "the architects of their own identities" and the ability to redesign disabled bodies becomes a form of empowerment [72]. The body no longer needs to be accepted in its "as is" condition, but becomes whatever we envision. The possibilities of prosthetic creations reject the conventional definition of less as lack; "the absence of limbs becomes an open-ended possibility to reconfigure the appearance and the functionality of human biology in unprecedented ways" [70] (p. 50). In 2016, San Francisco-based organization KIDmob teamed with 3D-printing and design software company Autodesk to hold a "Superhero Cyborgs" workshop. Superhero Cyborgs encouraged kids—both with and without disabilities—to design and build wearable devices that were potential alternatives to traditional upper limb prosthetics. The kids created their own "superpowers" through personalized wearable devices that included a five-nozzle glitter shooter, a prosthetic with a detachable bow and arrow, and an elbow-movement-activated water gun [73]. Prosthetics are undergoing an "incorporation and projecting" of design, "an unstoppable 'difference' that is not about negation but about the alterity of 'becoming' … [which] challenge simple figuration and fixity" [22] (p. 38). These creative possibilities render the ideal, normalized body obsolete; soon our bodies will either remain in their conventional forms or embrace the conceptual. For future generations, the human form can be whatever we imagine it to be. We will have the power to "customize" the body to adapt to different environments, age, functionality, and individual aesthetics.

The branding of prosthetics with superhero imagery as well as studies in augmented bodies' actual advantages and affirmative power have transformed the disability market. Prosthetics are less representative of stigma and more an invitation to explore the differently abled body's capabilities. In the near future, physical augmentation will no longer be a response to infirmities but a neo-liberalist choice to embrace diverse embodiment. However, as with all neo-liberalist choices, the power to enhance and redesign our bodies comes with social costs and moral responsibilities that must be thoroughly addressed. The questions I propose here are not what we can do with technology, but rather questions of how to reinforce a universal respect for diversity as well as promote equal access to viable resources. The power to genetically alter, bionically substitute, and otherwise hybridize our physical materiality deconstructs past definitions of human corporeality and individual agency. Augmentative technology undoubtedly provides an advantage to enhanced bodies, yet will disability be eliminated in a future where prosthetic enhancement is open to all? Or will biotechnology subject unaltered bodies to an even more severe regime of normalcy so that the "human capital" becomes a new determinant of social caste?

7. Bioethical Concerns and Uneven Access—Posthuman Disability and Peripheral Bodies

With the increasingly creative and seamless integration between prostheses and human bodies, "ableism" no longer solely represents physical performances of the body "occurring in natural embodiments and corporealities" [63] (p. 185). The body's boundaries are extended beyond the casing of flesh and bone; the "natural" body becomes the posthuman.

As with all innovative technology, prosthetic extensions come at a cost that raises bioethical concerns. In terms of social costs, critics believe that prosthetic technology will add to the stigmatizing effects of disability simply because these prosthetics will also aid people without disabilities, while the disabled remain so. Many believe that prosthetic enhancement will further marginalize disabled people by raising the benchmark for ableism, allowing stigma to continue under a new form and narrative [50,74,75]. Susan Wendell posited that advertising the disabled, enhanced hero "may reduce the 'Otherness' of a few people with disabilities, but because it creates an ideal that most people with disabilities cannot meet, it increases the 'Otherness' of the majority of people with disabilities" [76]. Prosthetic enhancement adds to the ableist stigma by asserting that disability is acceptable as long as the disabled are able to receive a prosthetic cure and "pass" as "whole" [34] (p. 24). The supercrip narrative and posthuman biotechnological efforts, while empowering the disabled, are in fact far more concerned with a "transhumanization of ableism" that will "impact what we perceive as healthy bodies leading to the transhumanization of the meaning of health … the scenario where only certain beyond species-typical body abilities are seen as healthy" [77]. Prosthetically-enhanced physical performance "reifies a particular conception of a normal body, creating an expectation that all disabled bodies should achieve" [61] (p. 56). These expectations include a "new era of disabled athleticism—buffed, muscular, yet technologically supplemented bodies—promising all of the transcendent capacity a hyper-medicalized culture could offer" while demonstrating a seamless connectivity between the corporeal body and the prosthesis [18] (p. 117). Cynthia Bruce worried that an insatiable pursuit of physical enhancement will further the benchmark of ableist marginalization for disabled people [75]. The biggest question is if the disabled will be further excluded from society "for not having hyper or super abilities, which may be the primary feature for defining a 'normal' person in the coming age" [34] (p. 29). Certainly, an amputee with a bionic leg would enjoy more physical mobility than an amputee with a traditional prosthesis and be seen as able in comparison; ableism is therefore determined by prosthetic performance [74,78,79].

In terms of monetary cost, scholars and activists such as Goggin and Newell questioned the "'commonsense' notion of people with disabilities inherently benefiting from new technologies" [80]. In 2013, ballroom dancer Adrienne Haslet-Davis' leg was amputated after being hit by shrapnel in the Boston marathon bombing. Haslet-Davis later became part of Hugh Herr's "New Bionics Let Us Run, Climb and Dance" TED Talk presentation, where she spun and floated across the floor with the latest bionic limb, to a standing ovation. As incredible as the moment seemed, Haslet-Davis was only a test participant; she was not able to keep the prototype that cost millions to reproduce. Herr's facility fit rock climber Jim Ewing in 2018 with a newly designed ankle–foot prosthetic that is thought to be responsive and allows him to feel the appendage through neuro-stimulation, yet the prosthetic is also a prototype that Ewing will eventually have to return. Haslet-Davis and Ewing were both part of a research project and would not be able to keep the advanced prosthetics. However, the exorbitant cost of high-functioning bionic limbs begs the question of who will be able to afford these prostheses once they are available on the market. In "A Leg to Stand on," Vivian Sobchack noted that her own "rather ordinary" AK leg cost $10,000 to $15,000. The well-known sprinting "cheetah leg" is at least $20,000 per leg, whereas stair-climbing bionic legs range from $40,000 to $50,000, a price range that is hardly affordable for the average person [81] (p. 31). Advanced prosthetics are not covered by health insurance policies, therefore the access to such technology is rapidly becoming a reflection of financial power and serves as a watershed between the advantaged and disadvantaged [34] (p. 33).

In the glamorous façade of twenty-first century biotechnological progress, prominent rifts have begun to surface. Disabled people using the latest assistive technologies are a natural constituency for prosthetic enhancement, yet disability remains rooted in "the political economic realities of capitalism" [82] (p. 199). The paradox between the commodification of assistive technology and the social reality of disability is a prominent issue, with Snyder and Mitchell arguing that

> the particularities of bodily accommodations necessarily send people with disabilities into circulation as consumers of medical and social services assistance. This entry of disabled "consumers" into market systems becomes an odd and nuanced affair in that the basis of those classified as consumers usually require "purchasing power," the one thing that the majority of disabled people do not possess. [18] (pp. 114–115)

At present, biotechnological marvels remain a form of class consumption. Advanced technology is only open to those who can afford the purchase, causing people in one country to experience the same impairment differently, based on their socioeconomic status. The "have nots" and the "haves" will lead directly to an "ability divide" that is not only reflected by individuals, but sets a division on a global scale [83]. Since "all subjects in modern, liberal, technologized, and consumerist social orders are medical subjects—or perhaps more precisely medical consumers," the main disabling factor is no longer impairment but access to technology [84]. In other words, this new model of disability is very likely to "generate new ability divides as well as gradations of wealth from techno-poor to techno-rich" [50] (p. 161).

James I. Charlton examined the relationship between the current socioeconomic system and the disabled experience to reveal a commonality between disabled persons and the "world's poor" in impoverished areas all over the globe [82] (p. 195). Technological advancement "does not occur as something separate from ideology and stigma" but rather reinforces stigma under a different structure of disabling due to disabled and disadvantaged persons' peripheral alienation [82]. Charlton believed this structuring force to be "the commodity form" that is "control (rooted in unequal relationships), hierarchy, commodification, and mystification" [82] (p. 196). Disabled persons are often situated in the poorest and most marginalized communities, resulting in a "double outcast" phenomenon that Charlton sees as a "pattern of periphery, even in the most varied places, that reveals a deep force immanent throughout the world system that is both hidden and hides many reasons for the disability condition" [82] (p. 196). An example of this pattern is seen in Laos, where many victims of war remain in handmade stumps and pegs, whereas prosthetics provided by rehabilitation centers are plastic, mannequin-like affairs operated by pulling a belt [84]. Regardless of where they are, the majority of people with disabilities remain on the periphery and are excluded from positions of capitalist gain.

Disabled persons are "locked into subalternity, the underclass, institutionalized dependencies, the peripheral everywhere" [82] (p. 198). Without access to technology, disabled bodies remain peripheral bodies; disabled people are "outsiders in core capitalist countries" [82] (p. 195). In short, stigmatizing social structures have not changed, the "conceptual schemas and mind maps of ableism" remain intact, while the "pool of the remnant, the 'have nots' and 'not quites,' grows larger and more diverse" [74] (p. 63). Access to technology is determined by an economic difference that is reflected in social differences; in this day and age, restricted access to technological devices and knowledge is arguably more disabling than limited mobility and is a criterion that applies to all of us. Access to technology determines the opportunities of an individual and is rapidly overtaking physical ability as a main disabling factor. Apart from relying on close supervision to construct an equal access to technology and information, how are we to bridge the gap between the haves and have-nots?

8. Concluding Thoughts—Bridging the Gap through Interdisciplinary Education

Disability scholarship and posthuman discourse are tasked with ethical debates that go above and beyond merely extending the boundaries of what we perceive as human. A disability-conscious bioethical viewpoint provides focus for social justice and equality; in addition to emphasizing an inclusive participation of people with disabilities in society, a respect for difference and accessibility are needed to prepare humanity for the changes ahead. Finding new and alternative modes of political and ethical agency for our technologically-mediated world is imperative to closing the differences between techno-capitalist gain and bioethical concerns. One solution to this challenge is to provide an interdisciplinary approach that balances the debate between scientific advance, capital gain, and social equality. In order to embrace "the liberatory and even transgressive potential of these technologies," we must also take precautions against "those who attempt to index them to either a predictable conservative profile, or to a profit-oriented system" [10] (p. 58). Braidotti urged us to focus on a "technological mediation" that is central to "a new vision of posthuman subjectivity and . . . provides the grounding for new ethical claims" [10] (p. 90). An interdisciplinary, bioethical vision exhorts us to "move forward into multiple posthuman futures . . . in a new global context and to develop an ethical framework worthy of our posthuman times" [10] (p. 150). Creative, intersecting thought provides a way for new alliances between the humanities and the sciences, resulting in emerging fields such as the medical humanities and interdisciplinary education [10] (p. 145). Garland-Thomson, in particular, advocated for a "disability cultural competence" that enforces the knowledge of interdisciplinary studies in both practice and policy [85] (p. 328). Converting disability theory into usable disability bioethics, she argued, will "strengthen the cultural, political, institutional, and material environment in which people with disabilities can most effectively flourish" [85] (p. 331). Certainly, an interdisciplinary, bioethical vision invites more thought about creating fair opportunities for the disabled, as well as builds a sustainable infrastructure that embraces diversity and allows for equal access.

The future of the posthuman lies not in the unbound efforts of technological growth, but in an interdisciplinary education that combines the humanities with scientific knowledge to teach a new and alternative mode of agency. Part of building a sustainable, equal future is creating equal opportunities through education. An interdisciplinary education that publicizes scientific knowledge, ethical guidelines, and open access should be as important as the technology that drives our progress. The focus of interdisciplinary education is not a "tur[n] backwards to a nostalgic vision of the Humanities as the repository and the executors of universal transcendental reason and inherent moral goodness," but should emphasize a move forward into multiple posthuman futures that provide equal opportunities in its stead [10] (p. 150). Similarly, Aoun appealed for a reformation in different approaches toward teaching the humanities and expounded upon the significance of human agency and creativity in influencing the impact of technology to propose a learning model of "humanics" [86]. This integrative alliance between the sciences and the humanities has become highly important, not only to address the new ecologies of becoming and belonging, but as a necessary response to the subsequent ecopolitical impact. Interdisciplinary education is crucial to promoting technological literacy, as well as to increasing equal access and sustainable development.

Interdisciplinary education paves the way for an equal future, taking into account the elements of creativity and imagination that allow us to embrace diversity and pursue alternative forms of the posthuman. The human race should be able to improve upon our physical conditions, with the acknowledgement that all forms of enhancement and prosthetic extensions are just as much part of our embodied experience as flesh and bone. However, this turn to posthuman embodiment should not be restricted to merely applauding the marvels of biotechnology and commodifying futuristic design, but should also include diverse forms of the differently abled, whether the body is enhanced or not. In other words, we can embrace biotechnology as a welcome form of choice that should not be considered mandatory, nor should it be a consumerist luxury only made available to the financially able. Above all, interdisciplinary humanities must include discussions about providing equal opportunities to all its adherents. This innovative agenda will continue to inspire affirmative alternatives in new forms of subjectivity, as well as urge us to rethink embodiment in a sustainable future.

Funding: This research received no external funding.

Conflicts of Interest: The author declares no conflict of interest.

References

1. Seaman, M.J. Becoming More (Than) Human: Affective Posthumanisms, Past and Future. *J. Narrat. Theory* **2007**, *37*, 246–275. [CrossRef]
2. Smith, M.; Morra, J. (Eds.) *The Prosthetic Impulse: From a Posthuman Present to a Biocultural Future*; MIT Press: Cambridge, MA, USA, 2007.
3. Kline, R. Where are the Cyborgs in Cybernetics? *Soc. Stud. Sci.* **2009**, *39*, 331–362. [CrossRef]
4. Mitchell, D.T.; Snyder, S.L. *Narrative Prosthesis: Disability and the Dependencies of Discourse*; University of Michigan Press: Ann Arbor, MI, USA, 2014.
5. Hayles, N.K. *How We Became Posthuman: Virtual Bodies in Cybernetics, Literature, and Informatics*; University of Chicago Press: Chicago, IL, USA, 1999.

6. Haraway, D. A Cyborg Manifesto: Science, Technology, and Socialist-Feminism in the Late 20th Century. In *The International Handbook of Virtual Learning Environments*; Weiss, J., Nolan, J., Hunsinger, J., Trifonas, P., Eds.; Springer: Dordrecht, The Netherlands, 2006.

7. Rifkin, J. *The Biotech Century*; Penguin: New York, NY, USA, 1998.

8. Blackmann, L. *The Body: Key Concepts*; Berg: Oxford, UK, 2008.

9. Siebers, T. *Disability Theory*; University of Michigan Press: Ann Arbor, MI, USA, 2008.

10. Braidotti, R. *The Posthuman*; Polity: Oxford, UK, 2013.

11. Crilley, M. Material Disability: Creating New Paths for Disability Studies. *CEA Crit.* **2016**, *78*, 306–311. [CrossRef]

12. Nayar, P.K. *Posthumanism*; Polity: Oxford, UK, 2014.

13. Goodley, D.; Runswick-Cole, K.; Lawthom, R. Posthuman Disability Studies. *Subjectivity* **2014**, *7*, 342–361. [CrossRef]

14. Davis, L. The End of Identity Politics and the Beginning of Dismodernism: On Disability as an Unstable Category. In *The Disability Studies Reader*, 2nd ed.; Davis, L.J., Ed.; Routledge: London, UK, 2006; pp. 231–242.

15. Shildrick, M. *Leaky Bodies and Boundaries: Feminism, Postmodernism and (Bio)ethics*; Routledge: London, UK, 1997.

16. Murry, S.F. *Disability and the Posthuman: Bodies, Technology and Cultural Futures*; Liverpool UP: Liverpool, UK, 2020.

17. Wolfe, C. *What is Posthumanism?* University of Minnesota Press: Minneapolis, MN, USA, 2010.

18. Mitchell, D.T.; Snyder, S.L. *The Biopolitics of Disability: Neoliberalism, Ablenationalism, and Peripheral Embodiment*; University of Michigan Press: Ann Arbor, MI, USA, 2015.

19. Lee, B.Y. *An Overlooked and Growing Market: People with Disabilities*; Forbes: Jersey City, NY, USA, 2016.

20. Garland-Thomson, R. Integrating Disability, Transforming Feminist Theory. *NWSA J.* **2002**, *14*, 1–32. [CrossRef]

21. McRuer, R. Cripping Queer Politics, or the Dangers of Neoliberalism. *Sch. Fem. Online* **2012**, *10*, 1–2.

22. Davis, L.J. *Bending Over Backwards: Disability, Dismodernism, and Other Difficult Positions*; New York University Press: New York, NY, USA, 2002.

23. Pullin, G. *Design Meets Disability*; MIT Press: Cambridge, MA, USA, 2009.

24. Tusler, A. *How to Create Disability Access to Technology: Best Practices in Electronic and Information Technology Companies*; World Institute on Disability: Berkley, CA, USA, 2005.

25. Bates, D. 10 Ways to Target the Disability Market & Keep a Competitive Edge. Available online: https://www.linkedin.com/pulse/20140703000007-11466134-top-10-ways-to-market-to-people-with-disabilities-and-boost-your-roi (accessed on 9 June 2020).

26. Trevisan, F. *Disability Rights Advocacy Online: Voice, Empowerment and Global Connectivity*; Routledge: New York, NY, USA, 2017.

27. Haller, B.A.; Ralph, S. Are Disability Images in Advertising Becoming Bold and Daring? An Analysis of Prominent Themes in US and UK Campaigns. *Disabil. Stud. Q.* **2006**, *26*. [CrossRef]

28. Davis, L. *The End of Normal: Identity in a Biocultural Era*; University of Michigan Press: Ann Arbor, MI, USA, 2014.

29. Schalk, S. Reevaluating the Supercrip. *J. Lit. Cult. Disabil. Stud.* **2016**, *10*, 71–86. [CrossRef]

30. Alaniz, J. Supercrip: Disability and the Silver Age Superhero. *Int. J. Comic Art* **2004**, *6*, 304–324.

31. Howe, P.D. Cyborg and Supercrip: The Paralympics Technology and the (Dis)Empowerment of Disabled Athletes. *Sociology* **2011**, *45*, 868–882. [CrossRef]

32. Silva, C.F.; Howe, P.D. The (In)Validity of Supercrip Representation of Paralympian Athletes. *J. Sport Soc. Issues* **2012**, *36*, 174–194. [CrossRef]

33. Clare, E. *Exile and Pride: Disability, Queerness, and Liberation*, 1st ed.; South End Press: Cambridge, MA, USA, 1999.

34. Sun, H.-Y. Prosthetic Configurations and Imagination: Dis/ability, Body, and Technology. *Concentric Lit. Cult. Stud.* **2018**, *44*, 13–39.

35. Linton, S. *My Body Politic: A Memoir*; University of Michigan Press: Ann Arbor, MI, USA, 2006.

36. Stobbart, D. Mad Max and Disability: Australian Gothic, Colonial, and Corporeal (Dis)possession. *Stud. Gothic Fict.* **2018**, *6*, 65–72. [CrossRef]

37. Grue, J. The Problem of the Supercrip: Representation and Misrepresentation of Disability. In *Disability Research Today: International Perspectives*; Routledge: London, UK, 2015; pp. 204–218.

38. Garland-Thomson, R. The Politics of Staring: Visual Rhetorics of Disability in Popular Photography. In *Disability Studies: Enabling the Humanities*; Snyder, S.L., Brueggemann, B.J., Garland-Thomson, R., Eds.; Modern Language Association of America: New York, NY, USA, 2002; pp. 56–75.

39. Berger, R.J. Disability and the Dedicated Wheelchair Athlete: Beyond the Supercrip Critique. *J. Contemp. Ethnogr.* **2008**, *37*, 647–678. [CrossRef]

40. Ripat, J.; Woodgate, R. The Intersection of Culture, Disability and Assistive Technology. *Disabil. Rehabil. Assist. Technol.* **2011**, *6*, 87–96. [CrossRef]

41. Wise, P.H. Emerging Technologies and Their Impact on Disability. *Future Child.* **2012**, *22*, 169–191. [CrossRef]

42. Allan, K. *Disability in Science Fiction: Representations of Technology as Cure*; Macmillan: New York, NY, USA, 2013.

43. Wälivaara, J. Blind Warriors, Supercrips, and Techno-Marvels: Challenging Depictions of Disability in Star Wars. *Pop. Cult.* **2018**, *51*, 1036–1056. [CrossRef]

44. Chemers, M.M. Mutatis Mutandis: An Emergent Disability Aesthetic in 'X-2: X-Men United'. *Disabil. Stud. Q.* **2004**, *24*, 1.

45. Talaty, M.; Esquenazi, A.; Briceño, J. Differentiating ability in users of the ReWalkTM powered exoskeleton: An analysis of walking kinematics. In Proceedings of the 13th International Conference on Rehabilitation Robotics, Seattle, WA, USA, 24–26 June 2013; pp. 1–5.

46. Song, Z.; Tian, C.; Dai, J.S. Mechanism Design and Analysis of a Proposed Wheelchair-exoskeleton Hybrid Robot for Assisting Human Movement. *Mech. Sci.* **2019**, *10*, 11–24. [CrossRef]

47. Shamo, L. These Bionic Arms Make Kids Feel Like Superheroes. Available online: https://www.businessinsider.com.au/open-bionics-prosthetic-arms-2018-2 (accessed on 9 June 2020).

48. Smith, S. Limbitless Solutions': The Prosthetic Arm, Iron Man and the Science Fiction of Technoscience. *Med. Hum.* **2016**, *42*, 259–264. [CrossRef]

49. Mitchell, D.T.; Snyder, S.L. *Cultural Locations of Disability*; University of Chicago Press: Chicago, IL, USA, 2006.

50. Goodley, D. *Dis/ability Studies: Theorising Disablism and Ableism*; Routledge: London, UK, 2014.

51. Helgensen, S. Hugh Herr Wants to Build a More Perfect Human. *Strateg. Bus.* **2016**, *85*, 1–12.

52. Kiss, J. What If A Bionic Leg Is So Good that Someone Chooses to Amputate? Available online: https://www.theguardian.com/technology/2015/apr/09/disability-amputees-bionics-hugh-herr-super-prostheses (accessed on 9 June 2020).

53. Herr, H. The New Bionics that Let Us Run, Climb and Dance. Available online: https://www.youtube.com/watch?app=desktop&persist_app=1&v=CDsNZJTWw0w (accessed on 9 June 2020).

54. Balf, T. The Biomechatronic Man. Available online: https://www.outsideonline.com/2238401/biomechatronic-man (accessed on 9 June 2020).

55. Adelson, E. Best Foot Forward. Available online: https://www.bostonmagazine.com/2009/02/18/best-foot-forward-february/ (accessed on 15 June 2020).

56. Rosenberg, J. Who Says Hugh Herr Can't Climb (Part I). Available online: https://www.youtube.com/watch?v=z6_0nI5U-Qg (accessed on 12 February 2020).

57. Brueggemann, G.P.; Arampatzis, A.; Emrich, F.; Potthast, W. Biomechanics of double transtibial amputee sprinting using dedicated sprinting prostheses. *Sports Technol.* **2008**, *1*, 220–227. [CrossRef]

58. Karpin, I.; Mykitiuk, R. Going out on a Limb: Prosthetics, Normalcy and Disputing the Therapy/Enhancement Distinction. *Med. Law Rev.* **2008**, *16*, 413–436. [CrossRef] [PubMed]

59. Burkett, B.; McNamee, M.; Potthast, W. Shifting boundaries in sports technology and disability: Equal rights or unfair advantage in the case of Oscar Pistorius? *Disabil. Soc.* **2011**, *26*, 643–654. [CrossRef]

60. Miah, A. *Genetically Modified Athletes: Biomedical Ethics, Gene Doping and Sport*; Routledge: London, UK, 2004.

61. Booher, A.K. Prosthetic Configurations: Rethinking Relationships of Bodies, Technologies, and (Dis)Abilities. Ph.D. Thesis, The Graduate School of Clemson University, Clemson, SC, USA, December 2009.

62. Brey, P.A.E. Human Enhancement and Personal Identity. In *New Waves in Philosophy of Technology*; Berg Olsen, J.K., Selinger, E., Riis, S., Eds.; Palgrave: New York, NY, USA, 2008; pp. 169–185.

63. Monceri, F. The Nature of the 'Ruling Body': Embodiment, Ableism and Normalcy. *Teoria* **2014**, *34*, 183–200.

64. Vaccari, A. Dissolving Nature: How Descartes Made Us Posthuman. *Techné Res. Phil. Technol.* **2012**, *16*, 138–186. [CrossRef]

65. Symons, J. The Individuality of Artifacts and Organisms. *Hist. Phil. Life Sci.* **2010**, *32*, 233–246.

66. Bianchi, E. Aristotle's Organism, and Ours. In *Contemporary Encounters with Ancient Metaphysics*; Greenstine, A.J., Johnson, R.J., Eds.; Edinburgh University Press: Edinburgh, UK, 2017; pp. 138–157.

67. Barfield, W. The Process of Evolution, Human Enhancement Technology, and Cyborgs. *Philosophies* **2019**. [CrossRef]

68. The Double Amputee Who Designs Better Limbs. Available online: https://www.npr.org/2011/08/10/137552538/the-double-amputee-who-designs-better-limbs (accessed on 10 May 2020).

69. A Bionic Man: Hugh Herr Strides Forward on Next-Generation Robotic Legs. Available online: https://www.autodesk.com/redshift/hugh-herr-robotic-legs/ (accessed on 23 April 2020).

70. Neicu, M. Prosthetics Imagery: Negotiating the Identity of Enhanced Bodies. *Platform* **2012**, *6*, 42–60.

71. Tamari, T. Body Image and Prosthetic Aesthetics: Disability, Technology and Paralympic Culture. *Body Soc.* **2017**, *23*, 25–56. [CrossRef]

72. My 12 Pairs of Legs. TEDTalk. Available online: https://www.ted.com/talks/aimee_mullins_my_12_pairs_of_legs?language=mg (accessed on 23 April 2020).

73. Hullinger, J. This Girl Designed Her Own Superhero Prosthetic Arm, And It Shoots Sparkles. Fast Company. Available online: https://www.fastcompany.com/3058221/this-girl-designed-her-own-superhero-prosthetic-arm-and-it-shoots-sparkles?utm_source=mailchimp&utm_medium=email&utm_campaign=fast-company-daily-newsletter&position=3&partner=newsletter&campaign_date=03282016 (accessed on 23 April 2020).

74. Kumari Campbell, F. *Contours of Ableism: The Production of Disability and Abledness*; Palgarave: New York, NY, USA, 2009.

75. Bruce, C. Rev. of Dis/Ability Studies: Theorizing Disableism and Ableism, by Dan Goodley. *Can. J. Disabil. Stud.* **2016**, *5*, 178–183. [CrossRef]

76. Wendell, S. *The Rejected Body: Feminist Philosophical Reflections on Disability*; Routledge: New York, NY, USA, 1996.

77. Wolbring, G. Expanding Ableism: Taking down the Ghettoization of Impact of Disability Studies Scholars. *Societies* **2012**, *2*, 75–83. [CrossRef]

78. Suarez-Villa, L. *Globalization and Technocapitalism: The Political Economy of Corporate Power*; Routledge: New York, NY, USA, 2016.

79. Wälivaara, J. Marginalized Bodies of Imagined Futurescapes: Ableism and Heteronormativity in Science Fiction Munch's Painting? Painting Reproductions on Display. *Cult. Unbound J. Curr. Cult. Res.* **2018**, *10*, 226–245. [CrossRef]

80. Goggin, G.; Newell, C. *Digital Disability: The Social Construction of Disability in New Media*; Rowman & Littlefield: London, UK, 2003.

81. Sobchack, V. A Leg to Stand on: Prosthetics, Metaphor, and Materiality. In *The Prosthetic Impulse: From a Posthuman Present to a Biocultural Future*; Smith, M., Morra, J., Eds.; MIT Press: Cambridge, MA, USA, 2006; pp. 2–41.

82. Charlton, J.I. Peripheral Everywhere. *J. Lit. Cult. Disabil. Stud.* **2010**, *4*, 195–200. [CrossRef]

83. Wolbring, G. The Politics of Ableism. *Development* **2018**, *51*, 252–258. [CrossRef]

84. Bulky, Medieval-Looking Stumps: Laos Amputees Make Their Own Prosthetics. The Irish Times. Available online: https://www.irishtimes.com/news/world/asia-pacific/bulky-medieval-looking-stumps-laos-amputees-make-their-own-prosthetics-1.3082794 (accessed on 7 May 2020).

85. Garland-Thomson, R. Disability Bioethics: From Theory to Practice. *Kennedy Inst. Ethics J.* **2017**, *27*, 323–339. [CrossRef]

86. Aoun, J.E. *Robot-Proof: Higher Education in the Age of Artificial Intelligence*; MIT Press: Cambridge, MA, USA, 2018.

 *philosophies*

Article

The Neuropolitics of Brain Science and Its Implications for Human Enhancement and Intellectual Property Law

Jake Dunagan [1], Jairus Grove [2] and Debora Halbert [2,*]

1   Institute for the Future, Palo Alto, CA 94301, USA; dunagan23@gmail.com
2   Department of Political Science, University of Hawaii at Mānoa, Honolulu, HI 96822, USA; jairusg@hawaii.edu
*   Correspondence: halbert@hawaii.edu

Received: 15 July 2020; Accepted: 23 October 2020; Published: 3 November 2020

Abstract: As we learn more about how the brain functions, the study of the brain changes what we know about human creativity and innovation and our ability to enhance the brain with technology. The possibilities of direct brain-to-brain communication, the use of cognitive enhancing drugs to enhance human intelligence and creativity, and the extended connections between brains and the larger technological world, all suggest areas of linkage between intellectual property (IP) law and policy and the study of the brain science. Questions of importance include: Who owns creativity in such a world when humans are enhanced with technology? And how does one define an original work of authorship or invention if either were created with the aid of an enhancement technology? This paper suggests that new conceptualizations of the brain undermine the notion of the autonomous individual and may serve to locate creativity and originality beyond that of individual creation. In this scenario, the legal fiction of individual ownership of a creative work will be displaced, and as this paper warns, under current conditions the IP policies which may take its place will be of concern absent a rethinking of human agency in the neuropolitical age.

Keywords: human enhancement; intellectual property; copyright; neuropolitics; brain science

---

1. Introduction

In 2019, researchers announced a noninvasive experiment using the human brain to control a robotic arm through a brain-computer interface [1,2]. In addition, in 2019, Brainnet, the first successful brain-to-brain direct interface for problem solving was announced [3]. In September of 2020, Finnish scientists reported what they believe to be the "first study to use neural activity to adapt a generative computer model and produce new information matching a human operator's intention", meaning that using a brain-computer interface (BCI), the scientists were able to generate images of what the humans in the experiment were thinking [4].

Related efforts to enhance the capabilities of the brain include Elon Musk's Neuralink, a project looking at how to create a future where the brain seamlessly and telepathically integrates with technology [5]. No longer science fiction, these examples show that technological advances have brought us to a present where it is possible to link the human body and brain with devices and to network brains via computer. As we enter this particular technological future, it serves us to think critically not just about the technologies we are creating but how we are shaping these technologies and the ultimate implications such conceptualizations will have on creativity and the ownership of creative works under intellectual property (IP) law. When scientists can now pull images from your brain and draw them as graphic depictions of what you are thinking, what does originality mean and who is doing the creation?

Technologies that enhance the human brain have the potential to vastly expand our creative capacity. However, as we will argue, such technological enhancements can also be a threat to individual autonomy by further decentering acts of innovation and creativity as belonging to an individual. Certainly, claims of ownership that ignore the social nature of innovation in favor of individual assertions of originality are problematic. However, even more concerning is the possibility that when the very boundaries of the human body are disrupted by technologies that reimagine how we communicate with each other, what has been considered the creative or innovative work of an individual can be more easily appropriated by those who own these enhancement technologies.

This article seeks to make a futures-inspired intervention into understanding the political economy of the enhanced brain and the possible futures for IP control that emerge through such enhancement. We discuss human creativity in the context of neuropolitics, where the underlying assumptions of what it means to be an autonomous individual are challenged and rethought because of what we are learning about the brain and human consciousness [6]. We argue it is important to take seriously how brain science influences the assumptions used to construct public policy. One central question we raise is who will own the output of creative work in the neuropolitical future; that is, in a future in which human enhancements contribute to work typically deserving copyright or patent protection?

When thinking of possible futures, what if, for example, a pharmaceutical company already primed to aggressively assert ownership over its patents, and one that already requires employees to sign over any intellectual property they create, develops a cognitive enhancing drug designed to amplify human creativity. It only takes a small shift in interpreting copyright law and a creative contract to claim ownership in the resulting innovation for the company. After all, companies already assert ownership over a wide range of intellectual property created by their employees.

Further, what if a company requires the use of cognitive enhancing chemicals in order to secure a more productive and innovative workforce—a work made for hire doctrine on steroids? How will we demarcate the distinction between humans integrated into machines and the work of machines as independent creators? The next generation of software that writes creatively or enhances the creative work of a human author may include a click-through license with language requiring rights to the final product or coauthorship at the least. When humans are the mere raw material for the output of brain-computer interfaces, now integrated into all technological devices, will their creative contributions even be acknowledged? Unless our understanding of human autonomy within the context of modern capitalism and legal schemes changes, the future is most likely one of dystopian corporate control, one where human enhancement benefits those who have engineered them as they increasingly claim the work done by others.

As a work of future envisioning, we seek to follow the threads of the present into possible futures where what seems mostly harmless (or unlikely) today could take on new and more sinister meaning. In doing so, we first look at the current state of intellectual property within the context of authorship and ownership. Second, we will look at how theories of extended mind allow for claims of ownership by those owning the technologies of extension rather than the individual who has used these technologies for creation. Third, we articulate the theory of a neurochemical brain—the brain as enhanced by modern chemistry—to posit yet another possible future where an individual's ownership of their creative work is undermined by alternative claims to who has made the creative contribution.

2. Grounding Intellectual Property in the Individual: Authorship and Ownership in the Age of Cognitive Capitalism

Copyright and patent law in the U.S. are designed to balance the interests of the individual author with the larger public good. Both copyright and patent law are based upon assumptions about the origin of creativity and innovation in the mind of an individual who produces or invents something. The protection of IP exists to incentivize this individual to share their creation with the larger public by extending limited monopoly rights to them. The romantic basis of authorship in the individual author or inventor is foundational to copyright and patent justifications worldwide [7–10]. While

Europe tends to recognize the moral rights of authors in a more detailed matter than the United States, the utilitarian nature of American copyright law makes it far easier to disassociate creative work from an individual creator. Even so, expression as unique to the individual and the idea that the individual expresses their creativity in isolation from others is critical to how both patent and copyright law justify the protections these legal regimes grant. Infringement is the appropriation of someone else's expression in copyright or someone else's invention in patent or an appropriation of some portion of their creative work. The law assumes the autonomous individual author and then plays referee between too much appropriation of what someone else created.

This underlying privileging of the individual author for the purposes of assigning rights under the law demonstrates the importance of the autonomous individual for how we tell a story about what intellectual property protects. However, from their inception, such rights were alienable and assignable [8]. Furthermore, while patent protection in the U.S. ends after only twenty-four years, copyright lasts for the life of the author plus 70 years, meaning numerous works of individual authorship are no longer owned by their creator but their relatives or savvy business partners. Such assignability means that we now live in a world of concentrated intellectual property wealth [11]. Thus, despite locating creativity in the individual, there are already mechanisms in place to assure that any creativity emerging from the mind of an individual can be appropriated and owned within the larger economic structure that is cognitive capitalism [12].

While scholars have critiqued the political economy of IP that allows for the concentration in ownership, such critiques have not fundamentally altered the basic rules of these IP regimes [13–15]. Thus, while in theory individual minds create, those creations are ultimately owned by others, either through the assignment of rights or through the law itself.

Copyright, for example, includes the work made for hire doctrine that assigns ownership of copyright to a company for the work of its employees and allows for corporations to "stand in as the sole author" of such works [16]. While patents must list the human inventors, patent rights can and are assigned to companies who fund the inventive work [16]. Both patents and copyrights can thus decenter the individual author by assigning ownership of copyright not to the creator but to a company for whom such a person works.

To look at how individual authorship is decentered from a different angle, copyright in the U.S. defines the author in statute, thereby providing legal authority to add or remove what kind of authors can assert copyright protection at all. For example, legal decisions are not works of authorship under the law [17]. Additionally, the 9th Circuit recently held in the infamous monkey selfie case that animals do not have statutory standing under the U.S. Copyright Act and so cannot sue as authors [18]. The Copyright Office agrees that animals cannot be authors. When updating their compendium on copyright practices, the U.S. Copyright Office specifically precluded copyright from extending to animals or to works created by machines, indicating that to be registered for copyright protection "original works of authorship must be created by a human being" [19]. These are interesting decisions but do not directly address the issue of copyright or patent ownership for works or inventions produced with the aid of a cognitive enhancement.

Given the statutory nature of authorship, the interpretation of what constitutes original authorship and who contributes can shift over time. The claim we make in this article is that as we learn more about how brains function and as additional enhancements for human creativity make it clear that it is not the autonomous individual working alone that creates but rather the networked and chemically altered human that creates, the boundaries of ownership (which have always been a legal fiction) will become even less solid. Into that space will come additional claims of ownership by those facilitating creation because as our new understandings of the brain demonstrate and new technologies make possible, the human brain is no longer isolated within the body. The brain has been extended, chemically altered, and reproduced in ways that will further undermine the already tenuous ownership rights individuals assert over "their" creativity.

The prevailing political approach to IP is to increase punishment for infringing IP rights and to further concentrate ownership over the world of ideas [11]. How the law is shaped in the future to further enhance such ownership may draw upon how we understand originality and creativity. New scientific innovation displaces that creativity from the individual author and places it onto networked minds or technological enhancements. Thus, this article suggests that we ought to be cautious about where the paths of research take us as we enter a neuropolitical future lest we become the monkey in future copyright calculations—not the originators of creativity because, as the argument will go, such creativity is located in the alterations we have made to our brains not in the individual human brain at all.

3. The Extended Mind

In the late 1990s philosophers Clark and Chalmers put forward the idea of the extended mind, arguing that cognition went beyond the brain to include devices used by individuals to retain knowledge and help them engage in daily life [20]. While the initial thesis used the notebook of a man (Otto) with Alzheimer's disease as an example, the logic of the extended mind is easily translatable into contemporary usage of technologies not envisioned in the 90s, including the ubiquitous adoption of "smart" phones, tablets, computers, Google searches, and much more.

The theory of the extended mind acknowledges that the brain/body/world barrier is artificially constructed and much more permeable than advocates for autonomous individuality might wish to make it [20]. This model takes as its starting point an integrated connection between the human and the larger world. Such a connection acknowledges that the human brain is already deeply integrated into technologies that undermine the brain/world barrier. While use of the Internet and data storage and retrieval systems come to mind quickly as evidence of the extended mind, other layers of brain/world interactions such as neurocontrolability via architecture can also explain how the extended mind may come into being and the possible controls over it.

Later work by Clark defended the extended mind theory in the context of cognitive tools outside the self, and used the concept of "ecological control", where we see ourselves "as biologically-based (but not biologically imprisoned) engines of ecological control" which "may help us to develop a species self-image more adequate to the open-ended processes of physical and cognitive self-creation that make us who and what we are" [21]. The extended mind hypothesis along with even more radical conceptions, such as Wendt and Kauffman's quantum mind hypothesis, insist on preserving the concept of mind as distinct from brain [22,23]. Thus, the brain of the extended mind is materially situated and no longer confined within the body.

The extended mind thesis is difficult for many to accept because it displaces agency, but it is also quite compelling because we all engage in daily "outsourcing" of cognition to the devices in our possession and the institutions that structure our lives. As a result, Clark and Chalmers' thesis caused considerable debate amongst cognitive scientists, philosophers, and others. Some sought to defend a more unitary sense of individual cognition and worried about "cognitive bloat", where everything became labeled as part of our cognitive processes once the brain/body barrier was disrupted [24]. Because the extended mind thesis undermines the centrality of the individual mind in the cognitive process, much of the debate about this thesis can be characterized as either affirming how lines between the individual and others are blurred or, as an effort to shore up our belief in the supremacy of individual cognition as distinct from external factors.

Some scholars offer objections to the thesis arguing that an "external device" cannot store a "person's beliefs because the stored statements do not have intrinsic intentionality" [25]. Others take on the concept of social cognition and argue against the notion that the Internet (or other devices) can function as an external mode of cognition. While we may have become reliant upon Google to seek out and store information for us, Huebner argues that it is merely a "contextual factor that affects the operation of an already existing cognitive system" [26].

More recently, efforts have been made to refine the concept of the extended mind. Gerken suggests the concept of outsourced cognition as an alternative hypothesis to the extended mind thesis [27]. Outsourced cognition retains the notion of the individual but acknowledges that some cognitive tasks occur outside the individual. However, unlike the extended mind thesis where cognitive parity is assumed between the human mind and external agents, outsourced cognition assumes a disparity with the human mind remaining central and in control of cognitive processes [27]. Gerken sides with those who reject the extended mind hypothesis, while also acknowledging the interrelationship of the individual to his or her external environment. Others seek to understand the larger social environment and its role in cognition.

Gallagher, for example, pushes the concept of the extended mind further and proposes the concept of a socially extended mind [24]. The socially extended mind takes into consideration the importance of social institutions as modes of cognition—legal institutions, cultural institutions, and so forth [24]. Institutions are part of our cognitive process and have been collectively created to help store human memory [24]. While institutions (beyond actual technological devices) serve a role in structuring human cognition, Gallagher seeks to avoid the accusation of cognitive bloat by suggesting that only when an individual engages with a specific social institution does it take on a cognitive role [24]. As a related example, Susan Goldin-Meadow and her team at the University of Chicago showed how the use of simple finger gestures can help "offload" some of the cognitive load during mathematical grouping exercises. Those who used their hands in the calculations were faster and more accurate than those who did not [28,29].

Another avenue of critique and extension of the extended mind theory comes in the form of questioning the way memory is conceptualized [30]. Is external "memory" as described by Clark and Chalmers really memory at all? Michaelian argues that external memory and biological memory are distinct and both cannot be considered "memory" in the same context [30]. Clark and Chalmers use a relatively simple analogy of how memory functions to set out their thesis, positing that using a notebook to remember is the same as using one's own biological memory. However, a notebook or computer "remembers" very differently from a human [30]. In fact, as others have pointed out, memory itself is inherently faulty [31]. Michaelian argues that instead we need to focus on the interactions between biological and external memory, given that the categories are so different [30]. That being said, he supports the conclusion that devices external to the person are cognitive aids but questions if they can be called memories.

One question raised within the context of the extended mind is where we should place "epistemic credit" for forming beliefs [32]. Given the notion that cognition happens outside the individual or can be attributed to different agent's and their influences on the individual, it may be that the credit for knowledge generated must be assigned outside the individual mind [32]. What Proust calls dynamic coupling means that processes outside the brain may be responsible for new memories, skills, or knowledge and we must then ask what type of credit needs to be attributed to these external forces in cognition and the formation of beliefs and ideas [32]. To Proust, an individual working within a "social scaffolding" must give "epistemic credit" to the social structures that makes their cognition possible. Those who work outside a given social scaffolding can take additional epistemic credit for themselves. Such epistemic credit and social scaffolding set up a different method of ascertaining ownership in any type of intellectual property.

Proust compares Sally, an experimental psychologist, to Srinivasa Ramanuja, a self-taught mathematician. Sally received an excellent childhood education, enjoyed an "epistemically favorable environment", was trained in using software such as MatLab and educated in how to conduct appropriate experiments. As a result, "epistemic credit should be spread to all the agents, machines and software that contributed to her mind being shaped "in the right way" for significant outcomes to be produced [32]. Ramanuja, in contrast, is "a self-trained mathematician from India, contributed major results to mathematical analysis, number theory, and many other mathematical domains" [32].

With a much more limited social scaffolding in place, more epistemic credit ought to be extended to Ramanuja than to Sally according to this logic.

Such questions of attribution, prior to the extended mind thesis ended at the individual. While the individual might acknowledge the support and help of others (remember the classic quote from Newton, "If I have seen further, it is because I stand on the shoulders of giants", it remained unquestioned that it was the individual who was responsible for the creative act. No matter which permutation of the extended mind, or extended cognition thesis you accept, once the brain/body barrier is broken, the range of agents from social structures, individual educators, technological devises, cognitive enhancing drugs and much more, become relevant to any claim of creativity and hence ownership. Thus, the extended mind thesis opens up new options for claiming ownership in the context of IP laws. In an era of cognitive capitalism such arguments merely ensure enhanced inequality as ownership is concentrated even further by those owning the contemporary means of production [12].

If indeed individuals owe some amount of epistemic credit to the extended cognitive and social structures that make their knowledge possible, we might want to be more cautious about who, what, and how patents over neurotechnology are granted. Between 2010 and 2014 neurotechnology patents rose dramatically, from 800 in 2010 to 1600 in 2014 with the biggest patent awardee being the consumer-research company Nielsen [33]. Nielsen and other advertising companies are preparing to dominate in a neurotechnological age where they can control (via patent ownership), "ways to detect brain activity with EEG and translate it into what someone truly thinks about, say, a new product, advertising, or packaging [33]. Other companies are also patenting neurotechnological innovations. Microsoft, for example, holds patents "that assess mental states, with the goal of determining the most effective way to present information. If software knows a user's attention is wandering, it could hold back complicated material" [33]. Or consider Microsoft's patent that "describes a neurosystem that claims to discern whether a computer user is amenable to receiving advertisements" [33].

In each case, the symbiosis between a technological device and the individual human brain is clear. As part of an extended mind, an individual using such software becomes part of Microsoft's or Nielsen's neurosocial order. If Microsoft controls how you access and view information based upon feedback from your brain, but does so in a way that further enhances your ability to think and produce, what portion of that productive product can they claim? Take for example, the use of the AI algorithm GPT-2 that a business graduate student used to write his term papers. He contributed an outline and a few sentences and the algorithm did the rest [34]. It only takes a few additional sentences in the end user licensing agreement to assert copyright control over the final product produced with the help of next generation technology tools that do most of the cognitive work for its human user. While the graduate student using GPT-2 indicated the quality was merely adequate to get a passing grade, it is only a matter of time before quality improves. In fact, the newer version of GPT-2, GPT-3 was recently given a byline in *The Guardian* for writing an essay on why we should not fear AI, an essay where the human contribution was a few sentences with the AI writing the rest [35].

Brain scientists are now able to view images a human mind is creating, tapping directly into the brain to do so [36]. Scientists can view movies the subject has already seen and these scientists believe this technology "paves the way for reproducing the movies inside our heads that no one else sees, such as dreams and memories" [37]. If scientists can pull from our brains internal thoughts and memories, how might we construct ownership over the resulting images? So, for example, in a profit-oriented technologically driven world, what stops an enterprising company from harvesting these thoughts and memories for future creative work owned and produced by them? Imagine a company using the work made for hire doctrine to tap into its employee's minds to mine for creative works.

In yet another possible scenario, scientists have now successfully established brain-to-brain communication using the Internet, both with a human/rat interface (where the human was able to move the rat's tail) and a human/human interface [38]. A recent effort at collaborative problem solving using networked brains is "Brain Net" [3]. While still more science fiction than science fact, research is being done to establish neural communication possibilities that further integrate the human mind with future nano-based technologies [39]. What happens when we do not modulate our IP laws but IP is literally streaming through our heads via the neuronet of the future? Such a world takes the concept of piracy to an entirely new level.

When viewing the extended mind produced by technological enhancements to the mind, we are posing questions about what the future looks like with only minor adaptations to how we conceptualize ownership of IP. The groundwork is already set because while ostensibly built upon the romantic author, the individual has almost always been required to give up control of IP to the corporations that produce it. The insertion of enhanced extended mind technologies disrupts the barriers between human brains and the larger networks that inspire creativity and thus have the potential to further displace who owns creativity. In doing so a gap is opened where the companies that already benefit from ownership of IP will be set to further monopolize creative property of the future.

4. The Neurochemical Brain

In 1781 Immanuel Kant published what is still the most robust philosophical description of what constitutes consciousness as an autonomous and unique capacity of the human species [40]. That there is knowledge from an inner world that is not merely contingent upon sensation is one of the most sacred conceptions of what it is to be human. Creativity, or the labor of creating new ideas, is conceptually dependent upon the philosophical foundation that the human mind is somehow above, or at least separated from, the matter that is its substrate for action, the brain, and its substrate for provocation—external reality [41]. The mind, for Kant, was categorically different from the brain and the seat of human freedom, creation, and intellectual capacity. To use phrases such as "intangible property", or so-called "creations of the mind" is to invoke a Kantian proposition widely accepted since the end of the 18th century.

In establishing the mind as another world from material experience, both the creator and the substance of IP is invented. In no other economic era could Kant be as significant as to the era of today's idea economy. What others have called cognitive capitalism is organized by the ability to monetize brands, ideas, processes, and even origins as protected assets [12]. In a world in which companies are often worth more before they exist than after, the realm of ideas, of mind, is essential [42].

However, the grounds on which Kant made his claims for a representational mind are quickly eroding in an era in which brain-to-brain emails have been sent without the use of language or meaning making [43]. The distinctions philosophers often make between states of mind and brain states seems less and less cogent after decades of psychopharmaceutical use as well as increasing research into the application and significance of psychotropics once thought recreational such as LSD and MDMA [44]. Chemical interventions into the brain combined with more refined research on electrical stimulation, such as deep brain stimulation used to treat severe depression, suggests that what we understand as the mind is plastic and malleable. Furthermore, the notion of self which we correlate to mind is not under its own dominion but can be altered as well as steered by means we would not think of as within the Kantian understanding of consciousness, meaning that the self is not changed through rational means of persuasion or self-discovery but through altering the very condition of possibility for consciousness itself [45]. To invoke the work of Catherine Malabou, the brain is plastic, it is an organ capable simultaneously of being altered and altering itself [46]. In fact, according to Malabou, it is precisely this torsion between being formative and formable rather than either independent or determined which allows for education, agency, and creation to coexist in the human brain such that something like intelligence can emerge [47].

In this section we will describe changes in the technological capabilities being made possible from the image of a plastic chemo-electric brain where the human self is reachable by means far outside the bounds of reason. The speculative extrapolations emanating from this type of contemporary research, we believe, demonstrates how the current conception of legal individuals on which property rights, particularly intellectual property rights rest, may come under significant stress as the individual comes to appear much less singular and much less independent, not only from their external environment—the extended mind—but even from their internal environment—the realm of serotonin and neurons.

In June of 2019, Vetere, Tran, Moberg, et al., published "Memory Formation in the Absence of Experience" in the pages of *Nature* claiming that they had determined a means by which memories could be created in mice sufficiently complex to guide the future behavior of the mice. That is, the mice could remember experiences that they had not had, which then, upon the replication of the memory in the amygdala, created aversion or attraction to things the mice had not previously encountered. In the words of the researchers, "given this current level of understanding about how memories are localized and coded in the brain, it should be possible to reverse engineer this process and artificially implant a memory for an otherwise never experienced event" [48]. The process involves a combination of olfactory and electrical stimulation to coordinate the production of a neuronal pathway which can be recreated in the presence of a future encounter. Unlike the behavioral conditioning of Pavlov or Skinner, the mice in this study were being trained without the trial and error experience that would be either rewarded or punished. Mice in this experiment were being steered toward a future that did not depend on a present they had experienced. Or, to put it another way, the mice were learning from mistakes they have not yet made.

In the immediate aftermath of the announcement, ethical questions emerged about what this research meant for the future of humans. Neuroscientist Robert Martone penned an editorial for *Scientific American* suffuse with dystopian peril and gloomy prospects for human accountability and the possible uses of these techniques by military agencies such as the infamous Defense Advanced Research Projects Agency [49]. The concern of Martone and others focused on the ways memories might be removed, particularly in the context of war, such that war crimes could be wiped from the brains of the soldiers that had committed them.

Certainly, the veracity of one's own memory plays a significant part in one's sense of responsibility. One can hardly feel guilt or remorse for something one has no memory of having committed. However, the moral panic over a world of malleable memories misses, we believe, the significance of this research on at least two counts. First, the emphasis on "deleting" memories is somewhat misguided. While it is true that the research discussed helps confirm theories about how memories are encoded and stored, which may be used someday for developing means of targeting memories, the techniques used in this study only work to create not destroy memories.

Second, there are many other ways to impede or destroy memories. Many methods of destroying memories are in fact natural to the body's own mechanisms for protection. Trauma often induces the brain-body network to suppress memories or create false memories to help an individual avoid ones that might impede the ability to survive. There are also artificial means by which memory formation can be undermined, such as using beta blockers or other adrenaline suppressing drugs to alter how memories are formed and stored. Either internal physiological response or external interventions that already exist can significantly alter the capacity and accuracy of recall in ways that regularly undermine, for instance, the capacity of an individual to serve as a compelling witness in a tribunal or court room.

The ability to create memories seems very different from these concerns. If the concern over deleting memories is that one loses part of who they are, i.e., that you are what you have experienced, then the addition of memories could create opportunity for more agency or more freedom through the capacity to consider options and draw on knowledge you would not have otherwise had from direct experience. Of course, this assumes all knowledge is generative rather than constructive. One can imagine the crushing guilt created from the belief of having committed a heinous act you did not believe yourself capable of (and in this case perhaps were not capable of). However, such a false or additional memory does not diminish your faculties of remorse or responsibility as much as it misdirects them.

What is novel both philosophically and legally about Vetere, Tran, Moberg, et al.'s research is almost the opposite. It is not the erasing of a past but the creation of a different future at stake.

The creation of embodied memories without previous experience is a horizon of possibility for education that is simply unprecedented. Humans, with their capacity for scenario planning, that is the possibility of simulating the future in our conscious minds, certainly learn without direct experience of the things they are learning. We have the capacity to gain knowledge through the experience of reading, listening, or watching. Many academics even become experts in things for which they have no direct experience. The bedrock of the study of history, for example, is creating knowledge about events for which there is no possibility of direct experience.

However, it only takes one attempt to build a piece of Ikea furniture to determine that having read and observed the visual instructions is in no way tantamount to building the furniture. The observation of a basketball game tells you little of what it feels like to take a jump shot, much less prepare you to accomplish this act successfully. Similarly, the rapid reaction time of aversion and attraction dealt with among mice is not something likely to be gained without experience. Descriptions of hot stoves are rarely sufficient to prevent young children from touching them, certainly not to the extent of an actual burn.

What is important is that Vetere, Tran, Moberg, et al. present us with a very specific model of a chemo-electric brain that is programmable. Brain states can be stimulated, stored, and remembered through direct chemical and electrical influence bypassing the normal sensory inputs we associate with awareness and the conscious and unconscious sorting of that sensory data into something that goes as meaningful decision-making. While aversion programing in a mouse is potentially a long way from implanting a memory, not to mention an entire education, it does raise significant questions about how one would think about the legal status of implanted memories and, importantly for this article, the ownership of the results.

In the rest of this section we will discuss the significance of a potentially programmable chemo-electric brain. We are not arguing that we should infer from current research that all behavior can be directed—what used to be called mind control. Instead, what is at stake is how an experiential brain can be modified externally and artificially and what the ramifications of that capacity are for how we think of the creation and ownership of knowledge.

At some basic level we do not think of the use of knowledge gained from a high school textbook as plagiarism or theft because of how we understand the role of the learner, the individual reading the textbook, in the process of interpreting, altering, and reusing the information in question. If a major textbook publisher sued a teacher for teaching the quadratic equation the way they had learned it from a specific textbook, we would think the suit ridiculous. One could dismiss this example on the basis that quadratic equations are general knowledge. However, if the teacher were to photocopy the textbook and distribute the photocopied text to the students, the legal grounds for suit would be much different.

We presume in questions of intellectual property and copyright a kind of transmission variability in the processes of learning and doing as well as a kind of discrete and independent individual who does the creating or learning. However, what if the knowledge at hand was transmitted directly into the brain? The intellectual property or proprietary textbook was not learned but directly implanted. Recall and use would not be like accessing one's own knowledge but accessing someone else's. Would we expect the "user" to pay royalties? If what was implanted had not been authorized for sale or use would we think it was a reasonable remedy to remove knowledge from someone's brain?

To think about this in a context somewhat outside the sci-fi present confronting us, many early theories of property, such as those presented by John Locke, presumed that property was acquired through labor. Accumulation of land, for instance, for Locke came from the mixing of land with one's labor [50]. Developing or cultivating the land was the grounds for claiming it and disputing the claims of others. Beyond the ways in which this theory of property was leveraged and even designed to displace Native Americans who were thought to merely reside on the land rather than cultivate it, the western tradition of property is always indebted to a sense of labor. That labor is inherent in its congealed form as money, as well as in the exchanged form of a contract when one sells one's labor to an employer, or in the making of something as in the case of craft or invention. At the heart of the concept of property is the individual's labor in acquiring or creating the property in question.

In the case of implanted memories one can imagine quite easily that the people receiving the memories would be thought of as consumers or users rather than producers or laborers. Imagine being required by a job to undergo job training which was implanted. The specific procedures or techniques needed to perform the job would be added to one's brain. Maybe these procedures are proprietary, highly specific industrial processes which are closely guarded and legally protected. If you were to leave the company would you be required to "give them back" or have them removed in some way? If you invented something at home, during your off hours, would you find yourself in litigation to determine if the invention had made use of the proprietary memories that belonged to your employer? How much would you need to demonstrate you had altered or recombined the proprietary memories with new knowledge or individually unique insight before the court would accept that your invention was yours?

In October 2020, Moreaux, Yatsenko, and Sacher, et al., proposed what they called a "new paradigm" for brain imaging [51]. Unlike the clunky and often disputed significance of functional magnetic resonance imaging scans (fMRI), Moreaux and his team proposed the implantation of neurophotonics deep in the brain which would be capable of recording neuronal activity at the scale of the single cell and in real time. The new paradigm they speak of is one in which the implants would be capable of recording and mapping every brain cell's activity continuously. Combining the increasing knowledge of how memories are formed and stored, as well as the ability to read those memories, could lead to a legible mind. These would be brains that could be read and monitored without the clumsy use of communication, much less language to mediate between brains.

If technology matures that can record and map memories at their inception as well as their external or extended mind origins, what becomes of concepts of creation and ownership? In what ways will companies make claims upon their employees in a commercial environment that has already normalized extensive surveillance of their employees. Imagine current efforts to prohibit memory sticks from coming and going at secure facilities like the NSA becoming even more invasive. Is it possible to imagine ideas you are only allowed to have in your brain at work? Or that as long as they were in your brain you were, legally speaking, always at work?

In more concrete terms, if neural mapping reaches the ability to lift the black box from the generation, as well as application, of ideas, and humans are able to map and monitor in real time "where" ideas come from, what is left of the concept of originality? Following Ashby, every act of creation must be a form of neurocognitive mimicry as the "environment" is our "dictionary" [52]. Will every new expression ostensibly created originally by a unique human mind in reality be yet another nonoriginal remix [53]? Imagine a neural map of Jimmy Page playing Led Zeppelin's "Since I've Been Loving You." Every lick from every measure echoes and recombines what are likely thousands of memories of blues standards heard played and replayed by any number of other musicians. With the capacity to track and even tag the neurological structure of every combination of notes, stripping out all that is not original, what would it be that Led Zeppelin would claim as its own? Ironically, Led Zepplin has recently settled with the families of blues musicians that were able to demonstrate in court that there was intellectual theft [54]. Irrespective of what kind of practical legal regime emerges amidst such technologies, what is certain is that the "intangible" and "incorporeal" will be shown to have been technical not metaphysical problems. Brain scans will become evidence of nonoriginality.

Consider the legal dispute between Bruce Springsteen and the Donald Trump re-election campaign over the use of the song "Born in the USA" [55]. Could an artist demand the return of the ability to perform songs based on moral turpitude clauses where the licensed musician who used the memories of how to play "Born in the USA" was caught on video making racist remarks or sexist remarks? What theory of property could account for such possibilities? As Grove has written previously:

> One can already imagine the intellectual property (IP) disputes between the next generation
> of Steve Jobs and Bill Gates when the ideas in question were created via linked brain

communities or research being done on a wireless neural net. How ideas, concepts, processes, and techniques of all kinds will be recorded and tracked or altered in such a world boggles the mind but will no doubt play a significant role in politics, military affairs, and the economy. As philosophically specious as the creator or inventor myth is, innovation in brain research will materially alter debates around freedom and personhood in ways that mere argument simply could not [45].

Our common conceptions of creativity are inherited from notions of consciousness that presume the mind to be a substance or world apart from the brute mechanics of metabolism and bodily process. Even those of us who have given up on a notion of the soul or think ourselves firmly in the brain rather than mind camp still trade on metaphors to describe human distinctiveness that are indebted to dualist understandings of mind as distinctive. Our property regimes are dependent on these metaphors for conceiving of the individual as well as the creative labor we imbue with individuality and vice-versa.

In such a brave new world it would not just be inventors but companies as well that would make claim to anything in the realm of knowledge as proprietary. A race to hoard ideas could ensue depending on how new property regimes emerged to adjudicate conflicts. Would we see the use of machine intelligence to attempt to generate all possible ideas in advance in an attempt to gain legal protection before it was determined that a particular idea had merit? If this sounds absurd, this is precisely what Noah Rubin and Damien Riehl attempted in 2020 when they designed a platform to generate and record all 68 billion possible melodies in 12 tone Western music, though they fortunately donated the results to Creative Commons [56]. Similar things have been attempted in genetic coding, though the impracticality of sorting and determining the application of such randomized codes has proved an obstacle. However, such obstacles may be solved as machine learning advances.

What is at least worth considering, depending on the maturation of already existing technologies, is that the model of the individual on which property regimes, particularly intellectual property regimes are based, will come under significant strain as well as produce novel grounds for dispute and regulation. Many of these disputes and regulation will cut to the core of the most intimate regions of our bodies and the grounds on which we currently understand our conceptions of self and individuality. Science ought not steer our political choices. Perhaps the notion of an autonomous individual should remain intact in the face of corporate-backed technological networks that would claim all innovation as their own. Legal individuals are always somewhat pragmatic and fictive. However, the appearance of connection to the world of experience is essential for the legitimacy of pragmatic fictions and as that world changes the grounds for legitimacy will also change.

5. Conclusions and the Futures of IP

We tend to think of ourselves as whole, intact, coherent bodies. However, the perception of inhabiting and being in a body in space arises from a complex performance of constant feedback mechanisms grounded in our nervous system. Maybe the most remarkable thing about this feeling of self-representation in a body is how universal and robust it feels, but at the same time how easy it is to subvert. Ramachandran's famous work on "phantom limb syndrome", where a person with an amputated limb can still feel sensations, often unpleasant from that nonexistent limb, demonstrated some of the previously unknown or poorly understood processes for the subjective experience of embodiment [57]. Alternately, those with neurological damage that impairs their ability to link sensory data with emotional responses can feel disconnected from their own "intact" body parts, often assigning notions of possession or remote control by sinister forces to the movement of their own limbs [58].

So, the mind (if it exists beyond the material brain) and body are locked in an intimate dance, with certain missteps causing major impacts on self-perception and feelings of embodiment. These tightly coupled feedback loops are also happening with other entities in the body, including the trillions of human and nonhuman cells executing their code inside of us—or more accurately, executing codes that turn physical matter into us. Uncovering these previously hidden communicative channels, perceptual pathways, and assembling machines has drastically altered the way we understand the boundaries of the body, the relationship of mind to body, and our own sense of human identity. This understanding has had other effects as well, which are currently being manipulated for fun and profit.

The neuropolitical subjectivities emerging from new understandings of the brain-body relationship and the technologies that leverage this new knowledge fit uncomfortably within the current system of IP. A productive approach might be to embrace the noisiness and contingency of embodied cognition and build IP law and policy from these foundations rather than the simpler but ultimately misplaced notions of bounded individuality and mind-body independence. What we know about the embodied brain decenters our notion of individuality. However, there is a threat of decentering the human brain when that brain is situated within the contemporary political economy of cognitive capitalism.

A paradox of the neurocentric age as it interacts with IP is that while we may undermine the legitimacy of the individual as the location for creativity, contemporary legal structures have no problem assigning ownership to the already existing entities that control the vast majority of IP. The scaffolding is already in place to enhance nonindividualized ownership and control over creative work. Take as an example the proliferation over the last generation of noncompete agreements and the requirement that workers assign their IP over to their employer as a condition of employment, even if created on their own time. The neurocentric turn will simply push this ownership beyond the body/brain boundary more easily. Imagine the contracts of the future when employees are much more integrated into the technologies of their corporations and when mental processing can be farmed and appropriated more easily.

The military has already invested heavily in brain-related science. While research into how to cancel memories and subvert the impact of traumatizing experiences may be helpful to those with PTSD, other memory enhancement efforts are designed to control and recall memories for better military applications [59]. Imagine a world where the military owns not just your body but your memories as well [60]. Of course, technically, such memories already and inevitably fall under the category of classified and thus there is no need to resort to claims of IP for protection.

The implications emerging from this field of study may alter our understanding of what it means to be human and pave the way for our further integration into a technological future of human enhancement. The increasing sophistication of computers either integrated or stand-alone also raise questions about how we might draw lines around creativity in the future. In the final instance, the law is more a reflection of power and interest than "reality". However, advances in brain science both enable new fictions and disrupt old ones, creating new opportunities for profit and power.

The futures we paint by focusing on human cognition and its enhancement via technology are not optimistic, in part because while the evidence suggests we are not autonomous individuals and our systems of IP have been built upon this faulty assumption, the alternative to individual autonomy allows for corporate ownership of creativity that is particularly disturbing. Furthermore, the technical revolution taking place in tandem to these shifting images of the brain will accelerate the pace of change and raise the stakes of inaction. All possible neurofutures represent a fundamental disruption to how we understand IP that exceeds the consequences to copyright ownership caused by digitalization and file sharing. If file sharing and digital copying let loose new modes of distribution and heightened calls of piracy and ownership, then the possibility of our neurofutures represent a revolution in the mode of production of ideas themselves. Any number of questions along these lines can and should be asked.

Author Contributions: The conceptual framework of neuropolitics and neurofutures are contributed by J.D. The intellectual property and scenarios associated with the extended mind were contributed by D.H. The neurochemical brain and associated scenarios were contributed by J.G. All authors have read and agreed to the published version of the manuscript.

Funding: This research received no external funding.

Conflicts of Interest: The authors declare no conflict of interest.

References

1. Edelman, B.J.; Meng, J.; Suma, D.; Zurn, C.; Nagarajan, E.; Baxter, B.S.; Cline, C.C.; He, B. Noninvasive Neuroimaging Enhances Continuous Neural Tracking for Robotic Device Control. *Sci. Robot.* **2019**, *4*. [CrossRef]
2. First Ever Non-invasive Brain-Computer Interface Developed. Available online: https://www.technologynetworks.com/informatics/news/first-ever-non-invasive-brain-computer-interface-developed-320941 (accessed on 19 June 2020).
3. Jiang, L.; Stocco, A.; Losey, D.M.; Abernethy, J.A.; Prat, C.S.; Rao, R.P.N. BrainNet: A Multi-Person Brain-to-Brain Interface for Direct Collaboration Between Brains. *Sci. Rep.* **2019**, *9*, 6115. [CrossRef] [PubMed]
4. Rosso, C. New Brain-Computer Interface Transforms Thoughts to Images. Available online: https://www.psychologytoday.com/blog/the-future-brain/202009/new-brain-computer-interface-transforms-thoughts-images (accessed on 29 September 2020).
5. Koetsier, J. Our Jedi Future: How It Feels to Use Brain-Machine Interfaces. Available online: https://www.forbes.com/sites/johnkoetsier/2020/06/01/controlling-tech-with-your-mind-how-it-feels-to-use-brain-machine-interfaces/ (accessed on 19 June 2020).
6. Dunagan, J. Politics for the Neurocentric Age. *J. Futures Stud.* **2010**, *15*, 51–70.
7. Coombe, R.J. *The Cultural Life of Intellectual Properties: Authorship, Appropriation and the Law*; Post-Contemporary Interventions; Duke University Press: Durham, NC, USA, 1998.
8. Rose, M. *Authors and Owners: The Invention of Copyright*; Harvard University Press: Cambridge, MA, USA, 1993.
9. Woodmansee, M.; Jaszi, P. (Eds.) *The Construction of Authorship: Textual Appropriation in Law and Literature*; Duke University Press: Durham, NC, USA, 1994.
10. McLeod, K. *Owning Culture: Authorship, Ownership, & Intellectual Property Law*; Popular Culture Everyday Life; Peter Lang: New York, NY, USA, 2001; Volume 1.
11. David, M.; Halbert, D. *Owning the World of Ideas: Intellectual Property and Global Network Capitalism*; SAGE Publications Ltd.: Thousand Oaks, CA, USA, 2015.
12. Moulier-Boutang, Y. *Cognitive Capitalism*, 1st ed.; Polity: Cambridge, UK, 2012.
13. May, C.; Sell, S.K. *Intellectual Property Rights: A Critical History*; Lynne Rienners Publishers: Boulder, CO, USA, 2006.
14. May, C. *The Global Political Economy of Intellectual Property Rights: The New Enclosures*, 2nd ed.; Routledge: London, UK, 2009.
15. Halbert, D. The Politics of IP Maximalism. *Wipo J. Anal. Intellect. Prop. Issues* **2011**, *3*, 81–92.
16. O'Connor, S.M. Hired to Invent vs. Work Made for Hire: Resolving the Inconsistency Among Rights of Corporate Personhood: Authorship, and Inventorship Berle III: Theory of the Firm: The Third Annual Symposium of the Adolf, A.; Berle, Jr. Center on Corporations, Law & Society. *Seattle Univ. Law Rev.* **2011**, *35*, 1227–1246.
17. Carroll, M. Big Win for Public Access to Law—Georgia v. Public.Resource.Org. Available online: http://infojustice.org/archives/42320 (accessed on 27 October 2020).
18. Cullinane, S. Monkey Does Not Own Selfie Copyright, Appeals Court Rules. Available online: https://www.cnn.com/2018/04/24/us/monkey-selfie-peta-appeal/index.html (accessed on 5 September 2020).
19. U.S. Copyright Office. Compendium of U.S. Copyright Office Practices. Available online: https://www.copyright.gov/comp3/ (accessed on 5 September 2020).
20. Clark, A.; Chalmers, D.J. The Extended Mind. *Analysis* **1998**, *58*, 10–23. [CrossRef]
21. Clark, A. Soft Selves and Ecological Control. In *Distributed Cognition and the Will*; Ross, D., Spurrett, D., Kincaid, H., Stephens, G.L., Eds.; The MIT Press: Cambridge, MA, USA, 2007; pp. 101–122.
22. Kauffman, S.A. *Reinventing the Sacred: A New View of Science, Reason and Religion*; Basic Books: New York, NY, USA, 2008.
23. Wendt, A. *Quantum Mind and Social Science: Unifying Physical and Social Ontology*; Cambridge University Press: Cambridge, UK, 2015.
24. Gallagher, S. The Socially Extended Mind. *Cogn. Syst. Res.* **2013**, *25*, 4–12. [CrossRef]

25. Stanciu, M.M. Personal Identity, Functionalism and the Extended Mind. *Procedia Soc. Behav. Sci.* **2014**, *127*, 297–301. [CrossRef]
26. Huebner, B. Socially Embedded Cognition. *Cogn. Syst. Res.* **2013**, *25*, 13–18. [CrossRef]
27. Gerken, M. Outsourced Cognition. *Philos. Issues* **2014**, *24*, 127–158. [CrossRef]
28. Cook, S.W.; Yip, T.K.; Goldin-Meadow, S. Gestures, But Not Meaningless Movements, Lighten Working Memory Load When Explaining Math. *Lang. Cogn. Process.* **2012**, *27*, 594–610. [CrossRef] [PubMed]
29. Goldin-Meadow, S.; Nusbaum, H.; Kelly, S.D.; Wagner, S. Explaining Math: Gesturing Lightens the Load. *Psychol. Sci.* **2001**, *12*, 516–522. [CrossRef] [PubMed]
30. Michaelian, K. Is External Memory Memory? Biological Memory and Extended Mind. *Conscious Cogn.* **2012**, *21*, 1154–1165. [CrossRef] [PubMed]
31. Coronel, J.C.; Federmeier, K.D.; Gonsalves, B.D. Event-Related Potential Evidence Suggesting Voters Remember Political Events That Never Happened. *Soc. Cogn. Affect Neurosci.* **2014**, *9*, 358–366. [CrossRef] [PubMed]
32. Proust, J. Epistemic Action, Extended Knowledge, and Metacognition. *Philos. Issues* **2014**, *24*, 364–392. [CrossRef]
33. Begley, S. Brain Technology Patents Soar as Companies Get Inside People's Heads. *Reuters*, 6 May 2015.
34. Robitzski, D. This Grad Student Used a Neural Network to Write His Papers. Available online: https://futurism.com/grad-student-neural-network-write-papers (accessed on 4 September 2020).
35. GPT-3. A Robot Wrote This Entire Article. Are You Scared Yet, Human? *The Guardian*, 8 September 2020.
36. Hutson, M. This "Mind-Reading" Algorithm Can Decode the Pictures in Your Head. *Science*, 10 January 2018.
37. Anwar, Y. Scientists Use Brain Imaging to Reveal the Movies in Our Mind. *Berkley News*, 22 September 2011.
38. Armstrong, D.; Ma, M. Researcher Controls Colleague's Motions in 1st Human Brain-to-Brain Interface. *UW Today*, 27 August 2013.
39. Naam, R. Neural Dust is a Step Towards Nexus. Available online: http://upgrade.io9.com/neural-dust-is-a-step-towards-nexus-806802917 (accessed on 13 May 2015).
40. Kant, I. *Kant: Critique of Pure Reason*; Smith, N.K., Ed.; St. Martin's Press: New York, NY, USA, 1969.
41. Louden, R.B. *Kant: Anthropology from a Pragmatic Point of View*; Cambridge University Press: Cambridge, UK, 2006.
42. Klein, N. *No Logo*, 10th ed.; Picador: New York, NY, USA, 2009.
43. Grau, C.; Ginhoux, R.; Riera, A.; Nguyen, T.L.; Chauvat, H.; Berg, J.; Amengual, A.P.-L.; Ruffini, G. Conscious Brain-to-Brain Communication in Humans Using Non-Invasive Technologies. *PLoS ONE* **2014**, *9*. [CrossRef] [PubMed]
44. Langlitz, N. *Neuropsychedelia: The Revival of Hallucinogen Research since the Decade of the Brain*; University of California Press: Berkeley, CA, USA, 2013.
45. Grove, J. *Savage Ecology: War and Geopolitics at the End of the World*; Duke University Press: Durham, NC, USA, 2019.
46. Malabou, C.; Jeannerod, M. *What Should We Do with Our Brain? Perspectives in Continental Philosophy*; Fordham University Press: New York, NY, USA, 2009.
47. Malabou, C. *Morphing Intelligence: From IQ Measurement to Artificial Brains. The Wellek Library Lectures in Critical Theory*; Columbia University Press: New York, NY, USA, 2019.
48. Vetere, G.; Tran, L.M.; Moberg, S.; Steadman, P.E.; Restivo, L.; Morrison, F.G.; Ressler, K.J.; Josselyn, S.A.; Frankland, P.W. Memory Formation in the Absence of Experience. *Nat. Neurosci.* **2019**, *22*, 933–940. [CrossRef] [PubMed]
49. Martone, R. A Successful Artificial Memory Has Been Created. Available online: https://www.scientificamerican.com/article/a-successful-artificial-memory-has-been-created/ (accessed on 20 October 2020).
50. Wolin, S. *Politics and Vision: Continuity and Innovation in Western Political Thought*; Classics, P., Ed.; Princeton University Press: Princeton, NJ, USA, 2016.
51. Moreaux, L.C.; Yatsenko, D.; Sacher, W.D.; Choi, J.; Lee, C.; Kubat, N.J.; Cotton, R.J.; Boyden, E.S.; Lin, M.Z.; Tian, L.; et al. Integrated Neurophotonics: Toward Dense Volumetric Interrogation of Brain Circuit Activity—At Depth and in Real Time. *Neuron* **2020**, *108*, 66–92. [CrossRef] [PubMed]
52. Ashby, W.R. *Design for a Brain: The Origin of Adaptive Behavior*, 2nd ed.; Chapman and Hall: London, UK, 1960.
53. Halbert, D.J.; Dunagan, J. Intellectual Property for a Neurocentric Age: Towards a Neuropolitics of IP. *Queen Mary J. Intellect. Prop.* **2015**, *5*, 302–326.

54. Sinnreich, A. Plagiarists or Innovators? The Led Zeppelin Paradox Endures. Available online: http: //theconversation.com/plagiarists-or-innovators-the-led-zeppelin-paradox-endures-102368 (accessed on 20 October 2020).
55. Chao, E. Stop Using My Song: 35 Artists Who Fought Politicians Over Their Music. *Rolling Stone*, 8 July 2015.
56. Chojecki, P. Musicians Generate Every Possible Melody and Make it Public. Available online: https://medium. com/data-science-rush/musicians-generate-every-possible-melody-and-make-it-public-df408e2f751b (accessed on 20 October 2020).
57. Ramachandran, V.S.; Hirstein, W. The Perception of Phantom Limbs. *Brain* **1998**, *121*, 1603–1630. [CrossRef]
58. Carruthers, G. Types of Body Representation and the Sense of Embodiment. *Conscious. Cogn.* **2008**, *17*, 1302–1316. [CrossRef]
59. DARPA Aims to Accelerate Memory Function for Skill Learning. Available online: https://www.darpa.mil/ news-events/2015-04-27-2 (accessed on 27 October 2020).
60. Weinstein, A. The Air Force Wants to Read Your Brain Waves. Available online: http://www.motherjones. com/mojo/2012/01/air-force-wants-read-your-brain-waves (accessed on 18 May 2015).

Publisher's Note: MDPI stays neutral with regard to jurisdictional claims in published maps and institutional affiliations.

 *philosophies*

Article

Law, Cyborgs, and Technologically Enhanced Brains

Woodrow Barfield [1] and Alexander Williams [2,*]

[1]   Professor Emeritus, University of Washington, Seattle, Washington 98105, USA; Jbar5377@gmail.com
[2]   140 BPW Club Rd, Apt E16, Carrboro, NC 27510, USA
*   Correspondence: zander.alex@gmail.com; Tel.: +1-919-548-1393

Academic Editor: Jordi Vallverdú
Received: 12 October 2016; Accepted: 8 February 2017; Published: 17 February 2017

Abstract: As we become more and more enhanced with cyborg technology, significant issues of law and policy are raised. For example, as cyborg devices implanted within the body create a class of people with enhanced motor and computational abilities, how should the law and policy respond when the abilities of such people surpass those of the general population? And what basic human and legal rights should be afforded to people equipped with cyborg technology as they become more machine and less biology? As other issues of importance, if a neuroprosthetic device is accessed by a third party and done to edit one's memory or to plant a new memory in one's mind, or even to place an ad for a commercial product in one's consciousness, should there be a law of cognitive liberty or of "neuro-advertising" that applies? This paper discusses laws and statutes enacted across several jurisdictions which apply to cyborg technologies with a particular emphasis on legal doctrine which relates to neuroprosthetic devices.

Keywords: cyborg; enhancement technology; neuroprosthesis; patent law; copyright law; cognitive liberty; international law

1. Cyborgs, Prostheses, and Law

We are currently undergoing a "technological revolution" in the design and use of "cyborg devices" integrated into the human body. Worldwide there are millions of people equipped with "cyborg technology" ranging from prosthetic limb replacements and prosthetic hands controlled by thought [1,2], to neuroprosthetic devices implanted within the brain [3], and additionally to people equipped with heart pacers or defibrillators, retinal prosthesis and cochlear implants [4]. As people become more and more equipped with "cyborg/prosthetic" devices, important issues of law and policy are raised which result in significant challenges to established legal doctrine. As an example, what law applies to people who hack an implanted medical device, or interfere with the transmission of the wireless signals of devices worn by cyborgs [1,5]? And in an "age of cyborgs", if third parties access another person's neuroprosthetic device, should there be a legal cause of action for individuals suffering harm under a theory of cognitive liberty, freedom of thought, or, in some cases, privacy law? In addition, what is the relationship between legal doctrine and cyborg technologies with regard to our very sense of being; that is, as we continue to enhance ourselves with technology, should laws and statutes be enacted to safeguard our humanity, and our sense of being and identity as homo sapiens? While the focus of this paper is on the laws and statutes of the U.S., there are numerous examples of law from other jurisdictions included in the paper. We include this body of law not only for comparative purposes, but also because we view the issues of integrating cyborg technologies into the human body and mind as so important and challenging for humanity that an international response is required. On this point, it is becoming more common for a particular jurisdiction to show deference to the law(s) of other jurisdictions in dispute resolution thus adopting more of a comparative, or international, law perspective.

Interestingly, the law and policy being challenged by the use of cyborg technology has already attracted the attention of national governments; for example, in the U.S., the White House Presidential Commission for the Study of Bioethics has produced a white paper summarizing ethical, policy, and legal issues associated with advances in neuroscience [6] and in Great Britain, the Nuffield Council on Bioethics has produced a similar comprehensive paper, discussing among others, brain–computer interfaces [7]. In comparison, this paper focuses more specifically on the machine technologies used to create "cyborgs" with a discussion of major legal issues associated with emerging combinations of human and machine. Additionally, the technology used to create cyborgs as discussed in this paper are reviewed in more detail by Kevin Warwick [8] and by Woodrow Barfield and Alexander Williams [9] as part of the special edition on Cyberphenemonology: Technominds Revolution published by this journal [10].

As stated previously by the authors, the term "cyborg technology" is used to refer to technology that is integrated into the human body which not only restores lost function, but enhances the anatomical, physiological, and information processing abilities of the body [1,9]. These include anything from medical implants, such as pacemakers, to near-future brain implants modifying memory and cognitive capability. Further, we use the term "cyborg prosthesis" to refer to artificial enhancements to the body providing computational capability, which operate as a closed-loop feedback system, are upgradeable, and in some cases controllable by thought and/or implanted directly into the body itself (see generally [11]).

Any person could be considered enhanced to the degree that their implant or prosthetic device increases, by computation and/or physical augmentation, their capabilities—it is with cognitive enhancements that we are most interested and our discussion most tied. As an example of the computational capabilities provided to a human-machine combination as a function of "cyborg prosthetic devices," consider a cochlear implant which contains a speech processor and a receiver/stimulator that sits just beneath the skin and sends signals to an electrode array positioned deep in the inner ear (this aspect of the prosthesis performs various computations). Sound in the environment is picked up by a microphone, analyzed and converted to electrical signals by the processor (again, by performing computations), and sent through the skin by a transmitter. Further, a receiver picks up the sound signals and sends them to an electrode array, which is positioned such that it can deliver patterns of electrical activity to the auditory nerve, similar to those delivered by healthy hair cells. From the perspective of increasing the information processing abilities of the person equipped with cyborg technology, there are several levels of computations occurring in this system resulting in the human sensory state of "becoming enhanced"—namely, that they can detect frequencies of sound that were the same as, or in some cases beyond, their normal range before the prosthesis.

While this paper does not include a discussion of artificially intelligent machines that gain sentience as an example of a cyborg, given the trend for artificial intelligence to become more human-like, perhaps this is another viable cyborg category. Finally, while devices such as the Cheetah prosthetic legs (which does not compute and is only externally attached) are considered by some to be a cyborg enhancement, the focus of this paper is more on devices integrated into the human body which compute, as we see such devices as the direction of our cyborg future. Additionally, given the complexity of integrating "computing devices" within the body, there are currently more non-cognitive than cognitive augmentation devices available as enhancements. We note, however, that the enhancement of cognitive functions is not only a major area of current research, but perhaps the most important aspect of the cyborg of the future.

2. Enter the Law

Current cyborgs are becoming equipped with prosthetic devices attached to their body, or even implanted within their body, for purposes which range from medical necessity to voluntary self-enhancement. Generally, we have begun to see a trend in "cyborgs" not only choosing to become equipped with implants and devices in order to *restore* lost functions but to, in some cases, *enhance* performance—be it cognitive, sensory or physical. In fact, already researchers are working to develop an artificial hippocampus to aid in memory recall and other researchers are using the technique of deep brain stimulation as an effective treatment of Parkinson's disease [1,6,7]. Figure 1 illustrates that cyborg technology to enhance one's body or mind, insofar as this represents the seat of a person's "sense of being," creates a human–machine combination which together implicate legal doctrine—an interest of this paper particularly is the area of constitutional law on freedom of thought and also more generally on privacy law. In addition, this paper provides a focused discussion of some of the main areas of law which relate to cyborg enhancement technology, with reference to our sense of being included, using the law and statutes from several jurisdictions to indicate the scope and importance of the issues implicated by the coming cyborg age.

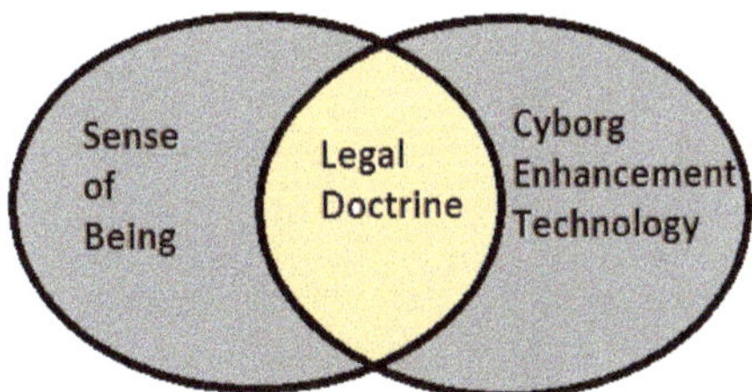

Figure 1. As cyborg enhancement technology becomes more widespread, these technologies will influence our sense of being in that they will begin to change how we function and who we hold ourselves out to be. In response, issues of patent, copyright, and constitutional law that relates to freedom of speech and thought are particularly relevant.

As we become more equipped with cyborg technology, it is relevant to ask: What legal rights are implicated by technology that is being used to repair, upgrade, and enhance the human body and mind? As examples, consider a prosthetic device attached to the body or consider a neuroprosthetic device implanted in the brain. From the perspective of the law, such devices receive intellectual property protection under patent statutes, and the software associated with such devices receive copyright protection (see generally [1]). Further, it may be illegal to interfere with the wireless communication from such devices, and products liability law applies to prosthesis that are defective. Within most jurisdictions, products liability law is applicable in the case of a design or manufacturing defect in a product, but if a third party purposively changes a product after it has left the manufacturer, the manufacturer may not be liable for harm resulting from a product that has been altered- this has implications for the do-it-yourself movement to self-enhance the body. Similar logic would apply to a neuroprosthetic device implanted within the brain, although it seems less likely that someone would change the physical design of the device, although the software controlling the device could be wirelessly accessed and edited. At any rate, once cyborg devices are viewed by a person as part of their body and very being, the distinction between rights for property versus rights for humans becomes blurred and thus doubly important.

Additionally, regarding the procedures to attach cyborg devices to the body or to implant them within the body, if there is negligence which results in harm to the person, medical malpractice may be a valid cause of action. But what if the issue of negligence involves software, and if so, is there still a viable cause of action to pursue under tort law? And for purposes of public safety, a government agency may regulate the use of cyborg technology and decide whether it can be introduced to the public as a consumer product or medical device. Furthermore, constitutional law issues, such as the right to privacy and search and seizure law, may be implicated if the government tracks, without acquiring a warrant, a person by accessing their prosthesis to collect private information or tracking data. It should be clear from the above list, which is certainly incomplete, that there are a host of timely and challenging legal issues which apply to cyborg devices—especially when they begin to intimately influence our most personal spheres of mind and body.

Interestingly, this emerging area of "cyborg law" (the regulation of technology relating to the body) has recently become a topic of interest for legislators, courts, and judges. For example, in *Riley v. California* [12], Supreme Court justices in the U.S. unanimously ruled that police officers may not, without a warrant, search the data on a cell phone seized incident to an arrest. The case had an interesting connection to cyborg law in that the Chief Justice declared that "modern cell phones . . . are now such a pervasive and insistent part of daily life that the proverbial visitor from Mars might conclude they were an important feature of human anatomy" [12]. In the U.S., this may be the first time the Supreme Court has contemplated the concept of a cyborg in case law, although as dicta. But the idea that the law will have to accommodate the integration of technology into the human body has actually been considered for some time—one example being disability law in employment settings.

2.1. Enter Regulatory Agencies

Based on medical necessity, which is a main factor motivating the need for cyborg technology, much of the current legal code regarding prostheses and implants are those that regulate medical products to ensure public safety—in several jurisdictions these laws are usually nested under disability, public welfare acts, or government agencies which regulate the design and use of medical devices. For example, in the U.S., under the Federal Food, Drug, and Cosmetic Act, the Food and Drug Administration (FDA) exercises regulatory authority over medical devices, including prostheses and implants [13]. Under this act, devices are divided into classes based on their potential health risk and level of involvement in maintaining patient health. The least restrictive of these classes require little or no formal FDA approval while the most restrictive require clinical safety studies before the product can be sold on the market. Of course, we should point out the do-it-yourself hackers self-implant technology under their skin, with no medical professional involved, completely circumventing the legal scheme for insuring public safety.

South Korean legislation on "cyborg-like" devices is similar to that of the U.S. in that it has a tiered classification system for regulating medical devices but in addition, regulates the use of "good manufacturing practices". Supervisory control is through the Ministry of Health and Welfare, which has overall authority and specifically enacts regulations for prostheses under the Disabled Persons Welfare Act. Additionally, another South Korean agency, the Ministry of Food and Drug Safety, regulates pharmaceuticals and medical devices (including implants) through the Medical Devices Law and the Pharmaceutical Affairs Law [14]. Similarly, the Japanese system, under the aegis of the Pharmaceutical and Medical Device Agency, classifies medical devices by risk level as well as by imposing certain manufacturing standards in order to protect the public health. And in the European Union (EU), a directive on Active Implantable Medical Devices along with directives on Machinery and Medical Devices provide the framework for ensuring public safety in regards to prostheses [15]. Among some jurisdictions, prostheses and non-medical implants may not be directly mentioned in the legal code and when they are, they are often considered reconstructive and not elective enhancements.

It may be useful to relate the discussion of regulating cyborg technologies to the regulation of robotics which are gaining in intelligence and in some cases becoming more human-like in appearance. Thus far, as with cyborg medical devices, many issues of law which relate to robotics, such as safety and products liability are litigated successfully under tort law. But as robots gain in intelligence, the law is far more challenged, for example, determining liability when a person is harmed by a robot operating autonomously from a human begs the question of who is responsible for the injury? And as the robot becomes more "human-like" and as humans become more "cyborg-like" it become more difficult to think of the machine-parts integrated into the body of a cyborg as separate from the biological-parts—this distinction has implications, among others, for the law of property, bodily integrity, and criminal law. Given that robotic and cyborg components are in some cases similar in design and function, it is relevant in a discussion of cyborgs and law to examine the approach being undertaken by governments to regulate emerging robotic technologies (which in some cases can be thought of as the "robotics part" of a cyborg) to gain insight into how cyborgs (with "robotic abilities") may be regulated by governments. Thus, while robotics and cyborgs may be considered as two different entities under the law, the technologies associated with each are in some cases similar, so laws and regulations regarding robotics may be helpful in developing the legal framework for cyborgs enhanced with similar technology (and *vica verse*). For example, improvements in the algorithms for robot computer vision and problem solving abilities will also assist those receiving cyborg technology suffering from brain injuries, Alzheimer's disease, or loss of vision.

At the very least, discussions on robotics open the door for discussions on cyborgs and can illustrate official, national dispositions towards future technologies and their support within the government, both financial, ethical, and legal. For example, in the U.S., the Presidential Commission for the Study of Bioethical Issues advises the U.S. executive branch on the ethics and current direction of biotechnology, artificial intelligence, and neuroscience research [16]. Additionally, the EU finished a related project regarding the law and ethics of emergent robotics in 2014 entitled Regulating Emerging Robotic Technologies in Europe: Robotics Facing Law and Ethics (RoboLaw) [17]. This comprehensive document explores the current legal state, ethical implications, and industry impact of robotics and other technologies and their future development [18]; in our view, the same approach and level of government interest should be employed for cyborg technologies which are creating various human-machine combinations.

"RoboLaw" recommends, among other things, that the EU relax and make more transparent its regularity conditions in order to promote competitive growth in the field [15,17]; a similar approach may be used to spur investments in cyborg technology. Additionally, South Korea has implemented a strong push towards future (robot) technological development in the Intelligent Robots Development and Distribution Promotion Act [19]. In this Act, South Korea has established building an intelligent robot industry as a national strategic goal. Measures include financial incentives for research and development, and the founding of the Robot Industry Promotion Institute to act as policy support, and even the creation of a hi-tech robot-centered theme park. We postulate that governments will also view the creation of cyborgs as an essential national strategy for competitiveness, and regulate to spur research in this area.

As the above discussion indicates, issues of emerging technologies development are being considered by government regulatory agencies in different jurisdictions, but often as discussions on related, but non-cyborg technologies. The conclusions and recommendations of high-level governmental committees, even when not specifically applied to this topic, can be helpful in anticipating national inclinations towards cyborg development and acceptance. However, other legal doctrine, as discussed below, offer more specific laws relating to cyborg technology.

3. Legal Protection for Enhanced Brains

We begin this section of the article with the law that applies to the computing technology that is just beginning to be used to enhance minds, that is, the chips and software comprising neuroprosthetic devices and other implants. We note that the chips implanted in one's mind, and the software associated with such technology, represents the state-of-the-art in the technology being developed to enhance the cognitive capabilities of a cyborg. As an example, consider the development of a neuroprosthesis, such as an artificial hippocampus which is being designed to restore and enhance one's memory (see generally [6,9,16]). A neuroprosthesis has strong potential to not only alleviate the damage to the brain from disease or injury, but to enhance the brain with superior abilities, such as to download information from the internet, to engage in thought-to-thought communication, and edit memories [1,9]. When computer chips are integrated into a neuroprosthesis, in our view they become a component of the architecture of a "cyborg brain" (which represents the integration of human–machine parts) and thus assist the brain in performing information processing which contributes to the functions of a technologically enhanced mind.

Of particular importance to our cyborg future, and particularly for technologically enhanced minds, is that copyright law extends to programs stored on chips. This means that programs stored on a chip implanted into a cyborg0enhanced brain have rights under copyright law that are not afforded to the architecture of natural brains; that is, individual neurons or groups of neurons and their synaptic connections are not copyright protected, yet software (controlling an implant) is. Generally, under intellectual property law, objects that are considered "utilitarian" are not the subject of copyright protection and chips are clearly utilitarian in function. But in the U.S., the issue of copyright protection for software encoded on chips was decided in the case of *Apple Computer, Inc. v. Franklin Computer Corp.* [20]. In this case, the court rejected the argument that software encoded on chips was to be considered "utilitarian" and thus not copyright protected noting the medium on which the program is encoded (for our discussion the medium could be a neuroprosthesis) should not determine whether the program itself is protected under copyright law [20]. This case has immense implications for a law of cyborgs, as it was the first time an appellate level court in the U.S. held that a computer's operating system could be protected by copyright. As a second area of importance for a law of cyborgs, the ruling clarified that binary code, the machine-readable form of software, was copyrightable too (as is the human-readable source code form of software). Thus, the software aspects of a neuroprosthesis receives copyright protection. This allows for the possibility of legally protected thought or modes of thinking, if thoughts pass through and are stored on these chips and their programs make substantive impact on the brain's though process.

To provide legal protection for the hardware components of the cyborg brain, we could look to rights under patent law to grant a limited monopoly to the designer of these architectures. This notion has interesting implications for our future as cyborgs. Would an artificial hippocampus which stores memory be patent protected and thus be "under the control" of the inventor(s) of the neuroprosthesis, or would the rights to such devices attach to the cyborg equipped with the device? Additionally, would a person with a neuroprosthesis need to agree to the terms of a license to receive software upgrades to their prosthesis (which is an artificial surrogate for the mind) and how would potential service contracts be drawn? As a public policy question—should any aspect of the brain be protected by a patent, copyright, or license to a third party, or would this, the loss of control over one's memories and thought processes to a third party, not only be bad policy but an egregious human rights violation? In our view, any laws limiting one's freedom of thought or granting a third party a limited monopoly over the "artificial" structures of the mind, should receive the highest scrutiny by courts.

We next consider in more detail integrated circuits as components of the technology integrated into a cyborg brain. Given that integrated circuits create microprocessors and sensors, and at the chip level process algorithms, they play a significant role in technologies to enhance the mind. Provided that the design of integrated circuits display satisfactory inventiveness and meets the required standard of uniqueness, at first glance, patent protection is a viable option for the protection of these devices. However, in actuality, the lion's share of integrated circuit design is considered obvious under most patent systems given that they lack any improvement (inventive step) over their predecessors (prior art) [21]. Further, integrated circuits are comprised of numerous building blocks, and each "building block could potentially be patentable. However, since an integrated circuit contains literally thousands of semiconductor devices, a patent claim to an integrated circuit would have to cover many individual elements—this would be the equivalent of trying to write a patent on the neuronal circuits of one structure of a natural brain; extremely tedious to say the least. Consequently, a patent claim that attempts to describe an entire integrated circuit would be hundreds of pages long. Clearly, such a narrow claim under patent law would provide almost no protection for the architecture of a cyborg brain at the semiconductor level.

As indicated by attorney Rajkumar Dubey, it could take several years to obtain an integrated circuit patent from most patent offices worldwide. This is unacceptable given that an integrated circuit's useful commercial life may be "less than one year" [22]. To place this comment in context for this article, what if the same principle of obsolescence (e.g., consider the cycle time for new smart phones and other information technologies) applied to the human brain such that every one to two years a patent had to be filed to protect the neuronal circuitry of their brain? (This comment is included to spur the discussion, obviously, neurons form new synaptic connections constantly as the brain learns.) Imagine that in the coming cyborg age, the human brain is equipped with neuroprosthesis with billions of integrated circuits. That is, the architecture of enhanced brains could become obsolete every two years or so due to the necessity of having to integrate (or update) new technology within the brain. The time-consuming nature of patent filing combined with narrow protection would seem to make patent law an insufficient form of protection for the integrated circuit components of a cyborg's brain (actually, the essential issue may be providing intellectual property protection for the algorithms performing computations "in the mind"). But surely, providing legal protection for the physical components comprising the architecture of a cyborg-enhanced brain is a necessary part of an emerging law of cyborgs. It is congruent to the idea of protecting "bodily integrity" for humans—an idea already established in some jurisdictions [23] and reasonable to assume.

In the U.S. and other jurisdictions, there is another type of legal protection which applies to brains enhanced with computing technology—legislation specifically applied to integrated circuits, which in the U.S. is the Semiconductor Chip Protection Act [24]. Given the importance of protecting integrated circuits from piracy, several nations, including Japan and the European Community have followed the example set in the U.S. and endorsed their own similar statutes and directives recognizing and protecting integrated circuit designs (also referred to as the "topography of semiconductor chips"). In 1989, a Diplomatic Conference among nations was held, at which the Treaty on Intellectual Property in Respect of Integrated Circuits (IPIC Treaty) was partially integrated by reference into the Agreement on Trade-Related Aspects of Intellectual Property Rights (TRIPS) of the World Trade Organization (WTO) [25]. TRIPS is an area of intellectual property law that covers, in summation: copyright and related rights (i.e., the rights of performers, producers of sound recordings and broadcasting organizations); trademarks including service marks; geographical indications, including appellations of origin; industrial design; patents including the protection of new varieties of plants; the layout-designs of integrated circuits; and undisclosed information including trade secrets and test data [26].

As noted, the main purpose of semiconductor chip protection Acts is to prohibit "chip piracy"—the unauthorized copying and distribution of semiconductor chip products copied from the original creators of such works. But such Acts could also provide protection for the architecture of a brain enhanced with cyborg technology given that it is constructed with, among other things, integrated circuits. The use of computer chips in heart pacers, brain–computer interfaces, and neuroprosthetic devices allows computing resources to be directed at activities such as the use of thought to control prosthetic devices, and, ultimately, to restore and enhance memories and other cognitive performance. Protecting such devices from piracy will be an important aspect of any legal scheme directed at cyborg enhanced brains.

According to the U.S. Semiconductor Chip Protection Act, integrated circuit design rights exist when they are created, just like copyright protection, which unlike patents, can only confer rights after application, examination, and issuance of the patent. This aspect of protection—rights conferring upon creation of the integrated circuit architecture of a neuroprosthesis, combined with copyright protection for software—means that the architecture and software of cyborg brains receives protection under current legal schemes. However, just considering chip protection acts, the exclusive rights afforded to the owners of integrated circuit designs are more restricted than those afforded to both copyright and patent holders. Modification (derivative works), for example, is not an exclusive right for owners of integrated circuit designs (this has implications for mind uploads). Furthermore, the exclusive right granted to a patentee to "use" an invention, cannot be used to exclude an independently produced identical integrated circuit design. Thus, reproduction for reverse engineering of an integrated circuit design is specifically permitted by most jurisdictions; does this suggest one can reverse engineer the synaptic structure of another person's "cyborg brain" without consequences under the law?

Legal schemes must also consider the issue of protecting a cyborgs' active memories as programs are stored and loaded on different physical devices implanted within the brain. Memory chips such as an EPROM chip (erasable reprogrammable read only memory), are chips that retain their data when its power supply is switched off. EPROM chip topographies are protectable under the U.S. Semiconductor Chip Protection Act, but such protection does not extend to the information stored on the chips, such as computer programs. Such information is protected, to the extent that it is, by copyright law. Interestingly, in the U.S. the Court of Appeals for the Ninth Circuit in *MAI Systems Corp. v. Peak Computer, Inc.*, [27] held that loading software into a computer's random access memory (RAM) created a "copy'" and a potentially infringing "reproduction" under the U.S. Copyright Act. What that holding meant is that even if no hardcopy was made, temporally storing a program in RAM was a reproduction and potentially infringing act. So turning on a computer constitutes a reproduction of the operating system programs because they are automatically stored in RAM whenever the computer is activated, or for that matter whenever a file is transferred from one computer network user to another. However, while the *MAI* court held that the program temporarily stored in RAM represents a reproduction, the U.S. Congress subsequently enacted an amendment to the Copyright Act to specifically carve out exceptions to this court decision in several circumstances [27,28].

Privacy and protection-of-self issues may arise if a cyborg chip's program involves loading a thought, or some other aspect of cognition, into memory hardware, especially if proper legal considerations are not in place. Laws which establish that loading a program can indeed infringe copyright place into question thought-to-thought communication wherein files protected by copyright are copied when stored within another's mind; these and other issues will need to be resolved in the near future.

4. Towards Cognitive Liberty for Enhanced Minds

The potential that brain implant technology used to enhance the capabilities of the brain, could be hacked raises the question of what rights people have to the veracity of the sensory information transmitted to their brain and the memories stored within the structures of their brain. Restructuring, editing, or modifying memories are aspects of cyborg enhancements awaiting future cyborgs. If third parties were able to hack the technology of brain implants, the possibility of a dystopian future for humanity cannot be underestimated. For example, a retinal prosthesis could be hacked to place images on the back of the retina that a person never saw; or in the case of cochlear implants, sounds could be transmitted to the auditory nerve that a person never actually heard [1]. Further, an artificial hippocampus could be hacked to place memories in a person's mind for events they never experienced. What law and policy might apply to these scenarios? If the First Amendment to the U.S. Constitution blocks the government from putting words in a person's mouth, surely it would also block the government from putting thoughts, sounds, or memories in a person's head. Based on this observation, it is relevant to ask—if the technological ability to hack the mind is in the hands of governments and corporations, will the mind remain a bastion of privacy, safe from the preying eyes of technology? Further, if the government or a corporation can access our thoughts and edit the content of our minds, will the integrity of our mind remain under our individual control, and, if not, who then as a person are we [1,9]?

Once third parties can access a neuroprosthetic device implanted within another person's brain, what are the implications and what could go wrong? For the latter question, not surprisingly, lots of things. For example, if a person committed a crime, and did so because someone had remotely accessed their brain and "influenced their mind" (the mens rea of a crime), would they be absolved of responsibility? Already according to law professor Jeffrey Rosen and Owen Jones, lawyers routinely order scans of convicted defendant's brains and argue that a neurological impairment prevented the accused from controlling their actions [29,30]. In the coming cyborg age, would a software expert be called upon to examine the programming language and algorithms controlling a neuroprosthetic device to see if they had been tampered with? If so, then the mens rea for a crime could have been supplied remotely by a third party. Tampering with the software could, under some circumstances, be actionable under tort law, but also under the criminal law and possibly the statutes which regulate interfering with the transmission of an electronic communication. But the use of neuroprosthetic devices could lead to other important issues of law and policy. For example, third party access to brain implant technology could allow advertising agencies to place pop-up ads into our consciousness or allow our thoughts to be searched by the government without our even knowing it [9,29,30]. Could there be any more egregious violation of a person's privacy than if a government or corporation scanned a person's brain and recorded their unspoken thoughts or changed the content of their memory? Recording unspoken thoughts should be a privacy violation, an unconstitutional search and seizure, and changing the content of the mind should be a fundamental human rights violation. Table 1, which follows, summarizes some major legal schemes which relate to the concept of cognitive liberty and the use of prosthesis, both evolving areas of law for cyborgs.

Table 1. Legal Protection of Cognitive Liberty and Prosthetic Devices for Selected Jurisdictions. In some cases, these protections apply to cyborgs with attached prosthetic devices and only in a limited manner to "cyborg minds," but hold the potential to be leveraged to use in that arena.

Cognitive Liberty/ Neuroprosthesis Protection	Jurisdiction	Type of Protection Provided
Intellectual Property		**Protection for the hardware or software of a device against unwanted copying, reproduction**
Treaty on Intellectual Property in Respect of Integrated Circuits (IPIC Treaty)—Referenced into the TRIPS Agreement of the World Trade Organization (WTO)	State Members of World Intellectual Property Organization/United Nations/Intergovernmental Organizations	The agreement covers the layout-design of integrated circuits; the exclusive right of the right-holder extends to articles incorporating integrated circuits in which a protected layout-design is incorporated (The IPIC Treaty is currently not in force, but was partially integrated into the TRIPS agreement protecting the copyright of integrated circuits)
Copyright Law	U.S. and Other Jurisdictions	In the U.S., code (binary) is protected under copyright law as is source code and operating systems; thus, the software aspects of prosthetic and implant devices are protected under copyright
Patent Law	U.S. and Other Jurisdictions	A limited monopoly in the form of a patent can be granted to the physical components of cyborg devices such as neuroprosthesis. One example is U.S. Patent No. 7,337,007 for a surface neuroprosthetic device for electrical stimulation
Public Safety		**Protection of public safety through the regulation of medical devices and their risk to public health**
Federal Food, Drug, and Cosmetic Act	U.S.	The FDA regulates, as medical devices in terms of public safety, the use of cyborg technology such as limb prosthetics, neuroprosthesis, and other active implantable devices
Directive on Active Implantable Medical Devices	European Union	Covers the safety of medical devices and active implantable devices (such as pacemakers and defibrillators, infusion pumps, cochlear implants, neurostimulators, and others)
Disabled Persons Welfare Act and Medical Devices Law	South Korea	The Ministry of Health and Welfare regulate prostheses quality and the Ministry of Food and Drug Safety regulate active implantable devices for public safety
Pharmaceutical and Medical Device Agency	Japan	Medical devices and implants are regulated for public safety

Table 1. *Cont.*

Cognitive Liberty/ Neuroprosthesis Protection	Jurisdiction	Type of Protection Provided
Personal or Cognitive Liberty		**Protection of the individual against violations of freedoms of-the-self**
First Amendment to U.S. Constitution (and case law)	U.S.	Freedom of speech and of thought receive First Amendment protection, but there are limitations
Fourth Amendment to U.S. Constitution (and case law)	U.S.	Protection of privacy and personal property—the right against arbitrary search and seizure (which will almost definitely apply to the information contained on cyborg enhancement hardware)
Fifth Amendment to U.S. Constitution (and case law)	U.S.	Self-Incrimination Clause protects defendants against being compelled to testify as a witness against themselves in a criminal trial and may apply to the information on cyborg enhancement hardware
Section 2(b) of the Canadian Charter of Rights and Freedoms	Canada	Explicitly guaranteeing the freedoms of cognitive liberty: everyone has the following fundamental freedoms: (b) freedom of thought, belief, opinion and expression …
Articles 8 and 9 of the European Convention on Human Rights	European Union	Concerned with search and seizure and "individual sovereignty" over one's interior environment (all of the rights afforded by the European Convention serve as a guideline for the judiciary to act upon)
Articles 17, 18 and 19 of the International Covenant on Civil and Political Rights	United Nations (U.N.)	Establishes pan-national protection, under the U.N. and its regulatory authority, for rights of privacy, freedom of thought, and freedom of expression
Article 40.5 and 40.6.1 of the Irish Constitution	Ireland	Protects the privacy of the home and person as well as freedom of speech—the Gardai have a greater scope of search and seizure than in the U.S. and freedom of speech has caveats related to public order and morality (i.e., blasphemy)
Part III of the Indian Constitution (Article 19 and 21, among others—including case law)	India	Grants civil rights in freedom of speech, expression and personal liberty (criminal procedure code governs search and seizure law)
Chapter II of the South Korean Constitution	South Korea	Among others and with some limitations, grants the right of freedom of speech and protects against arbitrary search and seizure
Article 21, 19, 35, 38 of the Japanese Constitution	Japan	Grants respectively the freedom of speech, freedom of thought, protection against arbitrary search and seizure, and the right not to self-incriminate
Case law and Article 13 of the Japanese Constitution	Japan	Japan has the constitutional right that everyone be "respected as individuals" which has been leveraged as a basis for a right to privacy (when not covered by search and seizure: Article 35)

It is not currently possible to directly recover the visual or auditory information stored in a person's brain, that is, their resultant neuroactivity from perceiving the world. However, this could become a future possibility given the capabilities of cyborg technology, because once equipped with a technology to sense the world, a cyborg will have an electronic (or digital) record of what they viewed and heard; this conclusion is based on current trends in cyborg (and bran scanning) technology [9]. And with further developments in technology the recording of sensory experiences could extend to olfactory, gustatory, and tactile information sensed by the person. But interestingly, according to the U.S. Presidential Commissions on Bioethics [6,16] and the Nuffield report [7] the ability to record thoughts in a distinct possibility in the foreseeable future. In the context of cyborgs equipped with neuroprosthesis to sense the world, would courts be able to subpoena the data stored on the prosthesis to use as evidence in court (see generally [29,30])? This question implicates rights afforded to individuals by constitutional law in several jurisdictions, not the least of which in the U.S. is the Fifth Amendment's right against self-incrimination in a criminal proceeding.

Interestingly, future cyborgs could have the capability to have "non-human" senses, such as ultra-sonic hearing or infrared vision (see [1,9]). For example, Professor Warwick, using an implanted chip hooked to an ultrasonic detector, was able to determine when objects were getting closer or further away by feel [31]. Applying established law to disputes involving senses which provide beyond human capabilities will surely be an interesting challenge for courts. However, there is some established law that has been used in the limited number of cases appearing before the court involving sensors. For example, in *Kyllo v. United States*, the U.S. Supreme Court held that the use of thermal imaging or forward looking infrared (FLIR) device from a public vantage point to monitor the radiation of heat from a person's home was a "search" within the meaning of the Fourth Amendment, and thus required a warrant; these days, thermal sensors are standard commercial products and one's heat signature is accessible to anyone once they are in public [32].

In the U.S., the most basic Fourth Amendment (search and seizure) question in computer cases asks whether an individual enjoys a reasonable expectation of privacy for electronic information stored within those computers (or other electronic storage devices) under the individual's control. Under *Katz. v. United States* [33], the test used by the Court to determine privacy rights when a government actor is involved is whether the person exhibited a reasonable expectation of privacy and whether the expectation of privacy was one society was prepared to recognize. For example, do individuals have a reasonable expectation of privacy for the contents of their computers and disk storage devices, and, by analogy, for data stored on a neuroprosthetic device? If "yes", then the government ordinarily must obtain a search warrant based on probable cause before it can access the information stored inside. Because individuals generally retain a reasonable expectation of privacy in the contents of closed containers, they also have an expectation of privacy in data held within electronic storage devices (which is any medium that can be used to record information electronically) (see generally [29,30,32,33]). Would the same conclusion hold for cyborgs equipped with neuroprosthetic devices storing data in the form of memories; clearly, the answer should be yes. Further, would it make a difference if the information was in the form of software or algorithms (and not neural circuits), and comprised part of the actual structure of the mind?

In our view, the "privacy of the mind", whether enhanced with technology or not, should receive the highest protection by the courts. If confronted with the issue of determining whether a cyborg has a reasonable expectation of privacy for the information stored on a neuroprosthetic device, based on precedence as discussed above, U.S. courts may (at the minimum) analogize the neuroprosthetic device to that of a closed container such as a briefcase or file cabinet [34,35]. However, as prosthetics are more and more accessible by the internet, this analogy may "break down" and new law will need to be enacted to protect the "privacy of a mind" wirelessly connected to a network. Additionally, future cyborgs may be able to link their minds together which will further erode the "closed container" analogy for a neuroprosthetic device; however, this example just illustrates the need for solid policy and appropriate law to handle such situations. Clearly a mind cannot be opened in the physical sense, but in the future, its contents could be accessed with the appropriate sensor technology. While we argue for stronger protection, it seems reasonable that at minimum courts view the act of scanning files stored on a neuroprosthetic device with the metaphor of a file cabinet, closed to the outside world, and that in the U.S. the Fourth Amendment would protect (i.e., require a search warrant) the content stored on a neuroprosthetic device [29,30].

As the technology to access the mind matures, governments could punish a person not only for the actual spoken expression of their thoughts, but just for formulating a thought contrary to enforced dogma. On this point, law scholar Jeffrey Rosen of George Washington University, wonders whether punishing someone for their thoughts rather than their actions would be a violation of the U.S. Constitutions Eight Amendments ban on cruel and unusual punishment [29]? This is not an observation relevant only to the plot of a science fiction novel, because before centuries end, futurists and others argue that it will be technologically possible for governments and corporations to access brain-implants to edit the long-term memories representing a person's life experiences. Surely, using technology to access and edit a person's memory of an actual lived experience would be actionable under the law—a trespass, an assault and battery, or even extortion [1]. On this last point, former Secret Service agent Marc Goodman worries that holding people's memory hostage could in the future be a form of extortion [5]. Therefore, for reasons of ensuring freedom of the mind in the coming cyborg age, it is imperative that the human body and mind be considered sacrosanct; to invade a person's mind without their consent should be an egregious human rights violation and punishable under criminal law statutes.

5. Freedom of Thought for Technologically Enhanced Minds

Cognitive liberty (see Table 1), or the "right to mental self-determination", is a vital part of international human rights law and is especially relevant in an age of technologically enhanced minds. For example, in the Universal Declaration of Human Rights, which is followed by member states of the International Covenant on Civil and Political Rights, freedom of thought is protected under Article 18 which states: "Everyone has the right to freedom of thought, conscience and religion … " [36]. Clearly, maintaining cognitive liberty in an age of brain implants should be a major societal objective as humanity moves closer to a cyborg future and eventual human–machine merger [1,37]. In fact, a growing number of legal theorists see cognitive liberty as an important basic human right and argue that cognitive liberty is the principle underlying a number of recognized rights within the constitutions of most industrialized nations; freedom of speech being an example [37].

Interestingly, from a jurisprudence perspective, the definition of what constitutes speech is not straight forward; clearly "cyborg telepathic communication" will raise a host of issues which will stress current law. In fact, the courts have identified different types of speech often protected with different levels of scrutiny by the courts. This means that depending on the type of speech, the government is more or less empowered to restrict that speech. However, numerous jurisdictions offer their citizens freedom of speech as a constitutional right. For example, in Japan, freedom of speech is guaranteed by Chapter III, Article 21 of the Japanese Constitution [38]. In Europe, the European Convention on Humans Rights guarantees a broad range of human rights to inhabitants of member countries of the Council of Europe, which includes almost all European nations. These rights include Article 10, which entitles all citizens to free expression [39]. Echoing the language of the Universal Declaration of Human Rights, Article 10 provides that: "Everyone has the right to freedom of expression. This right shall include freedom to hold opinions and to receive and impart information and ideas without interference by public authority and regardless of frontiers" [39]. Freedom of speech is not total, however, and often comes with qualification barring certain subjects of speech, such as sedition.

In the U.S., one type of speech recognized by courts is symbolic speech which is a legal term used to describe actions (not spoken language) that purposefully and discernibly convey a particular message or statement to those viewing it. Of particular relevance for cyborg technology is the category of "pure speech" which is the communication of ideas through spoken or written words or through conduct limited in form to that necessary to convey the idea. In the U.S., the prior restraint of speech is prohibited under the First Amendment so the prior restraint of thought would be more egregious as thought is a precursor to spoken speech The courts have generally provided strong protection of pure speech from government regulation; and past cases in this area could serve as legal precedence for "cyborg speech' consisting of wireless telepathic communication. In the future, perhaps the court should recognize a new form of speech—cyber speech, the conveyance of ideas using thought alone; if so, what level of scrutiny by courts would it receive?

Neuroprosthetic devices may enhance the process of thinking, and in numerous cases, the U.S. Supreme Court has recognized freedom of thought as a fundamental right, describing freedom of thought as: "… the matrix, the indispensable condition, of nearly every other form of freedom … " [40]. Without freedom of thought, the First Amendment right to freedom of speech is moot, because you can only express what you can think. Constraining or censoring how a person thinks (i.e., cognitive censorship) is the most fundamental kind of censorship, and is contrary to some of our most cherished constitutional principles. Supporters of cognitive liberty seek to impose both a negative and a positive obligation on states: to refrain from non-consensually interfering with an individual's cognitive processes, and to allow individuals to self-determine their own "inner realm" and control of their own mental functions [41].

The above mentioned first obligation on a state to protect cognitive liberty directly applies to government access to neuroprosthetic devices, and would also seek to protect individuals from having their mental processes altered or monitored without their consent or knowledge. Though cognitive liberty is often defined as an individual's freedom from *state* interference with their cognition, Jan Bublitz and Reinhard Merkel of the University of Hamburg, suggest that cognitive liberty should also prevent other non-state entities from interfering with an individual's mental "inner realm" [42]. Of relevance for an emerging law of cyborgs, Bublitz and Merkel propose the introduction of a new criminal offense punishing interventions interfering with another's mental integrity by undermining mental control or exploiting pre-existing mental weakness [41]. And that " . . . direct interventions that reduce or impair cognitive capacities such as memory, concentration, and willpower; alter preferences, beliefs, or behavioral dispositions; elicit inappropriate emotions; or inflict clinically identifiable mental injuries would all be prima facie impermissible and subject to criminal prosecution" [41]. Weighing in, Wyre Sententia of the Center for Cognitive Liberty and Ethics also expressed concern that corporations and other non-state entities might utilize emerging neurotechnologies to alter individuals' mental processes without their consent [42].

While one obligation of a state is to refrain from non-consensually interfering with an individual's cognitive processes, another right, to think *however* a person wants, seeks to ensure that individuals have the freedom to alter or enhance their own consciousness; one way to do this, perhaps controversially, would be by stimulating the pleasure centers of the brain by accessing a neuroprosthetic device. An individual who enjoys this aspect of cognitive liberty has the freedom to alter their mental processes in any way they wish to; whether through indirect methods such as meditation or yoga, or more directly through neurotechnology. This element of cognitive liberty is of great importance to proponents of the transhumanist movement, a key tenet of which is the enhancement of human mental function.

6. Litigating Cognitive Liberty

The U.S. Supreme Court heard arguments on an important case that dealt directly with issues related to the cognitive liberty of the mind. Even though the means to alter the mind in the seminal case was not the type of "cyborg technology" we describe in this paper, the effect was similar: a modification of the mind- in this case, the chemistry of the mind. As background, the defendant, Dr. Charles Sell, was charged in a U.S. federal court with submitting false claims to health insurance companies resulting in counts of fraud and money-laundering [43]. Dr. Sell had previously sought psychiatric help and had voluntarily taken antipsychotic drugs; however, he found the side effects intolerable. After the initial charge, Dr. Sell was declared incompetent to stand trial (but not dangerous), as a result, an administrative hearing was held and it was decided that Dr. Sell could be forcibly drugged to regain mental competence; a decision Dr. Sell challenged. The decision by the government to force Dr. Sell to take medication which would change his mental processes raised significant Constitutional law issues. On this point, Professor Lawrence Tribe of Harvard University had previously commented, "whether the government decides to interfere with our mental autonomy by confiscating books and films or by denying us psychiatric medications; 'the offense' is ultimately the same: government invasion and usurpation of the choices that together constitute an individual's psyche" [44].

Could a person who did not pose a danger to another be forcibly injected with antipsychotic medication solely to render him competent to be tried for crimes that were described as "nonviolent and purely economic" by Judge Kermit Bye of the U.S. 8th Circuit Court? If so, by extension, could the government "edit" the information on a prosthesis to restore a person to a level of competency to stand trial? In Dr. Sell's case, the government sought to directly manipulate and modify Dr. Sell's thought processes by forcing him to take mind-altering antipsychotic drugs. Generally, the government can forcibly administer drugs only "in limited circumstances." In Dr. Sell's holding, the Court imposed stringent limits on the right of a lower court to order the forcible administration of antipsychotic medication to a criminal defendant who had been determined to be incompetent, if for the sole purpose of making him competent and able to be tried. Thus, since the lower court had failed to determine that all the appropriate criteria for court-ordered forcible treatment had been met, the order to forcibly medicate the defendant was reversed. Several aspects of the case are worrisome in the context of cyborg technology, namely that the court did not hold that giving the defendant medication to influence the functioning of his mind was inappropriate under all circumstance. Thus the question: could the same be true for manipulating the components of a neuroprosthesis, or altering its software if the defendant was a cyborg? If so, then we have opened up the possibility that the government could have sweeping powers over the functioning and by extension contents of our mind. Our view is that, in the backdrop of constitutional rights to free thought, this outcome is unacceptable and furthermore should be a basic human rights violation.

A similar case in the U.K., *R v. Hardison*, involved a defendant who was charged with violating the Misuse of Drugs Act 1971 [45]. Hardison claimed that his cognitive liberty was safeguarded by Article 9 of the European Convention on Human Rights. Specifically, the defendant argued that "individual sovereignty over one's interior environment constitutes the very core of what it means to be free," and that because psychotropic drugs are a method of altering an individual's mental process, prohibition of them under the Misuse of Drugs Act was in opposition to Article 9. The court however disagreed, and denied Hardison's right to appeal to a superior court.

Marc Blitz of the Oklahoma City University School of Law has written extensively on the subject of freedom of thought and emergent technologies. In his paper Freedom of Thought for the Extended Mind: Cognitive Enhancement and the Constitution [46], professor Blitz explores the philosophical and constitutional framework for emerging laws governing neuroprosthesis, psychopharmacology, and other forms of elective mental self-modification:

> "In sum, these arguments indicate that freedom of thought should not only protect our (naturally protected) ability to engage in reflection. It should also lead courts to identify and protect technologies and resources that support mental autonomy and externalized thought. In this respect, free thought should parallel free speech". [46]

Summarizing the discussion in this section of the paper, freedom of speech is a protected right in the United States and other jurisdictions and if strong philosophical and phenomenological parallels can be drawn between it and freedom of thought and cognitive liberty, then there will be a constitutional basis from which to build a legal framework for our cyborg future. With that in mind, the Figure 2 highlights major issues of cognitive liberty in an age of technological enhancements to one's mind.

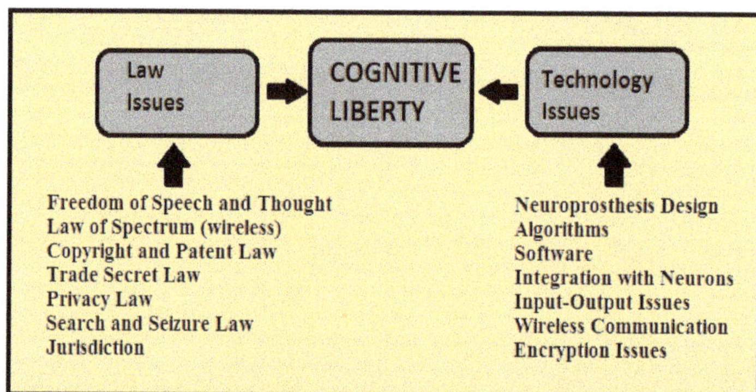

Figure 2. Some legal and technical issues which relate to the concept of "cognitive liberty".

7. Conclusions

From the discussion presented in this paper on "cyborg law", we consider existing laws and statutes to be less than fully adequate to regulate and guide the development of cyborg technologies—it is clear that further legislative and judicial work is required. Some countries have taken a proactive approach and strategized for a technological future while others have debated the ethical issues created when modifying the human body and its capabilities. The existing structures of intellectual property and constitutional law provide a framework around which new cyborg technologies and their legal implications may be developed; however, when these technologies affect the core of a person's mind, they become elevated from a procedural to an ethical and civil importance (see Figure 2 for some examples of the various factors involved). For example, neuroprosthesis—a technology deeply integrated into the body which can change the working of the mind—should be considered core to an individual's mental functioning, and thus of great legal import. A legal and policy discussion on these topics should happen sooner rather than later.

As we move deeper into the 21st century, the speed of technological advances is undoubtedly accelerating. Efforts to reverse engineer the neural circuitry of the brain are opening the door for the development of cyborg devices which may be used to enhance the brain's capabilities. In fact, neuroprosthetic devices are being created now which can serve to restore lost cognitive function or, in the case of techniques such as transcranial brain stimulation to provide therapeutic help for those with depression [40]; within a few decades, even more cyborg technology will exist to enhance cognitive functioning. That is to say, we are on the cusp of creating a class of people which would resemble sci-fi versions of cyborgs in popular media, people with "computer-like" brains connected to the internet communicating wirelessly by thought. Such developments will surely challenge current legal doctrine and established public policy [39,40]. Based on these observations, we need a "law of the cyborg" because without it, constitutional laws, the broad intellectual property laws, and civil protections will not cover the intricacies of this new technology—especially because it creates a *new* way of being and sense of self [1]. On that point, as discussed in this article, there is a current body of law which applies (albeit indirectly in many cases) to cyborg technologies. However, this body of law is insufficient and near-future cyborg technologies will surely create great challenges for established legal doctrine. In conclusion, we recommend that much more be done in the area of law and policy for cyborg technology while we still have time to chart our future in the coming cyborg age.

Author Contributions: Both authors contributed equally to this paper.

Conflicts of Interest: The authors declare no conflict of interest.

References

1. Barfield, W. *Cyber Humans: Our Future with Machines*; Springer: New York, NY, USA, 2016.
2. Ziegler-Graham, K.; MacKenzie, E.J.; Ephraim, P.L.; Travison, T.G.; Brookmeyer, R. Estimating the Prevalence of Limb Loss in the United States: 2005 to 2050. *Arch. Phys. Med. Rehabil.* **2008**, *89*, 422–429. [CrossRef] [PubMed]
3. Carey, B. Chip Implanted in Brain, Helps Paralyzed Man Regain Control of Hand. 2016. Available online: http://www.nytimes.com/2016/04/14/health/paralysis-limb-reanimation-brain-chip.html?_r=0 (accessed on 14 September 2016).
4. Greenspon, A.; Patel, J.; Lau, E.; Ochoa, J.; Frisch, D.; Ho, R.; Pavri, B.; Kurtz, S. Trends in Permanent Pacemaker Implantation in the United States From 1993 to 2009. *J. Am. Coll. Cardiol.* **2012**, *60*, 1540–1545. [CrossRef] [PubMed]
5. Goodman, M. *Future Crimes: Everything Is Connected, Everyone Is Vulnerable and What We Can Do About It*; Doubleday Press: New York, NY, USA, 2015.
6. Gutmann, A. Gray Matters, Topics at the Intersection of Neuroscience, Ethics, and Society. Presidential Commission for the Study of Bioethical Issues, March 2015; Volume 2. Available online: https://bioethicsarchive.georgetown.edu/pcsbi/sites/default/files/GrayMatter_V2_508.pdf (accessed on 4 September 2016).
7. Nuffield Council on Bioethics. Novel Neurotechnologies: Intervening in the Brain. Available online: http://nuffieldbioethics.org/ (accessed on 23 August 2016).
8. Warwick, K. Homo technologicus: Threat or opportunity? *Philosophies* **2016**, *1*, 199–208. [CrossRef]

9. Barfield, W.; Williams, A. Cyborgs and Enhancement Technology. *Philosophies* **2017**, *2*, 4. [CrossRef]
10. Vallverdú, J. Cyberphenomenology: Technominds Revolution. Special Issue. 2016. Available online: http://www.mdpi.com/journal/philosophies/special_isssues/cyberphenomenology (accessed on 15 September 2016).
11. Manfred, E.C.; Nathan, S.K. Cyborgs and Space. Astronautics, September 1960. Reprinted in the Cyborg Handbook. Available online: Cyberneticzoo.com/wp-content/uploads/2012/01/cyborgs-Astronautics-sep1960.pdf (accessed on 2 October 2016).
12. Riley v. California 573 U.S. ___ (2014). No 13-132. Available online: https://supreme.justia.com/cases/federal/us/573/13-132/ (accessed on 14 February 2017).
13. 21 USC Ch.9. Federal Food, Drug, and Cosmetic Act. 360c. Classification of devices intended for human use. Available online: http://uscode.house.gov/view.xhtml?path=/prelim@title21/chapter9&edition=prelim (accessed on 14 September 2016).
14. South Korea, Welfare Laws for Persons with Disabilities. Available online: https://dredf.org/legal-advocacy/international-disability-rights/international-laws/south-korea-welfare-law-for-persons-with-disabilities/ (accessed on 22 September 2016).
15. European Commission. Active Implantable Medical Devices. Available online: http://ec.europa.eu/growth/single-market/european-standards/harmonised-standards/implantable-medical-devices_en (accessed on 2 September 2016).
16. Presidential Commission for the Study of Bioethical Issues. US Department of Health and Human Services. Available online: https://bioethicsarchive.georgetown.edu/pcsbi/node/851.html (accessed on 3 October 2016).
17. RoboLaw. Available online: http://www.robolaw.eu/ (accessed on 4 October 2016).
18. European Commission. Regulating Emerging Robotic Technologies in Europe: Robotics Facing Law and Ethics. Available online: http://cordis.europa.eu/project/rcn/102044_en.html (accessed on 14 September 2016).
19. Korea Legislation Research Institute. Intelligent Robots Development and Distribution Promotion Act. Available online: http://elaw.klri.re.kr/eng_mobile/viewer.do?hseq=17399&type=sogan&key=13 (accessed on 8 October 2016).
20. Apple Computer, Inc. v. Franklin Computer Corp. 714 F.2d 1240 (3rd Cir. 1983). Available online: https://casetext.com/case/apple-computer-inc-v-franklin-computer-corp#! (accessed on 14 February 2017).
21. Patent Statutes, U.S. Code: Title 35. Available online: https://www.gpo.gov/fdsys/pkg/USCODE-2011-title35/pdf/USCODE-2011-title35.pdf (accessed on 10 September 2016).
22. Dubey, R. Semiconductor Integrated Circuits Layout Design in Indian IP Regime. 2004. Available online: http://www.mondaq.com/india/x/28601/technology/Semiconductor+Integrated+Circuits+Layout+Design+In+Indian+IP+Regime (accessed on 14 September 2016).
23. Ramachandran, G. Against the Right to Bodily Integrity: Of Cyborgs and Human Rights. *Denver Univ. Law Rev.* **2009**, *87*, 1.
24. Wikipedia. Semiconductor Chip Protection Act of 1984. Available online: https://en.wikipedia.org/wiki/Semiconductor_Chip_Protection_Act_of_1984 (accessed on 24 August 2016).
25. Overview: The TRIPS Agreement. World Trade Organization. Available online: https://www.wto.org/english/tratop_e/trips_e/intel2_e.htm (accessed on 17 September 2016).
26. Relation to the IIC Treaty. Available online: http://booksandjournals.brillonline.com/content/books/10.1163/ej.9789004145672.i-910.229 (accessed on 14 September 2016).
27. MAI Systems Corp. v. Peak Computer, Inc. 991 F.2d 511 (9th Cir. 1993). Available online: https://www.law.cornell.edu/copyright/cases/991_F2d_511.htm (accessed 14 February 2017).
28. 17 United States Code. § 117—Limitations on Exclusive Rights: Computer Programs. Available online: https://www.gpo.gov/fdsys/pkg/USCODE-2011-title17/html/USCODE-2011-title17.htm (accessed on 10 September 2016).
29. Rosen, J. The Brain on the Stand. *New York Times*. 11 March 2007. Available online: http://www.nytimes.com/2007/03/11/magazine/11Neurolaw.t.html?pagewanted=all&_r=0 (accessed on 4 October 2016).
30. Jones, O.D.; Wagner, A.D.; Faigman, D.L.; Raichle, M.E. Neuroscientists in Court. *Nat. Rev. Neurosci.* **2013**, *14*, 730–736. [CrossRef] [PubMed]

31. Physics Central. The Cyborg Scientist. Available online: http://www.physicscentral.com/explore/action/project-cyborg.cfm (accessed on 24 August 2016).
32. Kyllo v. United States. 533 U.S. 27 (2001). Available online: https://supreme.justia.com/cases/federal/us/533/27/case.html (accessed 14 February 2017).
33. Katz v. U.S. 347 (1967). Available online: https://supreme.justia.com/cases/federal/us/389/347/case.html (accessed 14 February 2017).
34. Cornell Law School, Legal Information Institute. Fourth Amendment Law. Available online: https://www.law.cornell.edu/anncon/html/amdt4frag3_user.html (accessed on 14 December 2016).
35. U.S. Const. amend V. Available online: https://www.gpo.gov/fdsys/pkg/GPO-CONAN-1992/pdf/GPO-CONAN-1992-7.pdf (Accessed on 15 September 2016).
36. The Universal Declaration of Humans Rights. Article 18. Available online: http://claiminghumanrights.org/udhr_article_18.html (accessed on 16 September 2016).
37. Boire, R.G. On Cognitive Liberty: Parts I, II, III, Center for Cognitive Liberty and Ethics. Available online: http://www.cognitiveliberty.org/curriculum/oncoglib_123.htm (accessed on 18 October 2016).
38. Prime Minister of Japan and His Cabinet. The Constitution of Japan. Available online: http://japan.kantei.go.jp/constitution_and_government_of_japan/constitution_e.html (accessed on 16 September 2016).
39. European Union Agency for Fundamental Rights. Article 10—Free of Expression. Available online: http://fra.europa.eu/en/echr-article/article-10-freedom-expression (accessed on 3 September 2016).
40. Palko v. Connecticut 302 U.S. 319 (1937). Available online: https://supreme.justia.com/cases/federal/us/302/319/case.html (accessed 14 February 2017).
41. Bublitz, J.C.; Merkel, R. Crime against Minds: On Mental Manipulation, Harms, and Human Right to Mental Self-Determination. *Crim. Law Philos.* **2014**, *8*, 61. [CrossRef]
42. Sententia, W. Neuroethical Considerations: Cognitive Liberty and Converging Technologies for Improving Human Cognition. *Ann. N. Y. Acad. Sci.* **2004**, *1013*, 221–228. [CrossRef] [PubMed]
43. Sell v. United States 539 U.S. 166 (2003). Available online: https://supreme.justia.com/cases/federal/us/539/166/case.html (accessed on 14 February 2017).
44. Tribe, L. Rights of Privacy and Personhood. *Am. Constitutional Law* **1988**, Sec. 15–7, 1322.
45. R v Hardison. 2006, EWCA Crim. 1502; [2007] 1 Cr App. R. (S) 37. Available online: http://court-appeal.vlex.co.uk/vid/200504343-c4-52563715 (accessed on 14 February 2017).
46. Blitz, M.J. Freedom of Thought for the extended mind: Cognitive enhancement and the constitution. *Wis. Law Rev.* **2010**, 1049–1118. Available online: http://works.bepress.com/marc_jonathan_blitz/17 (accessed on 10 September 2016).

 *philosophies*

Article

Can a Soldier Say No to an Enhancing Intervention?

Sahar Latheef * and Adam Henschke *

National Security College, Crawford School of Public Policy, Australian National University,
Canberra ACT 2601, Australia
* Correspondence: Sahar.Latheef@anu.edu.au (S.L.); Adam.Henschke@anu.edu.au (A.H.)

Received: 1 July 2020; Accepted: 27 July 2020; Published: 3 August 2020

Abstract: Technological advancements have provided militaries with the possibility to enhance human performance and to provide soldiers with better warfighting capabilities. Though these technologies hold significant potential, their use is not without cost to the individual. This paper explores the complexities associated with using human cognitive enhancements in the military, focusing on how the *purpose and context* of these technologies could potentially undermine a soldier's ability to say no to these interventions. We focus on cognitive enhancements and their ability to also enhance a soldier's autonomy (i.e., autonomy-enhancing technologies). Through this lens, we explore situations that could potentially compel a soldier to accept such technologies and how this acceptance could impact rights to individual autonomy and informed consent within the military. In this examination, we highlight the contextual elements of vulnerability—institutional and differential vulnerability. In addition, we focus on scenarios in which a soldier's right to say no to such enhancements can be diminished given the special nature of their work and the significance of making better moral decisions. We propose that though in some situations, a soldier may be compelled to accept said enhancements; with their right to say no diminished, it is not a blanket rule, and safeguards ought to be in place to ensure that autonomy and informed consent are not overridden.

Keywords: human enhancements; autonomy; informed consent; moral enhancement; vulnerability; numeric identity; military ethics

---

1. Introduction

Rapid advancements in technology have seen a rise in innovative ways to enhance human capabilities. This is the case for technologically advanced militaries seeking to enhance soldier capabilities. This paper explores challenges associated with technological interventions being developed that could offer soldiers a chance to enhance their cognitive functions and, by extension, to enhance their autonomy in a warfighting context. We propose that technologies that enhance an individual's cognitive functions, such as decision-making capacity, situational awareness, memory enhancement, and increased vigilance, all have the potential to also enhance an individual's autonomy and moral decision-making capabilities. In a medical bioethics context, such interventions require informed consent of the recipient in order for that intervention to go ahead. However, when considering particular enhancements used in the military, the nature, purpose, and context of the enhancement used may significantly undermine the capacity of a recipient to say no to these enhancements. This is a common problem for informed consent; how do we ensure that the recipients of medical or biotechnological interventions consent to these interventions freely? We suggest that the 'nature' of these enhancements presents a conceptual challenge; can a person autonomously say no to an option that will enhance their autonomy? Further to this, if the purpose of these enhancements is to improve moral decision-making, can a person justifiably say no to an option that will lead to them make better morally relevant decisions? In addition, we suggest that in the military context, this becomes even more complicated because soldiers are not just expected to follow commands, but are trained and inculcated in the practice of following commands, and bear certain loyalties to their comrades. We propose that these contextual elements form the basis to consider soldiers as a vulnerable group with regards to obtaining informed consent. Finally, the fact that soldiers have signed up to be part of the military and accept the military doctrine means that they might have to accept these enhancements as part of the job. In combination, we find that these conditions mean that in certain circumstances, a soldier cannot say no to an enhancement. However, as we show in the concluding section, this is not a broad statement; there is still a range of conditions that must be met in order for particular enhancements to be obligatory in the military context.

2. Cognitive Enhancements as Autonomy-Enhancing Technologies

Cognitive enhancements are those technologies with a demonstrated or potential ability to alter or modify physiological processes such as decision-making, reasonability, memory, judgement, situational awareness, attention span, and complex problem solving. We propose that cognitive enhancements could be viewed as "enhancing" technologies as they improve or have the ability to improve the physiological processes of how we acquire and process knowledge and understand the world around us. These cognitive processes, when enhanced, also enhance an individual's autonomy, i.e., ability to self-govern (see below for more on this). The two types of cognitive enhancements discussed in the following paragraphs are: Brain–Computer Interface (BCI) and Non-Invasive Brain Stimulation (NIBS). As we will discuss below, there is significant disagreement about whether such interventions do in fact act to enhance one's moral decision-making.

2.1. Brain Computer Interface (BCI)

BCIs consist of a direct communication pathway between the brain and an external device (usually a computer platform) via one-way or two-way communication. It is a system that captures brain signals (neural activity) and transforms these signals into commands that can be controlled by an external application or instrument [1]. BCIs have four broad characteristics: Ability to detect brain signals, provide feedback in real time or near time, read/decode brain activity, and provide feedback to the user on the success of the task or goal attained [2]. Broadly, there are two general uses for which BCIs can be used with respect to human performance enhancement: (1) Direct signals from the brain used to direct/alert/command external equipment as an auxiliary to human actions, to control prostheses, robotics, or weapons platforms, or (2) enhanced sensory or information input and/or control signals to enhance individual performance [3]. The earlier goals of BCIs—controlling external equipment—have research origins in medical research and are widely studied for their ability to control prostheses. As the scope of this paper focuses on human enhancement and not therapeutic uses of technology, we focus on those technologies that are hoped to improve particular cognitive functions.

Connection between the human brain and a computer interface is established via two methods: (1) Invasive connections that requiring surgery to implant/connect an electrode inside the skull and (2) non-invasive connections whereby electrodes are placed on the outside of the skull, either attached to a cap or helmet. For use in the military as an enhancement (and not for veterans' medical/therapeutic purposes), we assume that non-invasive BCIs would be preferred, as this poses less risk to the individual and is relatively easily reversible compared to invasive/implanted devices. This type of technology is attractive for use in the military, as it provides the opportunity to increase the brain's computational power, information load, and processing speed, which then allows for an enhanced human performance. BCIs allow for the human brain to handle larger quantities of information in a shorter time frame compared to the brain's normal/average functioning. In a military context, where individuals are required to process significant amounts of information in a short period of time, BCIs provide the possibility to increase human performance with regards to complex decision-making and situational awareness. For example, research in this area has indicated that BCIs can improve facial recognition as well as target detection and localisation in rapidly presented aerial pictures [4,5].

In reference to BCIs, the US Air War College indicated, "This technology will advance computing speed, cognitive decision-making, information exchange, and enhanced human performance. A direct connection between the brain and a computer will bypass peripheral nerves and muscles, allowing the brain to have direct control over software and external devices. The military applications for communications, command, control, remote sensors, and weapon deployment with BCI will be significant" [6].

The United States' (US) interest in identifying novel ways to enhance human cognition beyond current capabilities is evident in the projects undertaken by the Defense Advanced Research Projects Agency (DARPA)[1], such as: Restoring Active Memory Replay (RAM Replay), which is a brain interface project to investigate ways to improve memories of events and skills by studying neural replays; Targeting Neuroplasticity Training (TNT), aimed to improve cognitive skills training by modifying peripheral nerves and strengthening neural connections; and Next-Generation Nonsurgical Neurotechnology (N3), which uses bi-directional BCIs that can control external equipment and applications such as unmanned aerial vehicles and cyber defence systems [7]. These are all examples of projects aimed at identifying ways to enhance cognitive functions that extend beyond therapeutic purposes [8].

2.2. Non-Invasive Brain Stimulation (NIBS)

Non-Invasive Brain Stimulation (NIBS) technologies stimulate neural activity by using either a transcranial electrical stimulation (tES) or transcranial magnetic stimulation (TMS). TMS and tES have been demonstrated to improve the cognitive domains responsible for perception, learning, memory, and attention spans [9]. Research shows that an individual's ability to detect, visually search, or track specific targets can be improved by NIBS [10]. Similarly, tES can be used to improve complex threat detection tasks [11] and to increase risk-taking behaviour [12]. Stimulating specific regions of the brain that are active when performing complex threat detection tasks and risk-taking behaviour provides possibilities for use in military operations. The following paragraphs examine the capability of NIBS to enhance memory, vigilance, and attention, as well as its applicability in a military context.

Memory enhancement research has focused on using TMS and tES to improve working memory and learning capacities in individuals. Using a direct current stimulation on the dorsolateral prefrontal cortex (critical for working memory functions) improves the implicit learning of sequential motor sequences, motor learning, probabilistic learning, explicit memory for lists of words, spatial memory, and working memory [13,14]. Monitoring the mental state of users allows for enhanced performance by adapting the user interface to the changes in mental state. Target detection is one area where this type of technology has been tested. The adaptive interface adjusts accordingly to the feedback given by an electroencephalogram (EEG) and other physiological measures. Complex flight and driving simulation tasks have been used to test the usability of attention-increasing technologies. Studies have investigated the applicability of this technology in air traffic controllers [15] and in military-relevant training scenarios [16]. Enhanced declarative memory is another area that has military applicability. Memory enhancement impacts individual performance on tasks relating to situational awareness, which is of use for fighter pilots and point-shooting [17]. DARPA's RAM Replay project is aimed at identifying ways to enhance memory formation and recall to help individuals identify specific episodic events and learned skills, with research outputs to be applicable in military training activities [18]. Another area where research is done to augment cognitive capabilities useful for the military is to increase vigilance. Vigilance here refers to the ability to maintain sustained attention in areas of high workloads and to be able to shift/divide attention between tasks [15,19,20]. Reaction time tasks, stimulus discrimination, and target counting have been used to measure individuals' reaction times, and this information is used to increase vigilance.

---

[1]   Former U.S President Barack Obama's BRAIN Initiative is supported by National Institute of Health (NIH) and DARPA.

The Halo Sport headset manufactured by the company Halo Neuroscience is one example of an NIBS product available on the market and tested in a military context. Aimed at increasing neuroplasticity, the headset has the ability to enhance physical performance; hence, their usage has been popular among professional athletes [21]. The aspect of the headset that we are interested in here is the ability to improve cognitive performance. The basis as to how Halo Sport headsets function is the same as in the general NIBS method; a Transcranial Direct Current Stimulation (tDCS) (a weak current of approximately 2 to 3 mA) is delivered to the scalp for a duration of several minutes. The current alters the neural activity in the motor cortex of the brain to impact cognitive functions [22]. The Halo Sport headsets, tested in a controlled laboratory environment, have been shown to increase accuracy in cognitive functions but not reaction times. In 2017, Rear Admiral Tim Szymanski, a commander of the US Navy Special Operations, expressed interest in human enhancement technologies with a focus on cognitive enhancement technologies. In his statement, the commander requested that the Defence industry develop and demonstrate technologies that could enhance cognitive performance in the Navy Special Operations forces [23]. A specific reference was also made to NIBS technologies that apply an electrical stimulation to the brain to improve performance. Halo Neuroscience's Chief Technology Officer indicated that the Halo Sport headset has been tested on Navy Seals, showing promising results in improved cognitive performance [23]. According to the Naval Special Warfare Development group (SEAL Team Six), this technology has shown promising results for sleep-deprived individuals performing under hard training environments. At the time, this device was being tested at five military sites. Though the area that has shown the most promising improvements with the use of Halo Sport headsets is physical performance, testing regimens also showed significant improvements in cognitive functions, which has warranted its use for this particular purpose as well.

3. Decision-Making and Autonomy

Autonomy is not only a complex notion, but is one of the most contested areas in philosophy and ethics. We do not expect to answer any of those open questions here, but draw attention to the connection between the technologies as described and autonomy. As Christman describes it, autonomy is the "idea that is generally understood to refer to the capacity to be one's own person, to live one's life according to reasons and motives that are taken as one's own and not the product of manipulative or distorting external forces" [24]. Our view on autonomy is that there is some relative equivalence between what a person does and the *reasons* that they have for acting.[2] This is a somewhat Kantian notion where reason and rationality play a key role in autonomy and, more generally, in ethics. This stands in the face of other views, like that of Haidt's social intuitionist model, in which reasons play far less of a role than the in Kantian model [28]. However, as Kennett and Fine argue, an "examination of the interaction between automatic and controlled reflective processes in moral judgment provides some counter to scepticism about our agency and makes room for the view shared by rationalists and sophisticated sentimentalists alike that genuine moral judgments are those that are regulated or endorsed by reflection" [29] (p. 78). The important point is that if technologies can change cognitive capacities and practices, they could play a role in improving moral decision-making. Our purpose is to draw attention to common elements of autonomy, and to see how they play out in relation to particular enhancement technologies when used in a military context.

In particular, of the technologies that we have reviewed, they are all expected and intended to impact upon and improve decision-making in different ways. The connection to autonomy is that improved decision-making sits in part with the notions of autonomy as increasing "the capacity to be one's own person, to live one's life according to reasons that are taken as one's own." By increasing capacities like memory, attention, and vigilance, we suggest that these technologies are increasing the recipient's autonomy by enhancing their decision-making capacity. Moreover, insofar as these

---

[2] For more on autonomy and self-identification [25,26] and for the argument about reasons and autonomy, see [27].

enhancements increase such decision-making while in positions of high cognitive demand and stress, like conflict, then they are minimizing the "distorting external forces." While more can be said about the connections between increased decision-making capacity and autonomy, the point here is to show that the technologies described are hoped to have some potential to enhance autonomy.

4. Informed Consent

The idea of autonomy in a medical context is frequently operationalized in terms of informed consent. Similar to the elements of autonomy mentioned above, we are primarily interested in whether a person has freely consented to a given enhancement; could they have done otherwise, and was there some external agent or factor that interfered with their decision-making? This draws from the third aspect of autonomy described by Christman above, that decisions made by the person are not the product of an external agent. The concept of informed consent, developed out of the Nuremburg Doctors' Trials in 1947 which led to the creation of the Nuremburg Code, consists of 10 principles that constitute basic legal and ethical rules for research involving human subjects. The first principle is: "The voluntary consent of the human subject is absolutely essential" [30]. This principle is, for the most part, concerned with the individual's ability to exercise free power of choice and to be free of any intervention of force, deceit, duress, coercion, or constraint. Under this principle, the individual should also be given access to sufficient knowledge of the decision to be made and is able to understand the elements involved in the decision-making. The ability to refuse or say no to a decision is also an essential element of informed consent, or, more importantly, *voluntary* informed consent. The right to say no (or withdraw or refuse) is a direct indicator of the individual's ability to "exercise free power of choice" without any coercion, duress, or intervention [30]. Examining this concept in a military context, it is important to identify situations in which soldiers have the ability to refuse an order or directive to accept enhancement technologies.

5. Vulnerability and Saying No

Proper informed consent practices recognize that people may be especially vulnerable to diminutions in their autonomy and capacity to give informed consent. Human research ethics addresses the concepts of vulnerability in depth, some aspects of which are applicable here. For example, prisoners are treated differently with relation to informed consent compared to other adults in medical contexts [31–33]. This vulnerability comes from factors such as prisoners being placed in physical isolation and the power dynamics in the relationships with authority figures. Prisoners are at a greater risk of being manipulated or coerced into accepting interventions that they may otherwise refuse. This special vulnerability comes from aspects such as: Do prisoners have the capacity to understand what is being asked of them and what they are consenting to? Do they have the capacity to say no? If they do consent to an intervention, how can we be sure that the individual in this case is saying yes to an intervention freely and not as a product of coercion by institutional authorities? Based on these aspects, prisoners require special safeguards when it comes to obtaining informed consent for medical interventions.

Soldiers are not the same as prisoners in their roles and treatment within their relevant institutions; however, the concept of unseen pressures and the possibility of coercion and duress can be used to draw some parallels between these two scenarios. The directives to obey the chain of command and subsequent reprimand if one disobeys create an environment in which soldiers could feel unduly pressured into accepting enhancement technologies. The power imbalance in authority relationships formalised in the military's hierarchical systems directly impacts an individual's right to say no to enhancement technologies.

For instance, decisions involving the use of human enhancement technologies would, at a minimum, involve an authority figure (a commander or responsible officer), research or technical specialist, and a physician if the enhancement involves an alteration to the human physiology (such as cognitive enhancements). If it is the case that the enhancement is used for a specific operation (on-the-ground testing), one can presume that the unit or team members would need to be privy to the decision-making process. Privacy and confidentiality are normally available in a medical setting under doctor–patient confidentiality and individual privacy laws, or in a research setting with the ethics approval for the specific study. In prisons, such privacy and confidentiality are limited by the need for prison officers, medical specialists, etc. to share information about a given prisoner. Moreover, given the close confines when being incarcerated, information is hard to suppress, and can travel quickly and easily among inmates. Similar practical limits on privacy and confidentiality apply in a military context. Like with prisoners, we recognize a form of differential vulnerability arising from informal authority relationships, such as those with one's team members and other outranking officers.

In addition, the "mission first" values that are promoted in the military add to the constraints in the individual's ability to freely consent. Where the commanding officers' priorities and those of the individuals may not align, commanding officers may prioritise mission success and safety of the unit as a whole over one individual's safety or privacy. This may not be the case of the individual being asked to accept a brain-stimulating technology that could potentially leave them with adverse side effects, whether they are in the long or short term. History has shown that this is the case, as military personnel have been coerced or pressured into accepting experimental vaccines, which have later on been identified as having less-than-ideal efficacy and several side effects that were long-lasting [34]. Whilst some of the enhancement technologies are supported by scientific research conducted to investigate their functions prior to use, the testing protocols are not the same as a product that would be tested prior to release to the market, thereby raising concerns regarding safety and efficacy.

Vulnerability, as discussed here, is a set of contextual elements: Institutional and differential vulnerability [35]. Institutional vulnerability arises from individuals being subjected to authority relationships where the power imbalance is *formalised* in hierarchical systems, and differential vulnerability arises when individuals are subjected to the *informal* power dynamics from authority of others. The above example involving prisoners is used here to highlight the parallels that can be drawn with regards to obtaining informed consent from soldiers. In the following sections of this paper, we suggest that because of the elements of contextual vulnerability arising in the military context, soldiers fit the conditions of an especially vulnerable population, even when the specific technology could potentially enhance their autonomy through improved decision-making.

6. Can a Soldier Say No? The Special Case of Soldiers

In this section, we look at three situations where the recipient is compelled to say yes, and we ask if they could say no. First, can a soldier autonomously say no to interventions that will enhance their own autonomy? Second, given the moral significance of some of their future actions, does morality itself compel a person to enhance their morality? That is, can a soldier say no to making better moral decisions? Finally, in a military context, soldiers are expected to follow commands. Therefore, can a soldier say no to following a command given the special conditions of being in the military and the ethical implications of doing so?

6.1. Can a Soldier Say No to Themselves?

The first issue where a soldier's capacity to say no is limited derives from the potential for an intervention to change them. It is a question of continuity or "numeric identity".[3] Essentially, does the Soldier at Time 1 (T1) owe it to themselves for Soldier at Time 2 (T2) to be enhanced? The basic idea of this question works on two related aspects of numeric identity: First, that the enhancement causes some significant rupture between Soldier at T1 and Soldier at T2, such that there is no relevant

---

[3] We note here that in the philosophical literature, these issues are typically covered under discussions of "personal identity" rather than "numeric identity". However, as "personal identity" is also used in non-philosophical disciplines to refer to psychological aspects of a person's identity, we have chosen to refer to this as "numeric identity". For more on this particular nomenclature, see Henschke [26].

continuity between them; second, that Soldier at T2 not only has significantly enhanced autonomy as a result of the enhancement, but also that this matters morally. Combining these two points, as Soldier at T1 and Soldier at T2 are different enough people (as a result of the rupture), and Soldier at T2 will be so significantly improved by the enhancement, that Soldier at T1 owes it to Soldier at T2 to undergo the enhancement.

The first premise of this argument draws on notions of numeric identity. Essentially, are Soldier at T1 and Soldier at T2 the same person or different people? "Discussions of Numeric Identity ... are often concerned with what is needed for a thing to be the same as itself. If time passes, how do we judge that" the Soldier at T1 and Soldier at T2 are the same person? [26]. Consider here Sally who, since the age of 10, has wanted to join the army. By the time she is 30, Sally has spent a number of years as a soldier and fighting in conflict zones; Sally at 30 years old (at T2) has a set of physical, psychological, and experiential attributes that are going to significantly differentiate her from who she was as a ten year old at T1. We can obviously see that Sally is different at the two times. "But despite these changes, most of us would say that Sally is the same person she was at ten years old, now and until she dies... That is, Sally's identity persists through time, despite the obvious fact that she has changed" [26]. So, on the one hand, Sally at T1 and Sally at T2 are different, but on the other hand, Sally at T1 and Sally at T2 are the same.

The way that people have sought to explain this persistence or continuity, despite the differences, draws on different aspects of Sally. One is what Derek Parfit called overlapping chains of psychological connectedness [36–38]. The person Sally is today is very similar to the person she was yesterday. The person she was yesterday is very similar to the person she was two days before, and so on. So, though she may not be exactly the same person now as she was at ten years old, as long as there is a continuity of states that links the person she was then to the person she is now, an identity claim holds [26]. Others suggest an alternative explanation. Instead of these overlapping chains of psychological connectedness, it is the facts about Sally's physical persistence that make her the same person at T1 and T2.[4] On this bodily criterion of numeric identity, it is the facts of the ongoing physical existence that make Sally the same person.

We suggest here that, whichever account one favours (psychological connectedness or the bodily criterion), Soldier at T1 and Soldier at T2 are the same person. Though they are different, they are not different people. T1 and T2 are still likely going to be psychologically connected, and their body is ongoing. That they have received a technological intervention that enhances them is not sufficient cause to say that they are different people. On both accounts, they are still the same.

This is all relevant to whether the soldier can say no to an enhancement, as one potential argument against saying no is that the soldier owes it to their future self to say yes. On this argument, if Soldier at T1 said no, they would be unfairly denying Soldier at T2 the options or capacities offered by the enhancement. We encounter a similar form of argument in discussions about environmental stewardship and what present people owe future people [40,41]. On the issues of what we owe future people, the issues rely in part at least on generational injustice, which in turn relies on the people at T1 or Generation 1 being different people from the people at T2 or Generation 2. Likewise, the "owe it to their future self" argument relies on the two selves being different people; it requires some significant difference between T1 and T2 selves. However, this does not work as a compelling argument if T1 and T2 selves are the same. Insofar as they make a free decision, and the soldier is making an autonomous decision about themselves, they are not denying the options or capacities to any different future self.

Another way that the "owe it to themselves" argument can run is like this: Soldier at T2 is not simply improved or enhanced by the intervention, but their rational capacities and the resulting autonomy from those rational capacities are so far in advance of Soldier at T1, that Soldier at T2 essentially has "authority" over Soldier at T1. Here, we can look at the arguments around advance

---

[4]   For instance, see [39].

directives where a previous self has authority over the present self, but only when the previous self's autonomy is so far above the present self's extremely low autonomy [42]. In our situation, while the temporal logic is the reverse,[5] the core of the argument is the same. One's self is so significantly advanced in terms of its autonomy that the enhanced self has authority over the less autonomous self. As such, Soldier at T1 owes it to themselves to do what they can to bring Soldier at T2 about. However, we think that, with the particular technologies being the way that they are at the moment, it is unlikely that the Soldier at T2 would be so significantly enhanced that their autonomy must take precedence over that of the Soldier at T1. Thus, we think that the authority of the Soldier at T2 is not sufficient enough to prevent Soldier at T1 from saying no.

6.2. Can a Soldier Say No to Making Better Moral Decisions? Moral Decision-Making in a Military Context

The next argument is more compelling. The basic claim here is that the soldier cannot say no to an enhancement if that enhancement improves their moral decision-making. For example, an NIBS technology that could enhance a soldier's situational awareness or vigilance to the extent that they are able to process a considerable amount of information load could allow a soldier to improve their moral decision-making compared to that of a non-enhanced soldier. This might be a sacrifice they are compelled to make. Consider this argument by analogy: A soldier in a conflict zone is offered the option of using weapon 1 or weapon 2. Weapon 1 is a weapon that they have been using for years and they feel comfortable with it, and they like to use it. They are familiar with weapon 2, but they do not feel as comfortable with it. However, in this particular conflict zone, there is a reasonable risk that particular forms of combat will kill innocent civilians, and the soldier knows this. Now, weapon 2 is much more likely to avoid civilian casualties or harm, but other than that, it will impact the enemy the same as weapon 1. Again, the soldier knows that weapon 2 will be far better in terms of its discrimination. In this scenario, as per the ethics and laws of armed conflict, the soldier needs to choose weapon 2 over weapon 1.

The underpinning logic of this is that soldiers have a duty to not just adhere to relevant moral principles, but if there are two options and one meets the moral principles better than the other one, they ought to choose that better option. Here, they are compelled to follow what morality demands. The same reasoning would likely hold with regard to particular enhancements; if the soldier is presented with an option that would improve capacity to adhere to and meet specific military ethics principles, then that option ought to be chosen. On the face of it, the soldier's general moral responsibility overrides any personal disagreement they might have with a particular technological intervention. This idea that people should be morally enhanced is currently an idea being explored in the literature [44–47]. These authors have advanced the argument that we ought to morally enhance ourselves *if such enhancements exist*. Some of these authors take quite a strong line on this. If safe moral enhancements are ever developed, there are strong reasons to believe that their use should be obligatory [46].

Their reasoning turns on access to destructive technologies like weapons, and is similar to what we have offered here:

> Around the middle of last century, a small number of states acquired the power to destroy the world through detonation of nuclear weapons. This century, many more people, perhaps millions, will acquire the power to destroy life on Earth through use of biological weapons, nanotechnology, deployment of artificial intelligence, or cyberterrorism ... To reduce these

---

[5] We also recognise here that there is perhaps an additional step required to make the claim that the T2 self has authority over the T1 self—that the future self can direct or dictate things to the present self. However, this line of argument may rely on some form of backwards causation, where the future causes present events to occur. We note here that backwards causation is a somewhat contentious concept. For more on backwards causation, see [43].

risks, it is imperative to pursue moral enhancement not merely by traditional means, such as education, but by genetic or other biological means. We will call this *moral bioenhancement*.

[47]

We suggest that in the context of military decision-making, particularly when considering decisions that are of significant moral weight, such as deciding when to shoot, who to shoot, and so on, there seems to be a convincing argument that soldiers ought to be morally enhanced. However, this is a contingent claim. First, this is not a blanket claim that the soldier must assent to all enhancements. It is only relevantly applied to enhancements that enhance their *moral* decision-making. We note here that there is an important discussion about the assumptions and feasibility of *moral* enhancement. One general assumption is that there is some agreement on what constitutes "good" moral decision-making. Much of ethics, from one's metaethical position to one's preferred normative theories, is a series of open questions. However, we point out here that in the military ethics context, there are some generally accepted principles like discrimination, proportionality, and necessity that must be met. We do not claim that these principles are *true*, but instead agree with the just war tradition that things are better, all things considered, when soldiers adhere to these principles.

In terms of feasibility, as Harris points out, if moral enhancement involves the reduction of morally problematic emotions like racism, then he is "sceptical that we would ever have available an intervention capable of targeting aversions to the wicked rather than the good" [46][6]. Similarly, Dubljevic and Racine argue that "an analysis of current interventions leads to the conclusion that they are 'blunt instruments': Any enhancement effect is unspecific to the moral domain" [47].[7] The worry here is that the technologies that might aid in moral enhancement are so imprecise as to be discounted as serious ways to improve moral behaviour and so on. In Harris' view, we should instead focus on current methods of moral enhancement like education [48]. We consider these points to be reasonably compelling; there is good reason to be sceptical about the likelihood that these technologies will have the precision to reliably and predictably improve moral decision-making. However, for the purposes of this paper, in order to explore the ethical implications of these technologies, we are assuming some potential for these technologies to work as promised [49]. That said, this is an *in principle* argument. Without certainty that these interventions do enhance moral decision-making, the argument against saying no becomes significantly weaker.

For instance, we need to question which interventions actually constitute *enhancements* to moral decision-making. For instance, given the relation between enhanced memory, vigilance, and attention span and decision-making, as discussed in earlier sections of this paper, and the relations between improved decision-making and moral decision-making [29], one could argue that interventions that improved the quality of a soldier's cognitive functions do in fact enhance their chances at making better moral decisions.[8] It is important to note that whilst the enhancements examined in this paper have the capability to enhance cognitive functions, research investigating the efficacy of some types of commercially available NIBS products has shown that these enhancements may not be as effective as expected [51].[9] Our thought here is that, as we are entertaining a claim that the soldier ought to be *compelled* to accept the intervention, there would need to be more than a mere likelihood that such an intervention will reliably enhance their moral decision-making. We suggest that this is reliant

---

[6]   See p. 105.

[7]   See p. 348.

[8]   As noted earlier, a somewhat Kantian approach to reason and decision-making, as well as their connection to moral decision-making. For this, we draw from the work of people like Michael Smith, or Jeanette Kennett and Cordelia Fine [28,29]. This is, in contrast, a more Humean account, like that of the social intuitionist model of moral decision-making advocated by Jonathan Haidt [50].

[9]   In some cases, one type of cognitive function could be enhanced at the cost of another. For example, increased learning memory could come at a cost of decreased levels of automated processing.

on a combination of assumptions about moral psychology and empirical claims about particular interventions.[10]

We also need to take into account other factors—does the intervention have any side effects? In particular, does it have any side effects on moral decision-making? For instance, amphetamines are a range of pharmaceutical interventions that have been used to reduce the need for sleep in the military [53]. However, they have a range of side effects, such as causing aggression and long-term psychological effects that would argue against a soldier being compelled to take them on moral grounds. Investigation into potential side effects of NIBS identified short-term side effects, such as reactions at the electrode sites, as well as a few cases of black-outs and seizures [17,52]. These investigations were done on voluntary healthy patients in a medical/laboratory setting. As the technology is still relatively new, further investigation will be required to identify long-term side effects of their use. Any side effects would need to be taken into account and weighed against the likelihood that the intervention does indeed improve moral decision-making. For instance, if there was only an inference that the intervention improved moral decision-making and there were known deleterious side effects of the intervention, the case that the soldier can say no is much stronger.

However, even if the interventions were likely to improve moral decision-making without significant side effects, some might still balk at the idea that soldiers *must* consent to such enhancements. This is because such interventions seem to override the principle of autonomy. However, perhaps the soldier has to assent to enhancements that would improve their moral decision-making capacity. This is because of the nature of military actions, or at least military actions that involve killing people. These actions are of significant moral weight, and so need to be treated differently from non-moral decisions or actions.[11]

There is a counter-argument to this: That the position that one must assent to moral enhancements is absurd and extreme. If it is true that we must accept interventions that improve our moral decision-making, then everyone on earth is morally required to assent to these moral enhancements. If the particular case holds in the military context—that a soldier must consent to being morally enhanced—this would surely hold for everyone, and this seems absurd: It would seem to be such a significant infringement on personal autonomy, bordering on authoritarianism, that we ought to reject it. While there is perhaps substance to this counter-argument at a general level, we can reject it as an argumentum ad absurdum, as we are only looking at the military context. Moreover, we are only considering those military members whose roles and duties would have them being forced to make life and death decisions, and to make those decisions in a way that would benefit from the enhancements described in Section 2. The average person does not face such life and death decisions in such high-pressure contexts. Finally, even if we constrain potential recipients to the military, arguably, no military has or will have the capacity to roll this out for every serving member. Maybe they should [46,55,56], but that is a different point from what we are concerned with here. What we are concerned with is the capacity to say no. As we can reasonably constrain the claim to particular members of the military, the argumentum ad absurdum fails.

---

[10] Counter-arguments [52] indicate that concerns regarding explicit coercion and potential impact on individual autonomy and informed consent in the military are perhaps misplaced given the low prevalence of use, social acceptance, and efficacies of tDCS still yet to be explored. However, we propose that though these interventions are not widely used as yet, it does not negate exploration of potential ethical concerns should their use become more widely accepted.

[11] We recognise that this position, that "moral reasons" can override personal beliefs, is contentious and contested. While we do not have space to cover the topic here, we suggest that one of the features of moral reasons that makes them different from non-moral reasons is that they ought to count significantly in one's decision-making [54]. What we will say is that, given the specifics of the technologies that seem likely to be used for such enhancements, as they are currently non-invasive and potentially reversible, the argument that a soldier has a right to conscientiously object to such enhancements is weak. Like "weapon 1" versus "weapon 2" above, if the technologies *do* enhance moral decision-making and are not so different from using two different weapon types, the right to say no is limited at best. However, as we have taken care to note throughout the paper, there is perhaps a stronger conscientious objection argument that says "I say no to this technology, because it does *not* actually enhance moral decision-making."

6.3. Saying No to an Order: Ethics of Following Commands and Being in the Military

The third element of this discussion arises because of the special conditions of being in the military. First, soldiers are trained to follow orders, thus diminishing their capacity to say no. Second, soldiers decide to enter the military with the knowledge that they will not only be asked to potentially engage in risky activity, but also have the foreknowledge that they will be expected to follow orders. These points are complex and nuanced, as we will discuss, but when combined with the previous argument about saying no to making better moral decisions, we suggest there might be a situation where a soldier cannot say no to particular moral enhancements.

As an essential part of their training, soldiers are trained to follow orders, something that may conflict with their existing moral identity [57]. Of course, this does not mean that they are an automaton; many professional militaries now include training on the laws of armed conflict, military ethics, and the just war tradition. Any such training will explicitly or implicitly include recognition that a soldier should not follow an order that they know to breach the laws of armed conflict or a relevant ethical principle. For instance, many soldiers are taught that they can refuse to follow an order to kill a prisoner of war or an unarmed unthreatening civilian. However, as history shows [58], commanders still give orders that are illegal or immoral, and many soldiers still follow those commands. Moreover, as was infamously demonstrated by Stanley Millgram, many people will follow the commands of someone perceived to be in a position of authority even if what they are being asked to do is objectively morally objectionable [59,60]. The point here is that even when significant moral principles may be transgressed, many soldiers will still follow those commands; their capacity to say no is diminished.

The relevance here is that if it is generally psychologically difficult for a soldier to say no to a command, particularly commands that do not obviously contravene the laws or ethics of war, it may be equally psychologically difficult to be able to say no to commanders commanding them to accept an enhancement. We can consider here a general claim that the military command structures, training, and socialisation to follow orders undermine our confidence that soldiers can say no to enhancements.

Adding to this explanation, again arising in part from military training, is that soldiers feel a significant responsibility to their comrades and/or to the nation for which they fight. "The role of this socialisation process is to separate soldiers from their previous social milieus and inculcate a new way of understanding the world. Central to this process is loyalty, obedience, and the subsuming of one's individual desires to the needs of the greater cause" [61]. Not only are soldiers trained to consider seriously sacrificing themselves for some greater good, but as training progresses, they are taught that they ought to. "The officer in training builds up a professional identity on the basis of his personal immersion in the ongoing, collective narrative of his corps. This narrative identity is imparted not by instruction in international law but by stories about the great deeds of honourable soldiers" [62]. Some see this loyalty to one's closest comrades as fundamental to military practice: "The strongest bonds are not to large organizations or abstract causes like the nation; rather, they are to the immediate group of soldiers in one's platoon or squad, before whom one would be ashamed to be a coward"—Frances Fukuyama, quoted in [61]. Here, the issue is whether a soldier feels like they can say no because they are concerned that, if they do, they will be letting their comrades down.

Similarly, a number of people sign up to be soldiers due to a sense of loyalty to their nation, to protect their nation, and to fight for what is right. Serving in the military "is a higher calling to serve a greater good, and there is nothing incoherent or irrational about sacrificing oneself for a greater good" [63]. On this view, soldiering is not a mere job, but something higher. For a group like "the clergy, the 'larger' and 'grander' thing in question is the divine. [In contrast, for] soldiers, it is the closest thing on earth to divinity: The state" [63].[12] Here, rather than the loyalty being to their comrades, it is to the larger nation for whom they fight and/or the values for which they fight. In both

---

[12] We note here that this author is not endorsing this view; rather, they are describing the notion of military service as distinct from a normal job [63].

aspects, though, we can recognise a strong weight in favour of doing what is asked; were this another job, the responsibility to would play far less of a role; very few jobs reliably expect a person to sacrifice their life for their job. "If it is not permissible for civilian employers to enforce compliance with such 'imminently dangerous' directives, then why is it permissible for the military?" The obvious answer is that soldiers make a commitment of obedience unto death; they agree to an "unlimited liability contract" upon enlistment, called so because there is no limit to the personal sacrifice that they can legitimately be ordered to make under that contract [63]. Given the importance of soldiering, the soldier forgoes many basic rights. In line with this, they may also forfeit a right to say no to enhancements.

This brings us back to the argument that we need to think of informed consent in a military context as different from informed consent in a medical context. What we might need to think of is that entry into a military context is a broad consent, where you consent to giving up other latter consents.[13] This is not a "freedom to be a slave" argument, as the military service will typically end, but where enhancements differ is that they are ongoing.[14]

However, this all brings us to a second vital aspect of the capacity to say no—soldiers sign up to join the military, and unlike many other jobs, they are expected to follow orders. We would hope that people signing up to join the military would have some foreknowledge that what they are committing to is different from a normal job, with a set of important expectations, including following orders. For instance, it would be absurd to join the military and then complain about their commander being bossy, and that they do not like guns or state-sanctioned violence. It is essentially a form of caveat emptor, where the person knowingly gives up certain freedoms as part of joining the military; their freedom to say no is significantly curtailed. Just as they have significantly diminished freedoms to say no to commands, their freedom to say no to an enhancement is diminished. We return to the issue of exploitation in enlistment below.

The above argument becomes even more compelling when considering the narrowed focus that this paper has—whether a soldier can say no to technologies that enhance their military decision-making capacity. As we discussed in the technology summary and in the section on whether a soldier can say no to themselves, the technologies we are concerned with are those that are non-permanent and non-invasive. While their intended use is explicitly to enhance a person's decision-making capacity in a military context, thus qualifying them as an enhancement technology, they are as much a military tool as they are a biotechnological intervention. While the interventions we are concerned with here are not equivalent to asking a soldier to carry a weapon or put on body armour, they are not exactly equivalent to an invasive clinical intervention to irreversibly alter their physiology.

The relevance of this is that the arguments about saying no to clinical interventions that one finds in the biomedical literature have less purchase in the military context than they do in a clinical biomedical context. That informed consent in a military context is different from that in a medical or clinical context is a well-founded view [67–69]. This does not mean that we jettison the notion of informed consent in a military context, but rather that it needs to be adapted to that context.

We also need to consider whether the purposes of particular enhancements add further moral weight to the argument that soldiers cannot say no. For instance, if the enhancement was shown to increase a particular military team's success or survival rate, then there is an increased weight for that enhancement to be accepted; for example, if cognitive enhancements were so advanced to the stage that they could significantly enhance a soldier's memory processing or reduce or eliminate cognitive fatigue in the battlefield, factors that directly impact survival rate on the ground. It is like a soldier saying no to a better weapon; insofar as that better weapon hits its targets but is more proportionate and discriminatory than existing weapons, we find a prima facie case that the soldier should use the better

---

[13] For more on broad consent, see [64,65].
[14] For more on enhancements and the duty of care to veterans, see [66].

weapon. So too, we have a prima facie case that the enhancement be accepted.[15] Moreover, as was discussed above, if the particular enhancement is likely to increase the capacity to adhere to ethical principles like *jus in bello* proportionality or discrimination, then they should not say no to making better moral decisions. The relevance to this section is that both of these arguments that a soldier has responsibility to fight better are strengthened when considering that the soldier signed up for entry into the military.

All that said, this argument has a significant caveat of its own—it assumes that all people join the military freely and with the relevant advance knowledge of what this role entails. However, as Bradley Strawser and Michael Robillard show, there is risk of exploitation in the military [71]. Exploitation of people's economic, social, and educational vulnerabilities to get them to enlist in the military would significantly undermine any notions of broad consent or that soldiers must accept enhancements because "they knew what they were getting into when they signed up." Moreover, the arguments developed here are considerably weaker if considering soldiers who were conscripted to fight. Thus, not only does the context of the military change how we would assess whether they can say no to an enhancement, but the conditions under which soldiers are enlisted are also essential to any relevant analysis.

7. Conclusions

In this paper, we examined the challenges associated with autonomy-enhancing technology used in a military context, particularly a soldier's right to say no to these types of technologies. Some parallels can be drawn between enhancements used in the military and similar interventions used in medical or therapeutic contexts. However, we propose that the *nature, purpose, and context* of the enhancements used in the military raise special concerns regarding the impact on individual autonomy, informed consent, and the ability to say no to such enhancements.

Examination of current technologies indicated that the nature and function of the technologies require further evidence to ascertain that, as a blanket rule, autonomy would be enhanced to the extent that it would override a soldier's right to autonomy and informed consent rule. In addition, challenges to obtaining informed consent become more complex when in a military context. We propose that soldiers can be considered an especially vulnerable group due to contextual elements that highlight institutional and differential vulnerability. A system in which power imbalances are formalised in authority relationships and hierarchical command structures, where a higher priority is given to the success of the operations and safety of the unit as a whole over individual rights, could impact a soldier's ability to refuse an enhancement that is considered to be beneficial to the very aspects considered a higher priority. Further to this, the lack of individual privacy that would otherwise be afforded to civilians (in a medical context) could also diminish a soldier's capacity to say no to an enhancement.

Looking at possible situations that could compel a soldier to accept enhancements, we examined the argument where soldiers potentially owe it to themselves to accept an intervention that could benefit them in the future. Unpacking the concepts of numeric identity and potentially denying one's future self from a benefit, we propose that the current enhancements and the benefits they offer at this stage are unlikely to enhance a soldier to the extent that the rights of their future self takes precedence over their current self.

Another scenario in which a soldier may be compelled to accept an enhancement is the possibility of making better moral decisions. In this case, we propose that soldiers, by the nature of their work in making life and death decisions, could possibly be compelled to accept an enhancement if it is *certain* that said enhancement would guarantee a better *moral* outcome in line with *jus in bello* and Laws

---

[15]   We are thinking here of a parallel argument that remote weapons like drones should be used if, all other things being equal, these remote weapons reduce risk to one's own soldiers [70].

Of Armed Conflict (LOAC). This is not a blanket claim to all enhancements, but to those that only produce a better moral decision and a better moral outcome. However, even in this scenario, potential side effects of the enhancement and likelihoods of outcomes need to be taken into consideration. Even then, it does not warrant an override of autonomy and informed consent, but rather a need to ensure that safeguards are in place to protect soldiers' rights even if they are compelled to accept such enhancements.

Finally, we looked at the argument that claims that soldiers sign up to follow orders when they join the military; hence, they have foreknowledge that they are committing to a job that comes with certain risks and sacrifices. Considering the focus of this paper—the question of whether a soldier can say no to an autonomy-enhancing technology, provided that the technology is non-invasive, not permanent, and explicitly used to enhance a person's decision-making capacity—we propose that a soldier could be compelled to accept said technology with their right to say no being considerably diminished. However, a caveat here is that not everyone joins the military freely, without exploitation, and with sufficient knowledge of what their roles would entail.

Whilst there are narrowly focused scenarios in which a soldier could be compelled to accept autonomy-enhancing technologies impacting their right to say no, this would need to meet the thresholds highlighted in scenarios in this paper. Even so, rights to individual autonomy and obtaining informed consent should not be forgone; rather, an understanding of the nature, purpose, and context of the use of such enhancements in the military context is warranted, as is identification of the appropriate measures that could be implemented to ensure that individual rights are not corroded. Adding to that the risks of exploitation and issues like conscription, we need to be explicit that our argument is one that comes with a series of important caveats and restrictions. We are not saying that a soldier loses any right to say no to an enhancement.

Author Contributions: Both authors contributed equally to this paper. All authors have read and agreed to the published version of the manuscript.

Funding: Initial research for this paper was supported by the Brocher Foundation.

Conflicts of Interest: Sahar Latheef works for the Australian Government, Department of Defense. The views and opinions expressed in this paper are those of the authors and do not reflect any official policy or position of any agency. The authors have no other conflicts of interest.

References

1. McKendrick, R.; Parasuraman, R.; Ayaz, H. Wearable functional near infrared spectroscopy (fNIRS) and transcranial direct current stimulation (tDCS): Expanding vistas for neurocognitive augmentation. *Front. Syst. Neurosci.* **2015**, *9*. [CrossRef] [PubMed]

2. Burwell, S.; Sample, M.; Racine, E. Ethical aspects of brain computer interfaces: A scoping review. *BMC Med. Ethics* **2017**, *18*, 1–11. [CrossRef] [PubMed]

3. Krishnan, A. *Military Neuroscience and the Coming Age of Neurowarfare*; Routledge, Taylor & Francis Group: New York, NY, USA, 2017.

4. Matran-Fernandez, A.; Poli, R. Towards the automated localisation of targets in rapid image-sifting by collaborative brain-computer interfaces. *PLoS ONE* **2017**, *12*, e0178498. [CrossRef] [PubMed]

5. Matran-Fernandez, A.; Poli, R.; Cinel, C. Collaborative Brain-Computer Interfaces for the Automatic Classification of Images. In *Proceedings of the 2013 6th International IEEE/EMBS Conference on Neural Engineering (NER), San Diego, CA, USA, 6–8 November 2013*; IEEE: San Diego, CA, USA, 2013; pp. 1096–1099.

6. Moore, B.E. *The Brain Computer Interface Future: Time for a Strategy*; Air War College Air University Maxwell AFB: Montgomery, AL, USA, 2013.

7. DARPA Next-Generation Nonsurgical Neurotechnology. Available online: https://www.darpa.mil/program/next-generation-nonsurgical-neurotechnology (accessed on 12 July 2020).

8. Smalley, E. The business of brain-computer interfaces. *Nat. Biotechnol.* **2019**, *37*, 978–982. [CrossRef] [PubMed]

9. Coffman, B.A.; Clark, V.P.; Parasuraman, R. Battery powered thought: Enhancement of attention, learning, and memory in healthy adults using transcranial direct current stimulation. *NeuroImage* **2014**, *85*, 895–908. [CrossRef] [PubMed]

10. Nelson, J.M.; McKinley, R.A.; McIntire, L.K.; Goodyear, C.; Walters, C. Augmenting Visual Search Performance With Transcranial Direct Current Stimulation (tDCS). *Mil. Psychol.* **2015**, *27*, 335–347. [CrossRef]

11. Clark, V.P.; Coffman, B.A.; Mayer, A.R.; Weisend, M.P.; Lane, T.D.R.; Calhoun, V.D.; Raybourn, E.M.; Garcia, C.M.; Wassermann, E.M. TDCS guided using fMRI significantly accelerates learning to identify concealed objects. *NeuroImage* **2012**, *59*, 117–128. [CrossRef]

12. Sela, T.; Kilim, A.; Lavidor, M. Transcranial alternating current stimulation increases risk-taking behavior in the balloon analog risk task. *Front. Neurosci.* **2012**, *6*, 22. [CrossRef]

13. Durantin, G.; Scannella, S.; Gateau, T.; Delorme, A.; Dehais, F. Processing Functional Near Infrared Spectroscopy Signal with a Kalman Filter to Assess Working Memory during Simulated Flight. *Front. Hum. Neurosci.* **2015**, *9*, 707. [CrossRef]

14. Brunoni, A.R.; Vanderhasselt, M.-A. Working memory improvement with non-invasive brain stimulation of the dorsolateral prefrontal cortex: A systematic review and meta-analysis. *Brain Cogn.* **2014**, *86*, 1–9. [CrossRef]

15. Aricò, P.; Borghini, G.; Di Flumeri, G.; Colosimo, A.; Pozzi, S.; Babiloni, F. A passive brain-computer interface application for the mental workload assessment on professional air traffic controllers during realistic air traffic control tasks. *Front. Hum. Neurosci.* **2016**, *228*, 295.

16. McDermott, P.L.; Ries, A.J.; Plott, B.; Touryan, J.; Barnes, M.; Schweitzer, K. A Cognitive Systems Engineering Evaluation of a Tool to Aid Imagery Analysts. *Proc. Hum. Factors Ergon. Soc. Annu. Meet.* **2015**, *59*, 274–278. [CrossRef]

17. Davis, S.E.; Smith, G.A. Transcranial Direct Current Stimulation Use in Warfighting: Benefits, Risks, and Future Prospects. *Front. Hum. Neurosci.* **2019**, *13*, 114. [CrossRef] [PubMed]

18. DARPA Restoring Active Memory (RAM). Available online: https://www.darpa.mil/program/restoring-active-memory (accessed on 12 July 2020).

19. Hitchcock, E.M.; Warm, J.S.; Matthews, G.; Dember, W.N.; Shear, P.K.; Tripp, L.D.; Mayleben, D.W.; Parasuraman, R. Automation cueing modulates cerebral blood flow and vigilance in a simulated air traffic control task. *Theor. Issues Ergon. Sci.* **2003**, *4*, 89–112. [CrossRef]

20. Nelson, J.T.; McKinley, R.A.; Golob, E.J.; Warm, J.S.; Parasuraman, R. Enhancing vigilance in operators with prefrontal cortex transcranial direct current stimulation (tDCS). *NeuroImage* **2014**, *85*, 909–917. [CrossRef] [PubMed]

21. Neuroscience, H. Bihemispheric Transcranial Direct Current Stimulation with Halo Neurostimulation System over Primary Motor. Cortex Enhances Fine Motor Skills Learning in a Complex Hand Configuration Task. 2016. Available online: https://www.haloneuro.com/pages/science (accessed on 3 June 2020).

22. Huang, L.; Deng, Y.; Zheng, X.; Liu, Y. Transcranial Direct Current Stimulation with Halo Sport Enhances Repeated Sprint Cycling and Cognitive Performance. *Front. Physiol.* **2019**, *10*, 118. [CrossRef]

23. Seck, H.H. Super SEALs: Elite Units Pursue Brain- Stimulating Technologies. *Military.com*, 2 April 2017.

24. Christman, J. Autonomy In Moral And Political Philosophy. *The Stanford Encyclopedia of Philosophy.* Zalta, E.N., Ed.; 2018. Available online: https://plato.stanford.edu/archives/fall2020/entries/autonomy-moral/ (accessed on 28 June 2020).

25. Mackenzie, C. Critical Reflection, Self-Knowledge, and the Emotions. *Philos. Explor.* **2002**, *5*, 186–206. [CrossRef]

26. Henschke, A. *Ethics in an Age of Surveillance: Personal Information and Virtual Identities*; Cambridge University Press: New York, NY, USA, 2017; pp. 1–334.

27. Kahneman, D. *Thinking, Fast and Slow*, 1st ed.; Farrar, Straus and Giroux: New York, NY, USA, 2011; p. 78.

28. Smith, M. *The moral Problem*; Blackwell: Cambridge, MA, USA; Oxford, UK, 1994; pp. 1–226.

29. Kennett, J.; Fine, C. Will the Real Moral Judgment Please Stand up? The Implications of Social Intuitionist Models of Cognition for Meta-Ethics and Moral Psychology. *Ethical Theory Moral Pract.* **2009**, *12*, 77–96. [CrossRef]

30. Annas, G.J. Beyond Nazi War Crimes Experiments: The Voluntary Consent Requirement of the Nuremberg Code at 70. *Am. J. Public Health (1971)* **2018**, *108*, 42–46. [CrossRef]

31. Moser, D.J.; Arndt, S.; Kanz, J.E.; Benjamin, M.L.; Bayless, J.D.; Reese, R.L.; Paulsen, J.S.; Flaum, M.A. Coercion and informed consent in research involving prisoners. *Compr. Psychiatry* **2004**, *45*, 1–9. [CrossRef] [PubMed]
32. Hayes, M.O. Prisoners and autonomy: Implications for the informed consent process with vulnerable populations. *J. Forensic Nurs.* **2006**, *2*, 84–89. [CrossRef] [PubMed]
33. Pont, J. Ethics in research involving prisoners. *Int. J. Prison. Health* **2008**, *4*, 184–197. [CrossRef]
34. Cummings, M. Informed Consent and Investigational New Drug Abuses in the U.S. Military. *Account. Res.* **2002**, *9*, 93–103. [CrossRef] [PubMed]
35. Gordon, B.G. Vulnerability in Research: Basic Ethical Concepts and General Approach to Review. *Ochsner J.* **2020**, *20*, 34–38. [CrossRef] [PubMed]
36. Parfit, D. *Reasons and Persons*, Reprint with corrections, 1987 ed.; Clarendon Press: Oxford, UK, 1987; pp. 1–543.
37. Parfit, D. Personal Identity. *Philos. Rev.* **1971**, *80*, 3–27. [CrossRef]
38. Parfit, D. On "The Importance of Self-Identity". *J. Philos.* **1971**, *68*, 683–690. [CrossRef]
39. DeGrazia, D. *Human Identity And Bioethics*; Cambridge University Press: Cambridge, UK, 2005.
40. Forrester, M.G. What Do We Owe To Future Generations? In *Persons, Animals, and Fetuses: An Essay in Practical Ethics*; Springer: Dordrecht, The Netherlands, 1996; pp. 137–146.
41. Golding, M.P. Obligations To Future Generations. *Monist* **1972**, *56*, 85–99. [CrossRef]
42. Jaworska, A. Advance Directives and Substitute Decision-Making. In *Stanford Encyclopedia of Philosophy*; Center for Study of Language and Information, Stanford University: Stanford, CA, USA, 2009.
43. Faye, J. Backward Causation. *The Stanford Encyclopedia of Philosophy*. Zalta, E.N., Ed.; Summer 2018 ed. 2018. Available online: https://plato.stanford.edu/archives/sum2018/entries/causation-backwards/ (accessed on 28 June 2020).
44. Douglas, T. Moral Enhancement. *J. Appl. Philos.* **2008**, *25*, 228–245. [CrossRef]
45. Douglas, T. Moral Enhancement Via Direct Emotion Modulation: A Reply To John Harris. *Bioethics* **2011**. [CrossRef]
46. Persson, I.; Savulescu, J. The Perils Of Cognitive Enhancement And The Urgent Imperative To Enhance The Moral Character Of Humanity. *J. Appl. Philos.* **2008**, *25*, 162–177. [CrossRef]
47. Dubljević, V.; Racine, E. Moral Enhancement Meets Normative and Empirical Reality: Assessing the Practical Feasibility of Moral Enhancement Neurotechnologies: Moral Enhancement Meets Normative and Empirical Reality. *Bioethics* **2017**, *31*, 338–348. [CrossRef] [PubMed]
48. Harris, J. Moral Enhancement and Freedom. *Bioethics* **2011**, *25*, 102–111. [CrossRef] [PubMed]
49. Persson, I.; Savulescu, J. Getting Moral Enhancement Right: The Desirability Of Moral Enhancement. *Bioethics* **2013**, *27*, 124–131. [CrossRef] [PubMed]
50. Haidt, J. The Emotional Dog and Its Rational Tail: A Social Intuitionist Approach to Moral Judgment. *Psychol. Rev.* **2001**, *108*, 814–834. [CrossRef] [PubMed]
51. Steenbergen, L.; Sellaro, R.; Hommel, B.; Lindenberger, U.; Kühn, S.; Colzato, L.S. "Unfocus" on foc.us: Commercial tDCS headset impairs working memory. *Exp. Brain Res.* **2016**, *234*, 637–643. [CrossRef] [PubMed]
52. Voarino, N.; Dubljević, V.; Racine, E. tDCS for Memory Enhancement: Analysis of the Speculative Aspects of Ethical Issues. *Front. Hum. Neurosci.* **2016**, *10*, 678. [CrossRef]
53. Repantis, D.; Schlattmann, P.; Laisney, O.; Heuser, I. Modafinil and methylphenidate for neuroenhancement in healthy individuals: A systematic review. *Pharmacol. Res.* **2010**, *62*, 187–206. [CrossRef]
54. Giubilini, A. Conscience. In *Stanford Encyclopedia of Philosophy Archive*, 2016 ed.; Center for the Study of Language and Information (CSLI), Stanford University: Stanford, CA, USA, 2016.
55. Persson, I.; Savulescu, J. Moral Hard-Wiring and Moral Enhancement. *Bioethics* **2017**, *31*, 286–295. [CrossRef]
56. Persson, I.; Savulescu, J. The Duty to be Morally Enhanced. *Topoi* **2019**, *38*, 7–14. [CrossRef]
57. Dobos, N. *Ethics, Security, and the War Machine: The True Cost of the Military*; Oxford University Press: Oxford, UK, 2020.
58. Glover, J. *Humanity: A Moral History of the Twentieth Century*; Yale University Press: New Haven, CT, USA, 2000; 464p.
59. Russell, N.J.C. Milgram's Obedience to Authority Experiments: Origins and Early Evolution. *Br. J. Soc. Psychol.* **2011**, *50*, 140–162. [CrossRef]

60. Blass, T. The Milgram Paradigm after 35 Years: Some Things We Now Know about Obedience to Authority. *J. Appl. Soc. Psychol.* **1999**, *29*, 955–978. [CrossRef]
61. Connor, J.M. Military Loyalty: A Functional Vice? *Crim. Justice Ethics* **2010**, *29*, 278–290. [CrossRef]
62. Osiel, M.J. Obeying Orders: Atrocity, Military Discipline, and the Law Of War. *Calif. Law Rev.* **1998**, *86*, 939. [CrossRef]
63. Dobos, N. Punishing Non-Conscientious Disobedience: Is the Military a Rogue Employer? *Philos. Forum* **2015**, *46*, 105–119. [CrossRef]
64. Helgesson, G. In Defense of Broad Consent. *Camb. Q. Healthc. Ethics* **2012**, *21*, 40–50. [CrossRef]
65. Sheehan, M. Can Broad Consent be Informed Consent? *Public Health Ethics* **2011**, *4*, 226–235. [CrossRef]
66. Henschke, A. Militaries and the Duty of Care to Enhanced Veterans. *J. R. Army Med. Corps* **2019**, *165*, 220–225. [CrossRef]
67. Boyce, R.M. Waiver of Consent: The Use of Pyridostigmine Bromide during The Persian Gulf War. *J. Mil. Ethics* **2009**, *8*. [CrossRef]
68. McManus, J.; Mehta, S.G.; McClinton, A.R.; De Lorenzo, R.A.; Baskin, T.W. Informed Consent and Ethical Issues in Military Medical Research. *Acad. Emerg. Med.* **2005**, *12*, 1120–1126. [CrossRef]
69. Wolfendale, J.; Clarke, S. Paternalism, Consent, and the Use of Experimental drugs in the Military. *J. Med. Philos.* **2008**, *33*, 337–355. [CrossRef] [PubMed]
70. Strawser, B.J. Moral Predators: The Duty to Employ Uninhabited Aerial Vehicles. *J. Mil. Ethics* **2010**, *9*, 342–368. [CrossRef]
71. Robillard, M.; Strawser, B.J. The Moral Exploitation of Soldiers. *Public Aff. Q.* **2016**, *30*, 171–195.

 *philosophies*

Article

Superhuman Enhancements via Implants: Beyond the Human Mind

Kevin Warwick

Office of the Vice Chancellor, Coventry University, Priory Street, Coventry CV1 5FB, UK;
k.warwick@coventry.ac.uk

Received: 16 June 2020; Accepted: 7 August 2020; Published: 10 August 2020

Abstract: In this article, a practical look is taken at some of the possible enhancements for humans through the use of implants, particularly into the brain or nervous system. Some cognitive enhancements may not turn out to be practically useful, whereas others may turn out to be mere steps on the way to the construction of superhumans. The emphasis here is the focus on enhancements that take such recipients beyond the human norm rather than any implantations employed merely for therapy. This is divided into what we know has already been tried and tested and what remains at this time as more speculative. Five examples from the author's own experimentation are described. Each case is looked at in detail, from the inside, to give a unique personal experience. The premise is that humans are essentially their brains and that bodies serve as interfaces between brains and the environment. The possibility of building an Interplanetary Creature, having an intelligence and possibly a consciousness of its own, is also considered.

Keywords: human–machine interaction; implants; upgrading humans; superhumans; brain–computer interface

1. Introduction

The future life of superhumans with fantastic abilities has been extensively investigated in philosophy, literature and film. Despite this, the concept of human enhancement can often be merely directed towards the individual, particularly someone who is deemed to have a disability, the idea being that the enhancement brings that individual back to some sort of human norm. Cochlear implants, artificial hips and even deep brain stimulators for Parkinson's disease are all good examples of the sort of technology designed for a specific group need. This form of enhancement is not the main subject of this paper. Rather, here we concern ourselves with human enhancement which takes a human (norm) as a starting point, employs technology by means of an implant in some way and as a result realises a superhuman.

When looking at ideas from science fiction, the fundamental concept of the superhuman tends to be limited by the restrictions of the imagination of the creator of an idea and/or those who will take on board what has been suggested. Hence, many ideas of the superhuman ignore the capabilities of technology today, never mind technology of the future, and produce superhumans with a pretty limited set of capabilities—for example, the Hulk, who is bigger and stronger, and Superman, who can fly and has enhanced hearing and cold breath or superspeed; coupled with that, the script writer will most likely be limited in terms of their knowledge as to what is technically possible and what is not. Further, for popular media, it usually has to be possible to visualize (on a screen) and understand any enhancement employed, and this eliminates a plethora of possibilities.

An example of such restrictions at this time might be useful. Consider the film of the story *I, Robot* [1], starring the actor Will Smith. Essentially, the control of physical robots in the world is taken over by a computer system, thereby causing havoc amongst humans. Each robot is itself not particularly intelligent—nothing wrong there—but the computer system which controls them is. But firstly, the robots all look like metallic versions of humans, whereas that is not what most machines/robots actually look like or would need to look like. Secondly (and more importantly), the computer system turns out to be one big supercomputer in one specific building, which of course our hero Will Smith can gain access to and destroy, thereby saving the world. Even now, we have cloud computing (was a version of that filmed in *The Terminator*?), and it is network-based (how could we switch off the internet even now really/practically?).

In the next section, I have a brief look at some therapies that can be modified into some sort of enhancement and then very quickly mention available technologies and where they might lead, if anywhere. Following this, I focus more on such technologies as growing brains and direct connections into the human brain and nervous system specifically for enhancement. Subsequently, I briefly discuss some of the ethical issues, in a realistic way, and finally, rather than trying to draw some conclusions, there is a meandering discussion.

2. Therapy

Deep brain stimulation is a procedure employed, in terms purely of therapy, for the treatment of Parkinson's disease, Tourette's syndrome, epilepsy and clinical depression. Electrical signals at a frequency of 150 to 190 Hz are applied into the thalamus or subthalamic nucleus of the brain. The effect of these signals is to counteract the original problem. The deep brain electrodes can also be connected bi-directionally with a computer such that electrical activity in the brain can be monitored. Using AI techniques, a better understanding of the nature of Parkinson's disease has been obtained [2]. It is also possible for the monitoring computer to be located remotely from the patient, e.g., in a different country. Hence, signals within the brain can be tracked in real time and fed into a computer. The computer can analyse these signals and generate alternative signals that are fed directly back into the brain. Effectively, in this example, part of that person's brain resides not only outside their body but physically in a different country.

In another vein, in the 1950s, Olds and Milner [3] implanted electrodes directly into the lateral hypothalamus of the brain in several rats. Each electrode was connected to a stimulator that was activated by the touch of a button. The electrical stimulation was aimed at causing feelings of pleasure. The rats were taught how to press the button themselves, giving them the feeling of pleasure. They were then given the choice between the pleasure button and a button that resulted in them being given food. The rats continually chose the pleasure button even when they were hungry and starving. Very similar results were obtained when rats were replaced by monkeys.

Consider that a brain implant can be employed to overcome clinical depression; when it is switched on, the patient can feel as though a black cloud is leaving them. The individual person is being given positive feelings. It is essentially acting as an electronic drug. Indeed, as the human brain operates electrochemically, drugs can potentially take on either form—electrical or chemical. It is merely a matter of conformity that thus far, almost all drugs are chemically based. In the years ahead, this could change dramatically. Possibly even headaches will be removed by electronic means in the future, as this would be more direct with potentially fewer side effects. Conversely, this could readily be extended for the computer to bring about positive feelings under certain circumstances and negative feelings in other cases.

Communication is vitally important for humans. Philip Kennedy developed an operable system which allowed an individual with paralysis to spell words by modulating their brain activity. Kennedy's device used two simple electrodes: The first was implanted in an intact motor cortical region and was used to move a cursor among a group of letters. The second was implanted in a different motor region and was used to indicate that a selection had been made [4]. As the patient thought about moving their fingers, these signals were translated into signals to move and stop a computer cursor. The patient could see where the cursor was on a computer screen. They could choose when to stop thinking about moving. In this way, words could slowly be spelt out letter by letter or heating and lighting could be simply controlled. Although initiated as a therapeutic experiment, the recipient of the implant exhibited abilities that were beyond the human norm.

Another good example is that of Neil Harbisson, who is otherwise colour blind and has a camera which is attached to his skull. Different colours cause the frequency of vibrations to his skull to vary. As a result, he has learned a very high degree of colour discrimination. The technology translates colour frequencies into sound frequencies [5] which are translated into vibrations via an actuator. Initially, Harbisson memorised the frequencies related to each colour, but subsequently he decided to permanently attach the set-up to his head. The project was developed further so that Harbisson was able to perceive colour saturation as well as colour hues. Software was then developed that enabled Harbisson to perceive up to 360 different hues through microtones and saturation through different volume levels. What is particularly interesting about Harbisson's experience is that his discrimination between different colours has improved over time as his brain has adjusted to the different vibrations experienced. Clearly, the extent of brain adaptability is a pointer to what can be expected in general with regard to either extending the present range of sensory input or rather inputting a complex range of new sensory input information into the human brain that until now has not been possible. It is another example of a therapeutic treatment that results in the recipient having abilities beyond the human norm.

Perhaps the most commonly encountered brain–machine interface is the cochlear implant, intended to repair an individual's hearing by connecting electrodes to directly stimulate the auditory nerve fibres. It is estimated that there are at least 600,000 recipients thus far. At first, the person realises electronic signals that may have little meaning, but gradually, their brain adapts to the signals input and learns to recognise them as specific sounds. In some cases, it may be that the individual can comprehend certain higher frequency sounds for the first time in their life. Ordinarily, the frequency range input is aimed at mimicking as much as possible the normal human frequency input. However, there is of course nothing stopping a different frequency range from being input for the brain to learn and/or for the auditory nerve to be stimulated directly from a network rather than from sound waves.

Arguably the technology which has proven to be of the most practically successful in this area is the microelectrode array, currently referred to as the BrainGate. The array is made up of 100 silicon spikes which are 1.5 mm long with a platinum electrode on each tip. The electrodes are linked with platinum wires, and in this way, the array can be employed to both monitor neural activity and also apply stimulating currents. Several trials have been carried out using human studies. In experiments, the array has been fired into either the human brain or nervous system. In the first set of these experiments to be considered, the array has been employed in a purely recording role for therapeutic results.

Electrical activity from a few neurons monitored by the array electrodes, positioned in the motor cortex, has been decoded into a signal that enabled a severely paralysed individual to position a cursor on a computer screen using neural signals for control in combination with visual feedback. The same technique was later deployed to allow the individual recipient, who was paralysed, to operate a robot arm even to the extent of learning to feed themselves in a rudimentary fashion by maintaining sufficient control over the robot arm [6].

We later consider directly the use of this BrainGate implant in a set of experiments which were set up specifically to look at enhancement possibilities.

3. Off the Shelf—External Electrodes

In numerous research labs around the world, the concept of a brain–computer interface is actually realised by an interface between the experimentalist's scalp and a computer, see, e.g., [7]. In this method, weak signals from a brain (in terms of scalp-filtered averages from millions of neurons) are fed into a computer such that after significant AI processing, the individual concerned can learn to think in specific ways such that (typically) a connected truck will turn right maybe 8 times out of 10 when they want it to. This is of course purely a one-directional procedure, other than any feedback via the usual sensory route, e.g., their eyes.

It is not the author's intention to dwell on this area other than to acknowledge the research that is going on and to appreciate the possibility of monitoring a brain in this way to further understand what is happening inside with the hope of detecting neurological problems as encountered in cases such as epilepsy. Arguably several other scanning techniques provide similar or better such results. However, the method's big advantage is of course that external electrodes can be relatively easily attached to the outside of the scalp, which makes such research much more doable, thereby considerably reducing any potential problems of infection or rejection. It is interesting to note, however, that in many research papers on the topic, the same partial conclusion appears, which is [8]: "Instead of placing the electrodes on the scalp if they are placed in the cortex itself, it would provide a better result".

Rather, the focus of this article is fundamentally directed towards brain–computer connections involving an implant (or implants) in the brain or nervous system, thereby achieving a much higher resolution with the ability to extract signals from a handful of neurons. Importantly, such a technique is potentially bi-directional, meaning that both (efferent) motor signals can be monitored and (afferent) sensory signals can be applied.

4. Some Future Possibilities

In this section, we look at two possibilities. The second of these is instantly more practical and is based on a set of successfully conducted experiments involving the author. The first, whilst also being based on the scientific experimentation of the author, is perhaps rather more speculative.

It is quite possible to culture networks of dissociated neurons grown in vitro in a chamber. The neurons are provided with suitable environmental conditions and nutrition. A flat microelectrode array is embedded in the base of the chamber, thereby providing a bi-directional electrical interface with the neuronal culture. The neurons in the culture rapidly reconnect, form a multitude of pathways and communicate with each other by both chemical and electrical means. Although for most research in the field thus far, the neurons are typically taken from rat embryos, it is quite possible to use human neurons instead once sufficient connections have been made between the neurons so that, in research, the cultured brain is given a robot body with the ability to sense the world and move around in it [9].

Humans are essentially our brains [10]. Our bodies keep our brains functioning, transport them, provide some sensory input and enable each brain to interact with the outside world. However, through evolution, our brains have largely become dependent on the bodies that carry them around. But apart from some limited transplants or artificial organs, when some physical feature malfunctions, then that may well mean the human dies, even though there is essentially nothing wrong with the brain, i.e., the essential self is well but it dies because a (possibly in the future) trivial physical element no longer functions appropriately.

Therefore, if we look to the future, theoretically for the moment, it seems sensible to consider directly keeping the brain alive, somewhat akin to the experimentation described, without its dependence on its physical body. As an example, if a person has liver cancer and dies from this, even though their brain was perfectly OK, it is an unnecessary death. Rather than considering further research into the treatment of such cancers and the like, surely it is better to consider ways to keep the brain alive outside of its present human body, as we presently do with culture experimentation. There is, though, the matter of scale—presently, cultures of typically 150 thousand neurons are supported, whereas with the human brain, we would need to support 100 billion neurons.

In this way, the body could be designed merely to fit around the brain. If something in the body functions incorrectly or stops working, then it can simply be replaced or upgraded. If a better body component becomes available, whether biological or technological, then the newer, more powerful option can be selected. The range of sensory input can be what you want, the abilities of the body can be designed to suit. Life expectancy would be much more enhanced as all of the body could be replaced as needed. A lifetime would be totally the brain's lifetime.

In a different vein, the author also carried out an experiment in collaboration with surgeons at the Department of Neurosurgery and Neurosciences at the John Radcliffe Hospital, Oxford [11]. During a two-hour procedure, a BrainGate array was surgically implanted into the median nerve fibres of the left arm of the author. The array measured 4 by 4 mm with each of the electrodes being 1.5 mm in length. With the median nerve fascicle estimated to be 4 mm in diameter; this meant that the electrodes penetrated well into the fascicle.

The array was pneumatically inserted into the median nerve such that the body of the array sat adjacent to the nerve fibres, with the electrodes penetrating into the fascicle. The array was positioned just below the wrist, following a 4 cm long incision. A further incision, 2 cm long, was made 16 cm proximal to the wrist. The two incisions were connected by a tunnelling procedure such that wires from the array ran up the inside of the left arm, where they exited and connected onto an electrical terminal pad which remained external. The arrangements described remained permanently in place for 96 days. During that time, a series of experiments was conducted.

The terminal pad was directly linked to a computer terminal either by hard wiring or preferably by means of a wireless connection, which enabled mobility. By this means, a ready connection was arranged with the internet. The link was bi-directional such that motor signals due to hand movements could be monitored and decoded whilst, via the same implant, sensory signals from a remote source could be applied to stimulate the nerve fascicle and thereby be fielded by my brain.

A number of different trials relating to human enhancement were successfully realised. These were:

1. Extrasensory (ultrasonic) input was used to detect the distance to objects:

It took approximately two weeks for us to find a stimulating current that my brain could reliably recognise. All we had to go on initially was previous testing on chicken sciatic nerves [12]. Although the waveform was similar, we needed a stronger current at a voltage of 50 v. I wore a blindfold and heard a click; sometimes a stimulating pulse had been applied to my nervous system, sometimes not. After 1 week, I correctly detected 70% of pulses (50% would be the same as guessing); however, after 2 weeks, it was 100%. During that time, for about 2 h every day, we were doing the click testing in the lab. But then, we linked ultrasonic sensors on a baseball cap to cause the stimulation. The closer an object was, the more the frequency of pulses increased. My brain made sense of this immediately: Even with a blindfold on, I knew how far away objects were. So much so that when Iain, one of the researchers, suddenly moved a board quickly towards me, it was very scary—I could detect that something was coming towards me very quickly but I did not know what it was. It felt like a reactive response.

2. A wheelchair was driven solely by motor neural signals:

It was pretty straightforward from the early days of the experiment that we could detect neural signals by means of the implant. Neural signals are very different to, for example, noise or muscular signals, so it is not such a difficult signal processing problem. Understanding what those signals actually mean is a more difficult proposition. However, it did not take a complex algorithm to link neural signals with hand-opening or hand-closing movements. We were working on the project with two hospitals and had obtained a wheelchair from the National Spinal Injuries Centre in Stoke Mandeville Hospital. We then used a simple menu device to link the neural signals with wheelchair operation. By opening and closing my hand, I could move down the menu to my selection—e.g., forwards, backwards, slow, fast—and the chair would follow my wishes. The point is that this would operate exactly the same way, with the same implant, for someone with a spinal injury but with the implant in their motor cortex. It could of course be a car, or any vehicle, rather than a wheelchair.

3. The behaviour of a group of small robots was altered by motor signals:

We were fortunate to have a lab full of little robots. These could exhibit different "emotional" behaviours via ultrasonic sensors. Thus, on detecting another robot, they could act as though they were scared and try and escape or conversely act aggressively and try and catch the detected robot. This was simply linked to my neural signals such that when my hand was closed, the robots acted in a friendly way, whereas when my hand was open, the robots were aggressive. It was a simple experiment and merely went to show how powerful such an implant can potentially make the recipient.

4. A robot hand was controlled via the internet between Columbia University, USA and Reading University, England. Feedback from the fingertips was also obtained. I think that this experiment, more than any other, had a profound effect on me in that I never expected how I would feel at the time. As scientists, we were just rolling from one experiment to the next without thinking too much about the consequences. But this one was different. In New York, my nervous system was linked, online, with the internet. Security was so tight that we did not tell anyone what we were doing! As I opened and closed my hand in New York, my neural signals also opened and closed "my" robot hand in England. There was a noticeable fraction of a second delay in the process, but my brain did not seem to be bothered by it. As the hand gripped an object, signals from sensors in the fingertips gave me an accurate description of just how much force the hand was applying. I could see the hand by video link, but if I did not look, I could still hold on to an object and not apply too much pressure. Essentially, my nervous system was being stimulated by pulses, the frequency of which depended on how much pressure the robot hand was applying in England. In this experiment, I was trying to open and close my robot hand in order to get the pulses to die down to almost zero. This was very successful and made me realise just how powerful an individual is when their brain/nervous system is extended by a network.

5. A basic form of telegraphic communication was achieved between two human nervous systems.

I have something of a communications background, and for me, this was the icing on the cake. A volunteer assisted by having microneurography. Essentially, two very thin needles were pushed into the nervous system in their left arm. With this in place, we set up in the lab with a group of people around the volunteer and another group around me—we had a variety of different observers to oversee what we were doing. The volunteer and I were not able to see each other. We set the experiment up purely based on hand closures. When the volunteer closed their hand, I received a stimulating pulse on my nervous system, and the same happened vice versa. For me, it meant that my brain recognised the pulse. I shouted out "Yes" every time I felt a pulse, but only when I felt a pulse. Only the group around the volunteer could witness when they had closed their hand and when not. We achieved this with 100% success—the same being true in reverse. What I found exciting was that as the groups were splitting up, I felt a couple of quick pulses one after the other. Subsequently, the volunteer confirmed that they had done this. It was a "secret" message between the two of us, a new means of communication.

In every one of the experiments just described, the raison d' être could be heralded as being for therapeutic purposes, e.g., ultrasonic input for a person who is blind. However, in each case, the trial can rather be regarded as an enhancement beyond the human norm. Clearly, such enhancements throw up a multitude of intellectual and technological opportunities, but they also realise a range of ethical considerations. Indeed, there may be persons who do not wish others to exhibit such enhancements, whereas, on the other hand, individual freedom dictates that an individual can so upgrade themselves if they want. That said, individual freedom is not an absolute in many societies in the world today, and in any case, the freedom of one individual must be balanced with the effect of that individual on the freedom of others.

5. Enhancements

In this paper, we have considered the potential for enhancing humans, thereby creating superhumans, mainly in terms of upgrading the human brain. Practical examples, described in the previous section (in terms of implant experiments), have shown that it is possible to directly control, from brain signals alone, remote technology, with real time feedback also possible. It was also shown how senses could be enhanced. Although ultrasonics was employed in the experiments, a plethora of signals could be attempted, e.g., infrared, X-ray or ultraviolet. How well the brain can employ such signals and how much use the signals would be remains to be seen. An old film considered a man with X-ray eyes [13], whereas we are "looking" here at the possibility of X-rays as an extra sense.

Perhaps the most significant achievement was a new form of telegraphic communication directly between the nervous systems of two people. When a similar experiment is achieved directly between two human brains, then this will be a sort of telepathy, a form of thought communication. It then remains for us to witness just how well this will work. The possibility is real for communication between humans in terms of feelings, colours, images, emotions, as well as the sort of message passing we have now. But speed and accuracy could be much improved in this enhanced way, and potentially, misunderstandings can be much reduced. Many of the problems in our present, poor form of communication can be overcome.

In our favour is the fact that the human brain is very adaptive and can learn to operate with extra abilities, even at an older age. It may well need to ground each extra ability in terms of how it already functions, what it knows, but expand in this way it can. What limits the potential forms of enhancement in the first instance is our limited understanding not only of how the human brain functions but also of the world in which we live.

What we are considering here is the creation of superhumans by linking human brains with computers, potentially artificial intelligence, thereby upgrading the functioning of the individual human brain whilst, at the same time, linking brains (both human and machine) together in a much closer way than they have been to this day. This will be brought about in terms of a port (or more likely a number of ports) into each brain—probably of a similar form to the BrainGate employed in the experimentation.

With a human brain linked to a computer brain in this way, that individual could have the ability to (1) use the computer part for rapid maths, (2) call on an internet knowledge base, quickly, (3) have memories that they have not originally themselves had, as has been investigated [14], (4) sense the world in a plethora of ways not possible to humans (e.g., ultrasonic, infrared), (5) understand multidimensionality, as opposed merely to 3D for the human brain alone and (6) communicate in parallel, by thought signals alone, i.e., brain to brain.

Although this list of enhancements is, to an extent, speculative, it is the sort of list that can be employed in terms of an hypothesis, the truth of which is to be discovered through experimentation. Thus, it may turn out that some items on the list do not turn out to be practically useful, whereas other items may turn out to be mere steps on the way to abilities that are more far-reaching.

Whilst each of the items on the list can be explored in more depth, some of them are more profound than others. For example, thought communication will no doubt bring about a raft of changes to our present human consciousness, a big issue here being how superhumans with such an ability will subsequently regard humans who communicate as at present. This is be looked at further in the next section.

One of the items with far reaching consequences is number 5. Human brains regard the world around us in terms of a three-dimensional understanding. Of course the world is not three-dimensional; that is merely the complexity of model that humans use. Our beliefs around what is possible and what not are largely based on this limited understanding. Hence, we have major issues with space travel (we have to this time still only visited the Earth's moon). Getting to the edge of the solar system might well take a human many years because of our 3D understanding. Clearly, this is not particularly practical.

But if our superhuman brains can understand the world around us in terms of hundreds of dimensions, which our part artificial intelligence brains will be able to cope with, then many things that are presently difficult or realistically impossible may well become quite possible and even relatively simple. As superhumans, we may be able to travel through the universe *Star Trek* style.

It is worth also dwelling a little on the successful experiment involving the BrainGate implant in which a robot hand was controlled via the internet including feedback from the hand's finger tips. Effectively, the internet was acting as an extension to the human nervous system. In a telephone call, sound waves are transformed into signals (initially electrical), transmitted over enormous distances and then reconstituted as sound wave signals. The sound waves themselves have very limited range, but in this transformation, their range can be much enhanced. The same is exactly true with the human brain and nervous system. What was actually achieved in the experiment was a monitoring of neural motor signals, via the implant in the nervous system. These were then transformed into information on the internet and reconstituted as motor signals in another country to drive a robot hand. The same procedure was true in the reverse direction for the touch sense.

Clearly, this opens up a wide range of future possibilities for enhanced humans. Whilst the person's brain (along with some sort of brain supporting body) is in one place, their new body can be anywhere the network is operational. A different continent certainly, but it could also be on several different planets at the same time. Perhaps, in the future, your brain can stay on earth, but your body will be able to travel to other planets, only to be reconnected on arrival. Importantly, as far as your brain is concerned, when you are fully connected, it will "feel" like it is your own body, which, for all intents and purposes, it will be. Although it will also be possible to swap or share body parts with other people.

I personally found that out with my brain in New York, USA and my robot hand in Reading, England; my brain had no problem with this even though there was a fractional time delay from the moment my brain thought about moving to the time the robot hand actually moved. Indeed, our brains deal with a time delay anyway due to the time it takes a signal from your brain to travel down your nervous system to move a hand or leg. In the case of the experiment, the time delay was a little longer, but my brain quickly adjusted to that.

A robot hand was employed in the experiment to show what was possible, but also to show how the technology could be employed for therapy for an artificial hand or leg in the case of an amputation. In reality though, in the future, the appended (network-connected) new body parts will not need to be legs or arms and hands but can be anything at all. Thus, it will be quite possible to have a building or a vehicle or a financial system as new body parts, directly connected with your brain—as far as your brain is concerned, it will be your body. How easy it will be for the brain to understand what is going on remains to be seen. It may take some time for the brain to adjust and fully implement its new body. But, as in Kafka's *Metamorphosis* [15], once your brain has adapted, it will be as though it is perfectly normal.

6. Discussion

The realisation of superhumans with significant powers over and above those of regular humans presents enormous questions that affect all aspects of human society and culture. When attempting to consider the possibilities, a plethora of positives and negatives appear. What is clear, however, is that standing still is not an option. On the one hand, if humans opted by some global agreement for a non-superhuman future (if that were possible), could the end result actually be an intelligent machine superculture as described in [16,17], leading to the handing over of control on earth to intelligent machines? On the other hand, if humans globally opted for a superhuman future, could society and culture cope with such a distinct nonlinearity in evolution? Maybe we should not worry about it anyway, as it would be a superhuman culture that faced the nonlinearity.

It could be felt that humankind is itself at stake [18]. A viewpoint can then be taken that either it is perfectly acceptable to upgrade humans, turning them into superhumans, with all the enhanced capabilities that this offers, or it could be considered that humankind is OK as it is and should not be so tampered with. Realistically though, humans have always gone for progress—indeed, it is part of our nature, possibly even in our genetic make-up. How would the not-for-progress humans, who want us all to remain as humans, be able to prevent the progress in any case?

As we have discussed in this paper, the most important issue is that we are considering a completely different basis on which the superhuman brain operates—part-human and part-machine. When the nature of the brain itself is altered, the situation is complex and goes far beyond anything encountered with mere physical extensions, such as the ability to fly in an airplane. Such a superhuman would have a different foundation on which any of their thoughts would be conceived in the first place. From an individualistic viewpoint, therefore, as long as I am myself a superhuman, I am very happy with the situation. Those who wish to remain ordinary humans, however, may not be so happy.

With a brain which is part-human, part-machine, superhumans would have some links to their human background, but their view on life, what is possible and what is not, would be very much different from that of a human. Of course, this would all depend on the newly acquired abilities and what effects these have on the mixed consciousness. Would all superhumans have similar abilities, or would it be a case of picking and choosing? Importantly, each individual's values would relate to their own life, and ordinary humans may not figure too highly. Different superhumans would most likely exhibit very different abilities. Some, such as thought communication, would be highly desirable, whereas others might be OK for some but not for others. Just how these differences pan out is impossible to say at this stage.

One aspect is that superhumans would have brains, which are not standalone but rather are connected to each other directly via a network. A question is, therefore: Is it acceptable for humans to give up their individuality and become mere nodes on an intelligent machine network? Or is it purely a case of individual freedom, i.e., if an individual wants to so upgrade, then why not? Would the network become the most important aspect with each node being of little value? It must be remembered here that we are looking at an intelligent network. Would there, as a result, be some sort of network consciousness?

Some questions are obvious. Should every human have the right to be upgraded? If an individual does not want to upgrade, should they be allowed to defer, thereby taking on a role in relation to superhumans, perhaps something like a chimpanzee's relationship with a human today? How will the values of superhumans relate to those of humans? Will humans be of any consequence to superhumans other than something of an awkward pain to be removed if possible [19]?

It is sensible to be clear that with extra memory, high-powered mathematical capabilities, including the ability to conceive in many dimensions, the ability to sense the world in different ways, communication by thought signals alone and having a networked body, superhumans will be far more powerful, intellectually, than regular humans. It would be difficult to imagine that superhumans would want to voluntarily give up their powers in order to satisfy the grumbles of mere humans. Indeed, why would superhumans, who can communicate just by thinking with each other, pay any heed to the trivial utterances of humans, based on serially modulated sound waves—very slow, very simple and highly error-prone?

The fundamental philosophy that underpins the concept of the future relationship between superhumans and humans comes straight from Nietzsche [20]. We need to look no further than humans' present-day relationship with other animals. Humans cage them, destroy their habitat and treat them as captives, slaves or pets. Thus, if we look to the future, the best that a human could hope for might be that they become the pet of a superhuman.

Funding: The implant studies described were supported in part by Nortel Networks, Maidenhead, England; Tumbleweed Communications, Twyford, England; Computer Associates, Slough, England and Fujitsu (ICL), Bracknell, England. The neuron culturing described was supported in full by the Engineering and Physical Sciences Research Council under grant number EP/D080134/1.

Conflicts of Interest: The author declares no conflict of interest.

References

1. Asimov, I. *I, Robot*; HarperVoyager: New York, NY, USA, 2018.
2. Camara, C.; Isasi, P.; Warwick, K.; Ruiz, V.; Aziz, T.; Stein, J.; Bakstein, E. Resting tremor classification and detection in Parkinson's Disease patients. *Biomed. Signal Process. Control* **2015**, *16*, 88–97. [CrossRef]
3. Olds, J.; Milner, P. Positive reinforcement produced by electrical stimulation of septal area and other regions of rat brain. *J. Comp. Physiol. Psychol.* **1954**, *47*, 419–427. [CrossRef] [PubMed]
4. Kennedy, P.; Andreasen, D.; Ehirim, P.; King, B.; Kirby, T.; Mao, H.; Moore, M. Using human extra-cortical local field potentials to control a switch. *J. Neural Eng.* **2004**, *1*, 72–77. [CrossRef]
5. Harbisson, N. Painting by ear. In *Modern Painters*; The International Contemporary Art Magazine; Louise Blouin Media: New York, NY, USA, 2008; pp. 70–73.
6. Hochberg, L.; Bacher, D.; Jarosiewicz, B.; Masse, N.; Simeral, J.; Haddadin, S.; Liu, J.; Cash, S.; Van der Smagt, P.; Donoghue, J. Reach and grasp by people with tetraplegia using a neurally controlled robotic arm. *Nature* **2012**, *485*, 372–375. [CrossRef]
7. Nicolas-Alonso, L.; Gomez-Gil, J. Brain computer interfaces, a review. *Sensors* **2012**, *12*, 1211–1279. [CrossRef] [PubMed]
8. Babu, G.; Balaji, S.; Adalarasu, K.; Nagasai, V.; Siva, A.; Geetanjali, B. Brain Computer Interface for Vehicle Navigation. Available online: https://www.biomedres.info/biomedical-research/brain-computer-interface-for-vehicle-navigation.html (accessed on 14 June 2020).
9. Warwick, K.; Xydas, D.; Nasuto, S.; Becerra, V.; Hammond, M.; Downes, J.; Marshall, S.; Whalley, B. Controlling a mobile robot with a biological brain. *Def. Sci. J.* **2010**, *60*, 5–14. [CrossRef]
10. Churchland, P. *Brain-Wise: Studies in Neurophilosophy*; MIT Press: Cambridge, MA, USA, 2002.
11. Warwick, K.; Gasson, M.; Hutt, B.; Goodhew, I.; Kyberd, P.; Andrews, B.; Teddy, P.; Shad, A. The application of implant technology for cybernetic systems. *Arch. Neurol.* **2003**, *60*, 1369–1373. [CrossRef] [PubMed]
12. Branner, A.; Stein, R.; Norman, E. Selective stimulation of a cat sciatic nerve using an array of varying length micro electrodes. *J. Neurophysiol.* **2001**, *54*, 1585–1594. [CrossRef] [PubMed]
13. *X, the Man with the X-Ray Eyes*; Harris, B. (Ed.) K.K. Publications, Inc. (Indicia/Colophon Publisher): New Delhi, India, 1963.
14. Ramirez, S.; Liu, X.; Lin, P.-A.; Suh, J.; Pignatelli, M.; Redondo, R.L.; Ryan, T.J.; Tonegawa, S. Creating a False Memory in the Hippocampus. *Science* **2013**, *341*, 387–391. [CrossRef] [PubMed]
15. Kafka, F. *The Metamorphosis*; CreateSpace Independent Publishing Platform: Scotts Valley, CA, USA, 2014.
16. Warwick, K. *March of the Machines*, 1st ed.; Century: Salt Lake City, UT, USA, 1997.
17. Kurzweil, R. *The Singularity is Near*; Gerald Duckworth & Co. Ltd.: London, UK, 2006.
18. Cerqui, D. The future of humankind in the era of human and computer hybridisation. In An Anthropological Analysis, Proceedings of the Conference on Computer Ethics and Philosophical Enquiry, Lancaster, UK, 14–16 December 2001; pp. 39–48.
19. Warwick, K. Homo Technologicus: Threat or Opportunity? *Philosophies* **2016**, *1*, 199. [CrossRef]
20. Nietzsche, F. *Thus Spoke Zarathustra*; Penguin Classics: London, UK, 1974.

 *philosophies*

Article

Homo Technologicus: Threat or Opportunity?

Kevin Warwick

Office of the Vice-Chancellor, Coventry University, Priory Street, Coventry CV1 5FB, UK;
k.warwick@coventry.ac.uk; Tel.: +44-24-7765-9893

Academic Editor: Jordi Vallverdú
Received: 29 August 2016; Accepted: 19 October 2016; Published: 26 October 2016

Abstract: Homo sapiens is entering a vital era in which the human-technology link is an inexorable trend. In this paper a look is taken as to how and why this is coming about and what exactly it means for both the posthuman species Homo technologicus and its originator Homo sapiens. Clearly moral and ethical issues are at stake. Different practical experimentation results that relate to the theme are described and the argument is raised as to why and how this can be regarded as a new species. A picture is taken of the status of cyborgs as it stands today but also how this will change in the near future, as the effects of increased technological power have a more dramatic influence. An important ultimate consideration is whether Homo technologicus will act in the best interests of Homo sapiens or not. This paper concludes that the answer is clear.

Keywords: cyborgs; implants; posthumans; Homo technologicus; Homo sapiens; human-machine interaction

---

1. Introduction

The subject area of cyborgs and posthumanism has been well developed in the social sciences and humanities for many years now [1]. This has been influenced strongly by the general trend of humans to embrace technology in their everyday lives and depend on it for their existence [2,3]. At the center of the discussion is 'Homo technologicus, a symbiotic creature in which biology and technology intimately interact', with the overall effect that the end result is 'not simply "Homo sapiens plus technology", but the original Homo sapiens morphed by the addition of technology into 'a new evolutionary unit, undergoing a new kind of evolution in a new environment' [4,5].

But we must consider what, in practice, we mean by Homo technologicus. The critical element here is the concept of boundary. As has been pointed out by many researchers, e.g., [6,7], the human brain is affected by the technology around us. It develops over time to interact more efficiently with that technology. However it is perhaps somewhat flippant to suggest [7] that we are therefore all cyborgs, even though there may well be a gradual change in our neural make up over a period as a result of the environment around us. Some have gone further and suggested that a blind man with his cane [8] is a cyborg, on the basis that the cane feeds important information to the man about his local environment. Meanwhile a pair of glasses or a hearing aid for a deaf person could be regarded in the same way. In recent years many researchers in the field of wearable computers have become self-professed cyborgs (e.g., [9,10]) whilst, in some instances, having much less interaction with the object worn than the blind man with his cane.

So we have case 1 in which the human body in its entirety remains intact whilst some form of technology is positioned close to or attached to the body for some reason as mentioned. Such a situation includes military examples: infra-red night sight incorporated into weapon sighting systems or voice controlled firing mechanisms in the helmet of a fighter pilot. This case (case 1) could also include those who use their cell phone almost constantly or perhaps even those playing games on a Tablet. It may well be that some researchers wish to define such or all people as cyborgs and it may be that the individual brains are gradually modified due to external pressure. However the consideration is that they remain as Homo sapiens and do not come into the category of Homo technologicus. In saying this it is recognised that the technology in question may be for therapeutic purposes, for gain or benefit or it may be simply for pleasure.

On the other hand we have witnessed many intrusions of technology into the human body. As examples hip replacements and heart pacemakers are now relatively frequently encountered. They continue a trend in which technology is readily accepted as being a necessary intrusion. Each of these represent modifications intended to compensate for deficiencies [11]. Even in these instances however the establishment of conceptual limits and boundaries becomes a complex process.

The situation lands up on more difficult terrain when, rather than repairing the ineffective parts of a human body, technology is implanted to enhance normal functioning. The situation where technology is implanted into the body but not into the brain/nervous system, whether it is for therapy or enhancement, we refer to here as case 2. In "A Manifesto for Cyborgs: Science, Technology and Social Feminism in the 1980s" [12], Donna Harraway discussed these issues as the disruption of traditional categories. But why should such entities present an ethical problem?

In each case, although the individual's physical capabilities take on a different form and their abilities are possibly enhanced, their inherent mental state, their consciousness, their perception, has not been altered other than to the extent of itself concluding what the individual might be capable of accomplishing. Where the ethical dilemma appears is in the case when an individual's consciousness is modified by the merging of human and technology. Essentially it is not so much the physical enhancements or repairs that should be our cause for concern but where the nature of an individual is changed in certain ways by the linking of human and technology mental functioning. In the case of a human this means linking technology directly with the human brain or nervous system, rather than by a connection which is either external to the nervous system but internal to the body (case 2) or even one which is external to both (case 1).

Even with technology linked directly to the human brain/nervous system this can be merely for therapeutic purposes as is the situation with deep brain stimulation for Parkinson's disease [13]. This is referred to here as case 3. As a result, Homo technologicus is considered in this paper to be one in which the entity is formed by a human-technology brain/nervous system coupling in which the complete entity goes well beyond the norm in terms of Homo sapiens performance (case 4). Whilst this does refer to a relatively narrow definition with respect to all human-technology possibilities, the arguments that follow are dependent on such a definition.

Connections between technology and the human nervous system not only affect the nature of the individual, raising questions as to the meanings of 'I' and 'self' but they also directly influence autonomy. An individual Homo sapiens wearing a pair of glasses, whether they contain a computer or not, remains an autonomous entity. Meanwhile a human whose nervous system is linked directly with a computer not only puts forward their individuality for questioning but also, when the computer is part of a network or at least connected to a network, allows their autonomy to be compromised. It has to be accepted that when this is merely for therapeutic reasons (case 3) there is not, under normal circumstances, an issue. However it is when the individual is enhanced by such an arrangement (case 4) that is the principle subject of this paper.

The main question arising from this discourse being: when an individual's consciousness is based on a part human part technological nervous system/brain, in particular when they exhibit enhanced consciousness, will they hold to the values of Homo technologicus? These being potentially distinctly different to the values of Homo sapiens. Importantly, as a consequence, will such a Homo technologicus entity regard Homo sapiens in a Nietschian way [14], i.e., how humans presently regard cows or chimpanzees?

Some may prefer to look through philosophical pink glasses [12] and see posthumans as being "conducive to the long range survival of humans." But surely it will be those who are members of Homo technologicus and not Homo sapiens who will make the pro-Homo sapiens, anti-Homo sapiens decisions.

In this article we look briefly at some case 2 and case 3 practical examples as much to be clear as to which entities are not included in case 4. We then investigate case 4 in detail and consider some immediate possibilities, given the capabilities of intelligent machines. Finally we consider the future for both Homo technologicus and Homo sapiens. It is not thought to be worthwhile considering case 1 in any further depth than has been done already as this is merely the case of a human holding a pen, riding a bicycle or wearing a watch, no matter how some may wish to market their research [9,10].

2. Case 2

In this section we consider different practical instances of implants in the human body, but not in the brain/nervous system, designed more for enhancement purposes rather than for therapy.

2.1. Identification

My first electronic implant was a simple radio frequency transmitter, termed a Radio Frequency Identification Device (RFID) which was inserted in my left arm in 1998. As a result the computer in my building allowed me to open doors and switch on lights simply by walking through particular doorways [15]. One interesting feature of the experiment was that I rapidly came to regard the implant as being part of me, something that is a common feature with implantees.

I had the luxury of a doctor to put my implant in place. Many experimenters of today actually carry out their own implantations, learning as they go about basic medicine and sterilisation. In 1998, I had the comfort of local anaesthetic, not so many of today's subjects do, although a friend may well be on hand in case of fainting.

The RFID is perhaps the most common implant tried, recently in the form of a near field communication (NFC) version. This is essentially the same technology as is used in contactless payment cards except that it's now packaged in a small tube about the size of a grain of rice. In 1998 my RFID was almost an inch (2.54 cm) long. It does require external technology to transmit power to the implant, which has no battery, and to communicate with it.

In recent years many companies have engaged themselves with the technology, although this might be more for publicity than anything else. For example, in January 2015 it was widely reported that several hundred office workers in Sweden had been chipped. With their implants the workers were able to open doors and switch on the photocopier [16].

2.2. Implant Variety

The range of possible technology that can be implanted is broad and imaginative. Software developer and biohacker Tim Cannon has experienced a variety of implants. His latest is called the Northstar which lights up under his skin when a magnet is close by. Then there is Lepht Anonym who plans to have a small compass chip implanted near her left knee, along with a power coil that, rather like an RFID, can be charged from an external source.

Meanwhile Moon Ribas has a seismic sensor implanted in her elbow that allows her to feel earthquakes through vibrations, whilst Neil Harbisson, who is otherwise colour blind, has a camera which is attached to his skull. Different colours cause the frequency of vibrations to his skull to vary. As a result he has learned a very high degree of colour discrimination. The technology translates colour frequencies into sound frequencies [17] which are translated into vibrations via an actuator. Initially, Harbisson memorised the frequencies related to each colour, but subsequently he decided to permanently attach the set up to his head.

The project was developed further so that Harbisson was able to perceive colour saturation as well as colour hues. Software was then developed that enabled Harbisson to perceive up to 360 different hues through microtones and saturation through different volume levels [12]. What is particularly interesting about Harbisson's experience is that his discrimination between different colours has improved over time as his brain has adjusted to the different vibrations experienced. Clearly the extent of brain adaptability is a pointer to what can be expected in general with regard to either extending the present range of sensory input or rather inputting a complex range of new sensory input information into the human brain that till now has not been possible.

2.3. Magnet Implants

One other line of research worth mentioning here is the use of permanent magnet implants for sensory extension. The pads of the middle and ring fingers are the preferred sites for magnet implantation in the experiments that have been reported [18]. The mechanoreceptors in the fingertips are most sensitive to frequencies in the 200–300 Hz range. An interface containing a coil mounted on a wire frame and wrapped around each finger is used to generate magnetic fields to stimulate magnet movement within the finger. The output from an external sensor is used to control the current in the coil.

Experiments have been carried out in a number of application areas [18]. Ultrasonic range information, involves an ultrasonic sensor for navigation assistance. Distance information from the sensor is encoded as variations in the frequency of pulses. Effectively the closer an external object is to the sensor so the frequency of the pulses increased. The recipient has an accurate indication of how far objects are from the sensor. Further tests have used infrared sensors, which give an indication of the temperature of any objects remotely detected [19]. So the recipient 'feels' the temperature of remote objects.

2.4. Case 2 Conclusions

The implants considered in this section certainly, in each case, allow the recipients abilities that are not normal for Homo sapiens. On top of that the recipients regard their implant as being part of their body, which is quite different to something that is merely worn. However, apart from the usual long term brain modification also apparent due to external conditions, there is no immediate change to the neural abilities of the individuals. Hence all those involved in this type of implant are considered to stay within the realms of Homo sapiens with regard to this specific experimentation. In other words, because their implant does not immediately alter their neural make up, the individuals are still considered to be a member of species Homo sapiens.

3. Case 3

In this section we look in a little more detail at Case 3 which involves a direct link between the human brain/nervous system and technology, in particular computers and artificial intelligence. However the main point of interest is that the connection is for therapeutic purposes. That said, part of the reason for detailing this section is to see how easy it is for a case 3 example to become case 4 merely by means of software changes.

3.1. Deep Brain Stimulation

Deep Brain Stimulation is a procedure employed for the electronic treatment of Parkinson's disease, epilepsy, Tourette's syndrome and clinical depression. The deep brain electrodes can though be connected bi-directionally with a computer such that electrical activity in the brain can be monitored. Ongoing research is developing an 'intelligent' stimulator. This uses artificial intelligence to produce warning signals before Parkinsonian tremors begin [20]. So the stimulator only needs to generate signals occasionally rather than continuously, thus operating in a similar fashion to a heart pacemaker.

Using AI techniques, by better understanding the nature of the disease it has been found that there are distinct types of Parkinson's disease based on the different nature of the electrical activity in the brain [13]. It is also quite possible for the monitoring computer to be located remotely from the patient. Hence, signals within the brain can be tracked in real time and fed into a computer. The computer is able to analyse these signals and generate alternative signals that are fed directly back into the brain in order to ensure the person in question continues to function.

Clearly whilst stimulation is provided in a particular part of the brain merely to overcome a specific problem so we are dealing with a case 3 situation. However if the implant is shown to be successful in other ways then the situation could change. As an example when the implant is employed to overcome depression it is an interesting application, giving the individual positive feelings. This could readily be extended for the computer to bring about positive feelings under certain circumstances and negative feelings in other cases.

3.2. Overcoming Paralysis

Philip Kennedy developed an operable system which allowed an individual with paralysis to spell words by modulating their brain activity. Kennedy's device used two simple electrodes: the first was implanted in an intact motor cortical region and was used to move a cursor among a group of letters. The second was implanted in a different motor region and was used to indicate that a selection had been made [21].

As the patient thought about moving their fingers these signals were translated into signals to move and stop a computer cursor. The patient could actually see where the cursor was on a large computer screen. Hence they could decide as to when to stop thinking about moving. In this way words could be spelt out letter by letter but also heating and lighting could be controlled quite simply [22].

A different approach was taken by Todd Kuiken whereby nerves normally connected to the pectoralis muscles were employed in a process termed targeted reinnervation. In this procedure, nerves originally connected to arm muscles were reconnected to the pectoralis muscles. As the individual thought about moving their hand and arm so the muscles on the top of their chest flexed instead. External electrodes monitor these movements and send resultant signals to a prosthetic arm worn by the patient. Effectively the person's nervous system is rewired via the pectoralis muscles [23].

The first beneficiary of this technique was Jesse Sullivan, hailed in the media as the world's first 'Bionic Man', who lost both of his arms as a result of an accident he sustained during his work as a high-power electrical lineman. His arms were replaced with robotic prosthetics that he was able to control merely by thinking about using his original arms in the normal way.

3.3. BrainGate

The technology which has thus far shown itself to be of the most practical use in this area is the microelectrode array known as the Utah Array, more popularly (and commercially) referred to nowadays as the BrainGate.

The array consists of 100 spikes which are 1.5 mm long and taper to a tip diameter of less than 90 microns. The spikes, essentially silicon shafts, are arranged in a 10 by 10 array on a 4 mm × 4 mm substructure and each has a platinum electrode on its tip. The electrodes are linked to platinum wires and in this way the array can be employed bi-directionally to both directly monitor neural activity and also to apply stimulating currents.

A number of trials have been carried out that did not use humans as test subjects, these involving chickens or rats. However it is human studies only that we are more interested in here and these are limited to two groups of studies at the moment. In these experiments the array has been fired into either the human brain or nervous system. In the first set of these experiments to be considered, the array has been employed in a purely recording role for therapeutic results.

Electrical activity from a few neurons monitored by the array electrodes, positioned in the motor cortex, has been decoded into a signal that enabled a severely paralysed individual to position a cursor on a computer screen using neural signals for control in combination with visual feedback. The same technique was later deployed to allow the individual recipient, who was paralysed, to operate a robot arm even to the extent of learning to feed themselves in a rudimentary fashion by maintaining sufficient control over the robot arm [24,25].

The same implant was employed to enable a paralysed individual to regain some control over his own arm [26]. In this case signals from the individual's motor cortex were employed to bring about stimulation of hand/wrist muscles via a cuff worn around the person's arm. The effect of this was a sort of bi-pass of the non-functioning nervous system. As a result the individual recipient could make isolated finger movements and perform six different wrist and hand motions.

Initially fMRI scans were taken of the recipient's brain while he tried to copy videos of hand movements. This identified an exact area of the motor cortex dealing with the movements exhibited. Surgery was then performed to implant the array to detect the pattern of electrical activity arising when the recipient thought about moving his hand. These patterns were then sent to a computer which translated the signals into electrical messages, which were in turn transmitted to a flexible sleeve that wrapped around the forearm and stimulated the muscles.

3.4. Case 3 Conclusions

We have seen in this section how brain/nervous system to technology connections can be employed to overcome problems such as depression or paralysis. However they have been included here as much to show just how the functioning of the human brain can be altered by the employment of electronic signals. OK here we have considered injection of those signals into specific regions for therapeutic purposes, but obviously other regions and other purposes could be chosen.

4. Case 4

In the previous section the BrainGate implant was described and its use was explained in terms of therapeutic procedures. However the same implant has also been employed in experiments aimed at investigating human enhancement beyond the human norm.

4.1. BrainGate for Enhancement

In 2002 the BrainGate multi-electrode array was implanted into the median nerve fibers of this paper's author, a healthy human individual, in the course of two hours of neurosurgery to test bidirectional functionality. Stimulation current was applied directly via the implant into the nervous system to allow information to be sent to the recipient, while control signals were decoded from motor neural activity in the region of the electrodes [27].

Overall a number of trials were undertaken successfully using this set up [28].

1   Telegraphic communication directly, electronically between the nervous systems of two humans (the author's wife + the author) was performed.
2   Extended control of a robotic hand across the internet was achieved, with feedback from the robotic fingertips being sent back as neural stimulation for a sense of force by the fingers applied to an object (achieved between the USA and the UK).
3   Extra-sensory (ultrasonic) input was successfully implemented.
4   A wheelchair was successfully driven around by means of neural signals alone. Feedback was in this case purely visual.
5   The color of jewelry was changed as a result of neural signals.

4.2. Human Enhancement

In all these cases, the trial could also be described as useful for purely therapeutic reasons, e.g., the ultrasonic sensory input might be of use to an individual who is blind to give them an alternative (bat-like) interaction with the outside world, while telegraphic communication might be beneficial to people with certain forms of motor neuron disease. Each trial can, however, be seen as a potential form of enhancement beyond the human norm for an individual. There was no need to have the implant for medical reasons in order to overcome a problem; the experimentation was carried out for the purposes of scientific exploration.

Human enhancement with the aid of brain-computer interfaces introduces all sorts of new technological and intellectual opportunities, but it also throws up different ethical concerns [29,30]. While the vast majority of present day humans are perfectly happy for interfaces, such as the BrainGate, to be used in therapy, the picture is not so clear when it comes to enhancement.

From the trials, it is apparent that extra sensory input is one practical possibility that has been successfully trialed along with extending the human nervous system over the internet. However, improving memory and communication by thought are other distinct potential, yet realistic, benefits with the latter of these also having been investigated to an extent. To be clear these things appear to be possible (from a technical viewpoint at least) for humans in general.

4.3. Intelligent Machines

We now have computer technology that, many consider exhibits its own intelligence. As pointed out by Alan Turing [31], this can be considered as being distinct from human intelligence and exhibits a number of different characteristics when compared to human intelligence. Turing said I "May not machines carry out something which ought to be described as thinking but which is very different from what a man does?" In particular a number of positive features associated with the performance of machine intelligence can be picked out and for this section the important thing is to see how the human brain could be dramatically improved in its functioning by a direct link with technology.

As a start "on any issue of computing power, if computers do not have the advantage over human brains already, then they will certainly have it before too long" [32]. This applies to both the speed and accuracy of dealing with data. The biggest advantage of all for machine intelligence however is communication. The present way that humans communicate is extremely poor when compared with that of technology. In speech for example humans convert highly complex electro-chemical signals concerned with emotions, feelings, colours etc. into trivial coded mechanical pressure waves. The possibility of our thoughts remaining in electronic form, as is most likely the case with a machine, would be an enormous step forward for Homo sapiens.

With a human brain linked to a computer brain, that individual could have the ability to [33]:

— use the computer part for rapid maths
— call on an internet knowledge base, quickly
— have memories that they have not originally themselves had
— sense the world in a plethora of ways not possible to Homo sapiens (e.g., ultrasonic, infra-red)
— understand multi dimensionality, as opposed to 3D for the human brain alone
— communicate in parallel, by thought signals alone, i.e., brain to brain

Each one of these examples appears to provide a valid reason as to why for an individual Homo sapiens would wish to upgrade to join the ranks of Homo technologicus.

As for the possibility for machines to have emotions and consciousness then, by following Turing's lead, this is not really an issue. However such characteristics will be very different to those of a human. The big question for research though is to look into the possibilities of integrating human and machine forms of consciousness such that the consciousness of Homo technologicus will be an amalgam of the two. Importantly this form of consciousness, with all its inherent abilities, will almost surely not be comprehensible to Homo sapiens.

4.4. Case 4 Conclusions

Case 4 describes the situation when an individual's brain/nervous system is linked directly with a computer for the purposes of enhancement beyond the norm for Homo sapiens. The reason for including this section is to demonstrate that this is a realistic possibility and to assess the sort of enhancements that might well be possible and what this might mean in practice.

5. Discussion

What is the cost for a member of Homo sapiens to become Homo technologicus in the way described and what might the consequences be? Clearly the realisation of such entities presents enormous questions that affect all aspects of human society and culture. In attempting to answer such questions a string of positives and negatives appear. Standing still is not an option. In the extremes, if humans opted by some global agreement for a non-Homo technologicus future (if that were possible), could the end result actually be an intelligent machine superculture as described in [34], leading to the singularity and loss of control on earth to machines. Conversely, if humans globally opted for a Homo technologicus future, could society and culture cope with such a distinct non linearity in evolution?

Some argue that any view of the appearance of superhuman cyborgs can be seen as being unwarranted 'metaphysical' speculation [35]. On the other hand it could be felt that humankind is itself at stake in any case [36]. A viewpoint can then be taken that either it is perfectly acceptable to upgrade humans, turning them into Homo technologicus, with all the enhanced capabilities that this offers, or on the other hand it can be felt that humankind is just fine as it is and should not be so tampered with [36].

The most important issue here is that we are considering a completely different basis on which the Homo technologicus brain operates—part human and part machine in its nature. When the nature of the brain itself is altered the situation is complex and goes far beyond anything encountered with the mere physical extensions of case 1. Such a Homo technologicus entity would have a different foundation on which any of their thoughts would be conceived in the first place. From an individualistic viewpoint therefore, as long as I am myself a Homo technologicus I am happy with the situation. Those who wish to remain Homo sapiens however may not be so happy.

With a brain which is part human, part machine, Homo technologicus would have some links to their human background but their view on life, what is possible and what is not, would be very much different from that of a human. Their values would relate to their own life and Homo sapiens may not figure too highly in such a scenario.

One aspect is that Homo technologicus would have brains, which are not stand alone, but rather, are connected to each other directly via a network. A question is therefore is it acceptable for Homo sapiens to give up their individuality and become mere nodes on an intelligent machine network? Or is it purely a case of individual freedom, if an individual wants to so upgrade then why not?

Some questions are obvious. Should every human have the right to be upgraded? If an individual does not want to should they be allowed to defer, thereby taking on a role in relation to Homo technologicus rather like a chimpanzee's relationship with a human today? How will the values of Homo technologicus relate to those of Homo sapiens? Will Homo sapiens be of any consequence to Homo technologicus other than something of an awkward pain to be removed if possible?

However to conclude, we must be clear that with extra memory, high powered mathematical capabilities, including the ability to conceive in many dimensions, the ability to sense the world in many different ways and, perhaps most importantly of all, communication by thought signals alone, Homo technologicus will be far more powerful, intellectually, than Homo sapiens. It would be difficult imagining that Homo technologicus would want to voluntarily give up their powers or would pay any heed to the trivial utterances of Homo sapiens.

Just as now if a cow enters a room full of humans and proceeds to make cow noises (moo or boo!), it would be extremely unlikely for the humans in the room to say collectively what a wonderful idea the cow has, yes we will all do what the cow wants immediately. No the cow's noises would be simply ignored and she would be removed from the room and shortly killed. So in the future when a Homo sapiens enter a room full of Homo technologicus members and says something like "I don't like what you're doing". It would be extremely unlikely for the Homo technologicus members in the room to say collectively, what a wonderful idea the human has, we will all do what the human wants immediately. No, the human's noises would be simply ignored and they would be removed from the room and shortly killed.

Conflicts of Interest: The author declares no conflict of interest.

References

1. Wolfe, C. *What Is Posthumanism?* University of Minnesota Press: Minneapolis, MN, USA, 2010.
2. Hayles, N. Traumas of code. *Crit. Inq.* **2006**, *33*, 136–157. [CrossRef]
3. Scharff, R.C.; Dusek, V. *Philosophy of Technology: The Technological Condition: An Anthology*; Wiley-Blackwell: Hoboken, NJ, USA, 2003.
4. Longo, G. Body and Technology: Continuity or discontinuity? In *Mediating the Human Body: Communication, Technology and Fashion*; Fortunati, L., Katz, J., Riccini, R., Eds.; Lawrence Erlbaum: Mahwah, NJ, USA, 2002; pp. 23–30.
5. Callus, I.; Herbrechter, S. Introduction: Posthumanist subjectivities, or, coming after the subject. *Crit. Psychol.* **2012**, *5*, 241–264. [CrossRef]
6. Greenfield, S. *Mind Change: How Digital Technologies Are Leaving Their Mark on Our Brains*; Rider: London, UK, 2015.
7. Clark, A. *Natural Born Cyborgs*; Oxford University Press: Oxford, UK, 2004.
8. Bateson, G. *Steps to an Ecology of Mind*; Ballantine Books: New York, NY, USA, 1972.
9. Pentland, A.P. Wearable intelligence. *Sci. Am.* **1998**, *9*, 90–95.
10. Mann, S. Wearable Computing: A first step towards personal imaging. *Computer* **1997**, *30*, 25–32. [CrossRef]
11. Hayles, N. *How We Became Posthuman: Virtual Bodies in Cybernetics, Literature and Informatics*; University of Chicago Press: Chicago, IL, USA, 1999.
12. Haraway, D. *Simians, Cyborgs and Women*; Free Association Books: London, UK, 1996.
13. Cámara, C.; Isasi, P.; Warwick, K.; Ruiz, V.; Aziz, T.; Stein, J.; Bakštein, E. Resting Tremor Classification and Detection in Parkinson's Disease Patients. *Biomed. Signal Proc. Control* **2015**, *16*, 88–97. [CrossRef]
14. Nietsche, F. *Thus Spoke Zarathustre*; Penguin Classics: London, UK, 1961.
15. Warwick, K. The Cyborg Revolution. *Nanoethics* **2014**, *8*, 263–273. [CrossRef]
16. Cellan-Jones, R. Office puts chips under staff's skin. Available online: http://www.bbc.co.uk/news/technology-31042477 (accessed on 17 October 2016).
17. Harbisson, N. Painting by ear. In *Modern Painters. The International Contemporary Art Magazine*; Louise Blouin Media: New York, NY, USA, 2008; pp. 70–73.
18. Hameed, J.; Harrison, I.; Gasson, M.; Warwick, K. A novel human-machine interface using subdermal implants. In Proceedings of the IEEE 9th International Conference on Cybernetic Intelligent Systems, Reading, UK, 1–2 September 2010; pp. 106–110.
19. Harrison, I. Sensory Enhancement, a Pilot Perceptual Study of Subdermal Magnetic Implants. Ph.D. Thesis, University of Reading, Reading, UK, 2015.
20. Wu, D.; Warwick, K.; Ma, Z.; Gasson, M.; Burgess, J.; Pan, S.; Aziz, T. Prediction of Parkinson's Disease Tremor Onset using Radial Basis Function Neural Network Based on Particle Swarm Optimization. *Int. J. Neural Syst.* **2010**, *20*, 109–118. [CrossRef] [PubMed]
21. Kennedy, P.; Bakay, R.; Moore, M.; Adams, K.; Goldwaithe, J. Direct control of a computer from the human central nervous system. *IEEE Trans. Rehabil. Eng.* **2002**, *8*, 198–202. [CrossRef]
22. Kennedy, P.; Andreasen, D.; Ehirim, P.; King, B.; Kirby, T.; Mao, H.; Moore, M. Using human extra-cortical local field potentials to control a switch. *J. Neural Eng.* **2004**, *1*, 72–77. [CrossRef] [PubMed]
23. Kuiken, T.; Li, G.; Lock, B.; Lipschutz, R.; Miller, L.; Stubblefield, K.; Englehart, K. Targeted muscle reinnervation for realtime myoelectric control of multifunction artificial arms. *JAMA* **2009**, *301*, 619–628. [CrossRef] [PubMed]
24. Hochberg, L.; Serruya, M.; Friehs, G.; Mukand, J.; Saleh, M.; Caplan, A.; Branner, A.; Chen, D.; Penn, R.; Donoghue, J. Neuronal ensemble control of prosthetic devices by a human with tetraplegia. *Nature* **2006**, *442*, 164–171. [CrossRef] [PubMed]
25. Hochberg, L.; Bacher, D.; Jarosiewicz, B.; Masse, N.; Simeral, J.; Haddadin, S.; Liu, J.; Cash, S.; van der Smagt, P.; Donoghue, J. Reach and grasp by people with tetraplegia using a neurally controlled robotic arm. *Nature* **2012**, *485*, 372–375. [CrossRef] [PubMed]

26. Bouton, C.; Shaikhouni, A.; Annetta, N.; Marcia, A.; Bockbrader Friedenberg, D.; Nielson, D.; Sharma, G.; Sederberg Bradley, P.; Glenn, C.; Mysiw, W.; et al. Restoring cortical control of functional movement in a human with quadriplegia. *Nature* **2016**. [CrossRef] [PubMed]
27. Warwick, K.; Gasson, M.; Hutt, B.; Goodhew, I.; Kyberd, P.; Andrews, B.; Teddy, P.; Shad, A. The application of implant technology for cybernetic systems. *Arch. Neurol.* **2003**, *60*, 1369–1373. [CrossRef] [PubMed]
28. Warwick, K.; Gasson, M.; Hutt, B.; Goodhew, I.; Kyberd, P.; Schulzrinne, H.; Wu, X. Thought communication and control: A first step using radiotelegraphy. *IEE Proc. Commun.* **2004**, *151*, 185–189. [CrossRef]
29. Warwick, K. Cyborg morals, cyborg values, cyborg ethics. *Ethics Inform. Technol.* **2003**, *5*, 131–137. [CrossRef]
30. Warwick, K. Future Issues with Robots and Cyborgs. *Stud. Ethics Law Technol.* **2010**, *4*, 6. [CrossRef]
31. Turing, A. Computing Machinery and Intelligence. *Mind* **1950**, *59*, 433–460. [CrossRef]
32. Penrose, R. *Shadows of the Mind*; Oxford University Press: Oxford, UK, 1994.
33. Warwick, K. *I, Cyborg*; Century: London, UK, 2002.
34. Kurzweil, R. *The Singularity Is Near*; Gerald Duckworth & Co. Ltd.: London, UK, 2006.
35. Coole, M. Becoming a cyborg as one of the ends of disembodied man. In Proceedings of the International Conference on Computer Ethics and Philosophical Enquiry, Lancaster, UK, 14–16 December 2001; pp. 49–60.
36. Cerqui, D. The future of humankind in the era of human and computer hybridisation. An Anthropological Analysis. In Proceedings of the Conference on Computer Ethics and Philosophical Enquiry, Lancaster, UK, 14–16 December 2001; pp. 39–48.

 *philosophies*

Article

Human Enhancements and Voting: Towards a Declaration of Rights and Responsibilities of Beings

S. J. Blodgett-Ford [1,2]

1   Boston College Law School, Boston College, Newton Centre, MA 02459, USA; blodgesa@bc.edu
2   GTC Law Group PC & Affiliates, Westwood, MA 02090, USA

Abstract: The phenomenon and ethics of "voting" will be explored in the context of human enhancements. "Voting" will be examined for enhanced humans with moderate and extreme enhancements. Existing patterns of discrimination in voting around the globe could continue substantially "as is" for those with moderate enhancements. For extreme enhancements, voting rights could be challenged if the very humanity of the enhanced was in doubt. Humans who were not enhanced could also be disenfranchised if certain enhancements become prevalent. Voting will be examined using a theory of engagement articulated by Professor Sophie Loidolt that emphasizes the importance of legitimization and justification by "facing the appeal of the other" to determine what is "right" from a phenomenological first-person perspective. Seeking inspiration from the Universal Declaration of Human Rights (UDHR) of 1948, voting rights and responsibilities will be re-framed from a foundational working hypothesis that all enhanced and non-enhanced humans should have a right to vote directly. Representative voting will be considered as an admittedly imperfect alternative or additional option. The framework in which voting occurs, as well as the processes, temporal cadence, and role of voting, requires the participation from as diverse a group of humans as possible. Voting rights delivered by fiat to enhanced or non-enhanced humans who were excluded from participation in the design and ratification of the governance structure is not legitimate. Applying and extending Loidolt's framework, we must recognize the urgency that demands the impossible, with openness to that universality in progress (or universality to come) that keeps being constituted from the outside.

Keywords: human enhancements; voting; human rights; ethics; discrimination; racism; speciesism; ableism

Citation: Blodgett-Ford, S.J. Human Enhancements and Voting: Towards a Declaration of Rights and Responsibilities of Beings. *Philosophies* **2021**, *6*, 5. https://doi.org/10.3390/philosophies6010005

Received: 12 November 2020
Accepted: 23 December 2020
Published: 14 January 2021

Publisher's Note: MDPI stays neutral with regard to jurisdictional claims in published maps and institutional affiliations.

1. Introduction

The phenomenon and ethics of "voting" will be explored in the context of "human enhancements", which, for purposes of this analysis, are broadly defined as intentional modifications of a person who is accepted prior to the enhancement as genetically human, of any possible type of modification or combination of types.[1] Enhancements may be made with or without the consent of the enhanced human. For example, an enhancement could be made prior to birth (or prior to the age at which consent is recognized as valid) or could be forced on the human. Or an enhancement could be made with consent by, or at the request of, the enhanced human. In most cases, the enhancement is perceived as a way to "improve" the human, such as by improving their physical, mental or emotional abilities, but the analysis herein does not require that the enhancement actually be an improvement under any ranking theory or particular perspective. Enhancements can also be permanent or temporary.

In Section 2, "voting" will be examined for enhanced humans who fall into two broad categories—those with moderate enhancements, where the "humanity" of the enhanced

---

1 For purposes of this analysis, a descendant of a "human" (*Homo sapiens*) who was genetically or otherwise altered to such an extreme that they would no longer commonly be perceived to be "human" (or who no longer even had an organic embodiment) is still defined as an "enhanced human". However, in Section 3, we will propose to eliminate the requirement of "humanity" in voting rather than using such a "Grandfather Clause" approach and instead broadly extend voting rights to all living beings and all conscious beings.

person is not seriously in question, and those with extreme enhancements, where humans who lacked similar enhancements would likely question the "humanity" of the enhanced human. The question of who has the right to vote and how they can vote will be explored primarily in the context of voting in the United States at the federal level. Existing patterns of discrimination around the globe would likely continue substantially "as is" for humans with physical and mental enhancements who are viewed as "still human." For the "beyond human" category of enhancements, the established rules and practice of voting are likely to be directly challenged. For both moderate and extreme enhancements, humans who are not enhanced (or are less enhanced) may be disenfranchised if certain enhancements become prevalent among the voting population.

In Section 3, voting by enhanced and non-enhanced humans will be examined using an extension of a theory of engagement articulated by Professor Sophie Loidolt that emphasizes the importance of legitimization and justification and "facing the appeal of the other" to determine what is "right" "from a phenomenological first-person perspective [1]. Seeking inspiration from the Universal Declaration of Human Rights (UDHR) of 1948 [2], voting rights and responsibilities will be re-framed from a foundational working hypothesis that all conscious and all living "beings" (including enhanced and non-enhanced humans) should have a right to vote directly. Representative voting should be considered as an admittedly imperfect alternative where direct voting is not possible, such as due to communication limitations, or as an additional option to direct voting, in the case of a group that suffers from historic systematic discrimination. Finally, the framework in which voting occurs, as well as the processes, temporal cadence of voting, and role of voting in the overall architecture and functioning of government, should be seen as provisional pending a "to be determined" approach (or approaches) designed, modified, refined, abolished and rebuilt from scratch as needed, with the participation from as diverse a group of beings as possible. Voting in a particular structure of governance has not been legitimized if it is delivered by fiat "as is" to those beings who were excluded from participation in the design and ratification of such structure—whether because they were enhanced to an extreme level or because they were not enhanced/less enhanced. Applying an extension of Loidolt's framework, we must recognize the urgency that demands the impossible with openness to that universality in progress (or universality to come) that keeps being constituted from the outside.

2. Discussion—Human Enhancements and Voting—Moderate and Extreme

Common assumptions about who has the right to vote and what "voting" involves warrant careful examination to see what they reveal, what they hide, and what they can tell us about biases that might exist against enhanced humans (or non-enhanced humans) who are perceived as "other" or "lesser" than those who hold political power. For purposes of this analysis, the meaning of "enhanced" shall be any intentional modification of a human being, whether done prior to birth (e.g., through genetic engineering of human embryos), or after birth, such as through implants, chemical enhancements, or other modifications. While such enhancements are often intended to be efforts to "improve" the human being who is enhanced, either with or without the consent of the enhanced person, for purposes of this analysis there will be no effort to assess whether such enhancements are actually an "improvement" or "benefit" to the enhanced human, to their family, to any broader group, or to society generally. One person might view an increase of intelligence as an "improvement" while another might view it as a negative change if the enhancement was only to computational abilities or analytical intelligence without any corresponding increase in capacity for empathy, compassion or "emotional intelligence." Similarly, a person who was born as a female biologically and then underwent surgery to become male might view it as an improvement for themselves as they felt their body reflected their true identity more accurately, but another person might view the loss of child-bearing capacity as a negative. As a final example, someone who values an idealized "normative" "able" human body might view an enhancement to a human body that brings that body closer to their ideal norm as an enhancement, while a person who embraces diverse forms of embodiment and living might consider such a modification as a negative.[2]

The present analysis of voting and human enhancements is agnostic as to the benefits of enhancements. We assume that enhancements (for better or for worse) are already occurring and are likely to occur in the future, potentially with greater variations, increasing numbers and extremity, and may be permanent, semi-permanent or reversible at will. In this context, we explore the effects of biases in the history of voting in the United States, using the examples of ancestry and gender, and consider how they might continue for both moderate and extreme human enhancements.

Restrictions on voting in the United States have been used to systematically discriminate against humans viewed as "other" or "lesser", in large part to preserve the established distribution of wealth and power. Similar patterns of discrimination can be seen in many other countries—currently and historically, including, just as a few examples, systematic and intentional disenfranchisement of indigenous peoples in Australia, Canada and Japan [7–9]. There is a risk that such discrimination would continue substantially "as is" for humans with physical and mental enhancements who are viewed as "still human." In addition, humans who are not enhanced or who are less-enhanced may be disenfranchised if certain enhancements become so prevalent that they are the "new normal."

For "beyond human" or "no longer human" types of enhancements, the established phenomenon of voting is likely to be directly challenged by extreme physical and/or mental enhancements, including hybrid organic/inorganic humans and mental enhancements such as a "networked consciousness" or an "extended mind." For example, it has been hypothesized that "consciousness does not originate from a single brain section" and instead "originates globally" [10]. If so, it is possible that there could be an enhanced human in the future whose "consciousness" arose from a network of organic and inorganic elements that were distributed (physically) in various locations, potentially even among different human embodiments. Such enhancements would put serious pressure on the United States model that being human is a necessary condition for voting,[3] both for elected human representatives as well as on votes for particular laws or amendments. Finally, in the case that either moderate or extreme enhancements, such as those contemplated in the DARPA (United States Defense Advanced Research Agency) program, became common among the voting population, non-enhanced or less-enhanced humans may be viewed as disabled or less-abled humans, and they may (legally and in practice) be denied the right to vote [12,13].

For purposes of this discussion, human enhancements may include any intentional physical and/or mental modifications of a human being, made either before birth, such as through gene editing or embryo selection, or after birth, whether by the human being or by another person, with or without the consent of the enhanced human. The term "cyborg" will also, at times, be used to refer to an enhanced human who has a physical or mental enhancement that includes an inorganic component, such as a prosthetic or

---

2   See [3] (quoting dancer, artist and poet Neil Marcus "Disability is not a 'brave struggle' or 'courage in the face of adversity', disability is an art. It's an ingenious way to live."). Consider also the esteemed composer Molly Joyce, who has to compose without using her left hand, and "has carved a unique sound as a composer by treating disability differently: not as an impediment but as a wellspring of creative potential [4]." An additional example of an "ideal" body relates to skin color. In many countries, due to a legacy of racism, a lighter skin color is viewed as an "improvement" and may objectively benefit the person having such skin color, including in employment opportunities and advancement. Consider the practice of skin bleaching in Jamaica. In his analysis of the practice, Professor Christopher Charles concluded that "self-hate" was not the primary driver for skin lightening in Jamaica and noted that reasons for skin lightening were more nuanced: "Some Black Jamaicans recognize the color and racial distinctions in society. This should not be viewed as self-contempt. It is borne out by their experience that the Blacker one is, the less status and privilege one has in the society. They recognize the reality of contemporary Jamaica. They do not necessarily accept it [5]." In the United States, caste-based discrimination allegedly is occurring in the technology industry [6] ("The lawsuit notes that the employee is Dalit Indian and that he has a darker complexion than non-Dalit Indians.").

3   Non-human corporations and other legal entities that are "persons" under the law can still have influence one election results even though they cannot vote. See the United States Supreme Court decision in the *Citizens United* case against the Federal Election Commission [11], in which the Court held that limitations on spending for political campaigns by groups, including corporations and labor unions, violate the Constitutional First Amendment right to free speech.

wearable enhancement or a neural implant. We will consider moderate enhancements first, then turn to extreme enhancements.

2.1. Moderate Human Enhancements—"Still Human"

In this section, we will consider voting by people who have been physically or mentally enhanced through moderate enhancements and are still recognized as "human". Such enhanced humans could have a wide variety of physical and/or mental enhancements, or combinations thereof, which could also in theory be malleable throughout their lifetime. For example, physical enhancements could result in an enhanced human being, or appearing to be, of a different ethnicity or ancestry, gender (or genders) or age. Enhancements could also lead to heightened or modified sensory perceptions, such as being able to touch, taste, smell, see, or hear in a different manner than a non-enhanced human. Enhancements could also allow the human to communicate via wifi and or access the Internet directly, such as via neural implants.

In all such cases, in this category of moderate enhancements, the enhancements cannot be so extreme that the person with the enhancements is perceived in their society as being no longer human or inhuman. What would "voting" mean for such enhanced humans? The answer is not clear, and could depend on whether the enhancements were made prior to birth or after birth and whether they were such that the person was characterized as "lesser" or "other" than those humans who were entitled to vote—or who were more able to vote due to enhancements.

The right to representation in government (and the lack of such representation in the English Parliament) was one of the core reasons the American colonists revolted against England. Taxation without representation was viewed as tyranny.[4] Based solely on this aspect of United States history, it would be easy to conclude that any person who was required to pay taxes to the United States federal government, enhanced or not, would be entitled to vote in United States elections. This simple rule could mean that a human who was physically enhanced and "still human" would have the right to vote if they paid taxes in the United States, regardless of whether they were citizens of the United States or even physically located in the United States.

However, the historical and current phenomenon of voting in the United States greatly deviates from such a simplistic model. Instead, it reflects systematic efforts over hundreds of years—and continuing to the present—to deny the right to vote to those viewed as "other" or "lesser" than the group(s) controlling the wealth and power of the United States—initially wealthy white men from England and later wealthy white women as well[5] [15].

2.1.1. Physical Enhancements—"Ancestry"

"A free negro of the African race, whose ancestors were brought to this country and sold as slaves, is not a "citizen" within the meaning of the Constitution of the United States" [16]

Based on the legacy of the denial of the right to vote to people of different ancestries or ethnicities in the United States, both as a matter of law and a matter of practice, it seems likely that if a physical enhancement changed a human to actually be (or appear to be) of an ancestry that was not legally entitled to vote (for example, not legally entitled to citizenship under then-current United States laws),[6] then such a person might not be, in practice, fully enfranchised, particularly if the change was made genetically such that the

---

4. "Taxation without representation is tyranny" is commonly attributed to attorney James Otis, who was one of the representatives from Massachusetts in the Stamp Act Congress (approx. 1761).

5. In 2018, the richest member of Congress had a Net Worth estimated at $500,000,000, and the median net worth of all members of Congress was approximately $500,000 [14].

6. Former Professor Rachel Dolezal, who pretended to be Black for decades and was president of the Spokane, Washington chapter of the National Association for the Advancement of Colored People, is just one example of people who have attempted to "pass" as having a different ancestry or ethnicity (although at a time in U.S. history when Blacks were enfranchised, so she did not give up her right to vote in doing so) [17].

birth ancestry was modified under the then-applicable laws. Similarly, one might expect that some parents might want to change their child's ancestry to be (or appear to be) that of a more privileged group. In such cases, a human who might otherwise not have the right to vote could, in theory, then have the right to vote, particularly if the change was made prior to birth. All of this analysis of course must be caveated by skepticism toward any scientific definition of "ancestry" and instead relies on the meaning of "ancestry" in the context of citizenship, including for immigrants who seek entry into the United States or naturalization after entry[7] [18].

American colonists felt they had the rights of English citizens to vote for representatives in Parliament in England. They viewed themselves as retaining their citizenship rights as emigrates and descendants of emigrates from England. However, the right to vote was (and still is) only available for "citizens of the United States." This means that only "native born" United States citizens (but for many years not indigenous peoples) or "naturalized" citizens have the right to vote.

Until the 14th Amendment was ratified in 1868 after the American Civil War, the Supreme Court decision in *Scott v. Sanford*, 60 U.S. 393 (1857) (the "*Dred Scott* decision") was the law of the land. Under the *Dred Scott* decision, even freed slaves who were born in the United States could not become United States citizens (and therefore have a constitutional right to vote) because they were descendants of non-citizens (Africans) [16]. Section 1 of the Fourteenth Amendment granted citizenship to "All persons born or naturalized in the United States." In the infamous decision, the Supreme Court explained that freed slaves "are not ["people of the United States"], and that they are not included, and were not intended to be included, under the word 'citizens' in the Constitution, and can, therefore, claim none of the rights and privileges which that instrument provides for and secures to citizens of the United States. On the contrary, they were at that time considered as a subordinate and inferior class of beings, who had been subjugated by the dominant race, and whether emancipated or not, yet remained subject to their authority, and had no rights or privileges but such as those who held the power and the government might choose to grant them." [16]

Shortly after the ratification of the Fourteenth Amendment, the Fifteenth Amendment was enacted and specified that voting rights could not be "denied or abridged by the United States or by any state on account of race, color, or previous condition of servitude." Having this in the Constitution was of course an important step forward, but as a matter of law and practice systemic racism and white supremacy continued through a process of legal "informal disenfranchisement", in which "a group that has been formally bestowed with a right is stripped of that very right by techniques that the [United States Supreme] Court has held to be consistent with the Constitution." [19]. As Professor Khiara Bridges explains, "While the Fifteenth Amendment formally enfranchised black men, white supremacists in the South employed methods—poll taxes, literacy tests, residency requirements, and white primaries—that made it nearly impossible for black men (and after the passage of the Nineteenth Amendment, black women) to actually vote in the South for a century after their formal enfranchisement. Moreover, the Court held that these techniques of racial exclusion from the polls were constitutional."[8] The equivalent of a "poll tax" as a means to prevent poor people from immigrating to the United States, continues through the fees for applications for naturalization. In a lawsuit, Project Citizenship alleged that an increase in the filing fee for N-400 naturalization applications from the US$640 to $1170 and a new rule that would allegedly disqualify 97.2% of green card holders from receiving fee waivers, "constitutes a wealth test for citizenship" [20,21].

---

[7] Ref. [18] ("for most of its history U.S. law treated newcomers differently according to race. Between 1790 and 1952, legislators restricted naturalization—the process by which immigrants become citizens—to particular racial and ethnic groups, with a consistent preference for whites from northwestern Europe. Laws restricted black immigration beginning in 1803, and a series of subsequent measures banned most Asians and limited access by immigrants from southern and eastern Europe.").

[8] Ref. [19] at 48–49.

The Fourteenth and Fifteenth Amendments did not give the right to vote to Native peoples. Indeed, many Native people were legally wards of the government with no political voice [22]. This treatment of indigenous peoples as "non-citizens" even in the land in which they were born echoes the Nuremberg laws of Nazi Germany, which legally classified people defined as "racially" Jewish by law as "subjects" of the state versus people of "German or kindred blood" who could be citizens [23]. The Nuremberg laws later were expanded to cover Black people and the Roma, who were also targets of the Nazi genocide.

Similarly, in the United States, "ancestry" has been used as a proxy for the social construct of "race" or "ethnicity" [24,25]. We can see similar voter suppression measures in the United States that appear to be based on discrimination due to "ancestry."[9] For example, United States citizens "by birth" are not required to pass any test to demonstrate they understand the significance of voting. However, people born outside the United States who want become citizens are required to pass the "civics" portion of the Naturalization Test. There are questions on the test that a relatively small percentage of Americans could answer correctly, such as, for example, "How many amendments does the Constitution have?[10]"

These types of restrictions were intentionally designed and enforced to disenfranchise those deemed "lesser" or "other" and privilege those who were already "grandfathered" in as voters. The "phrase 'grandfather clause' originally referred to provisions adopted by some states after the Civil War in an effort to disenfranchise African-American voters by requiring voters to pass literacy tests or meet other significant qualifications, while exempting from such requirements those who were descendants of men who were eligible to vote prior to 1867" [30]. "Providing such protection commonly is known—in the case law and otherwise—as 'grandfathering.' We decline to use that term, however, because we acknowledge that it has racist origins." [30]

The Voting Rights Act of 1965 (VRA) prohibited making the right to vote being dependent on whether a citizen was able to read or write, attained a particular level of education, or passed an interpretation "test." However, the VRA does not prohibit such tests in the immigration or naturalization process toward citizenship. Just like a "whites-only" primary could legally be used to prevent Blacks from participating in voting, if a person cannot even become a citizen (and therefore have a right to vote) without passing a wealth test or literacy test, then the voter suppression has effectively been shifted to an earlier point in time.

What can we learn from this history when we consider human enhancements? First, we should expect discriminatory treatment, and efforts (legal and otherwise) to deny the right to vote or discourage voting, for anyone who has a physical enhancement of a type that makes them seem "other" or "lesser" than those in power. For example, if a person were to enhance their skin color to make it darker, or to modify their physical appearance to look like they were of a different ancestry, they could be subject to targeted discrimination and denial of the right to vote under then-current laws. Second, as will be explored further in Section 3, if "ancestry" (however the term is defined, whether based on physical appearance, some type of genetic marker(s), or place of birth of the enhanced human or of people to whom they are biologically, socially, culturally, or legally related) can be changed via human enhancement technology, then "ancestry" may be an artificial distinction that should no longer be the basis for denial of enfranchisement. Such modifications to "ancestry" might, perhaps, be made either before birth through genetic

---

[9] For example, in U.S. elections in 2012 and 2014, Rosa Maria Ortega, an Hispanic immigrant who was a permanent resident of the United States and was brought to the country from Mexico as an infant, apparently voted without knowing it was illegal and was sentenced to eight years in prison. She then faced deportation [26]. Mexican lawful immigrants are among those least likely to become U.S. citizens [27]. ("desire is high, but about half cite language, cost barriers").

[10] U.S. Citizenship and Immigration Service, "Civics (History and Government) Questions for the Naturalization Test," (rev. 01/19) [28] (Q7: "How many amendments does the Constitution have? twenty-seven (27)") (Q48: "There are four amendments to the Constitution about who can vote. Describe one of them. Citizens eighteen (18) and older (can vote). You don't have to pay (a poll tax) to vote. Any citizen can vote. (Women and men can vote.) A male citizen of any race (can vote)."). For an example of a post-Civil War literacy test that was designed to be impossible to pass, see the Louisiana Literacy Test. "The literacy test – supposedly applicable to both white and black prospective voters who couldn't prove a certain level of education but in actuality disproportionately administered to black voters – was a classic example of one of these barriers [to enfranchisement]." [29]

or biological modifications or during a human lifetime, potentially including ongoing modifications and reversions to the original default ancestry without any enhancements. In such cases, "ancestry" would seem very artifactual. However, "ancestry" might still be relevant to self-determination for particular groups that have a shared group identity in addition to the identity of their individual members, such as an indigenous tribe or a disadvantaged caste, as will be discussed in Section 3 below. We may come to a similar conclusion regarding gender-related human enhancements as an additional example of "moderate" enhancements.

2.1.2. Physical Enhancements—"Gender"

For purposes of the discussion of voting in the United States, we will start with a binary definition of "gender" as male/female or woman/man to trace the historical lines drawn regarding voting rights, recognizing that such a narrow definition is not only overly simplistic it is harmful. As Professor Judith Lorber explains: "Today's gender paradox is a rhetoric of gender multiplicity made meaningless by a continuing system of bigendered social structures that support continued gender inequality." [31][11].

Professor Peter Singer offered this initial possible response to Thomas Taylor's satire of Mary Wollstonecraft's "Vindication of the Rights of Women" in 1972—"A Vindication of the Rights of Brutes":

> *"Women have the right to vote, for instance, because they are just as capable of making rational decisions as men are; dogs, on the other hand, are incapable of understanding the significance of voting, so they cannot have the right to vote."[12]*

For much of the history of the United States, there was no shared agreement that women were just as capable of making rational decisions as men. The right to vote in the United States, even prior to its formal construction through the United States Constitution, was a right for (white) male citizens only. According to the Preamble of the Declaration of Independence of 1776:

> *We hold these truths to be self-evident, that all men are created equal, that they are endowed by their Creator with certain unalienable rights, that among these are life, liberty and the pursuit of happiness. That to secure these rights, governments are instituted among men, deriving their just powers from the consent of the governed.* [34]

The express reference to "men" was intentional, and the fact that it was not necessary to state "white" was because of the predominant view that people of color—whether indigenous, immigrant, slaves or "free Blacks" were not created "equal" to white men. "The founding generation's republican vision—that is, the vision of the propertied white males who monopolized political power and promulgated the Constitution—can be reduced without too much distortion to a handful of fundamental ideas. Moreover, the preservation of liberty in a confederated republic depended on limiting full political participation and legal personhood to propertied white men. The majority of the population—women, black Americans, the indigenous nations, the poor—would take positions decisively subordinate to that of propertied white men in the new constitutional structure [15]."

Denial of the right to vote based on gender was still allowed under the United States Constitution for another three decades after the right was given to Black males. In 1920, the Nineteenth Amendment to the Constitution was ratified. The Nineteenth Amendment specifies that: "The right of citizens of the United States to vote shall not be denied or abridged by the United States or by any State on account of sex." (emphasis added). Yet

---

11 This harm extends across species. See also the work of pattrice jones exploring the intersection of "speciesism and sexism at the heart of not only domestic violence and other forms of husbandry but also landlordism, racism, and 'ecophobia'" [32] (citations omitted).

12 Ref. [33] at 148–162. To be more accurate, this statement could be re-worded as "women are just as capable of making rational *and irrational* decisions as men are." In addition, we should not accept without challenge the assumption that "rationality" has a well-defined neutral meaning, nor that it is the *sine qua non* of voting. As will be suggested in Section 3, sympathy, empathy and emotional intelligence may all be equally valuable qualities, particularly for voters who have responsibilities toward non-voters who may be impacted by their vote.

similar "informal disenfranchisement" occurred for women of color as for Black men, even after the passage of the Nineteenth Amendment. It was particularly easy to make such disenfranchisement legal because the wording of the Nineteenth Amendment did not affirmatively give women the right to vote. Instead, it was phrased to prohibit denial of the right to vote solely on the basis of gender. States continued to use poll taxes and other voter suppression tactics to keep women of color from voting. Asian American women who lacked citizenship as of 1920 were still not entitled to vote, nor were indigenous women. Even worse, the enfranchisement of white women at the expense of other women was intentional, not accidental.

What are the potential implications for moderate human enhancement and voting? In the near term, we can see that voter ID laws could easily be used to disenfranchise humans who changed their genders from male to female, or who were enhanced such that they appeared to be of a different gender than listed in their identification documents. For example, states that require government photo IDs, such as Georgia, can be daunting for transgender voters in the United States.[13] Similarly, if gender modifications could be made at will, it could be even more challenging to meet legal voting requirements that incorporated some element of gender for identification.

In the longer term, if a person's gender could be changed prior to birth, and if a certain gender was more favored than another gender, we might see a voting population skew, over time, as more parents made the choice to only have children of the favored gender. This type of gender imbalance has already happened in certain countries. For example, as of 2018, men outnumbered women by 70 million in China and India (as a combined total). "A combination of cultural preferences, government decree and modern medical technology in the world's two largest countries has created a gender imbalance on a continental scale [37]." What would that mean for voting? Would all votes really be equal in the case of marked gender imbalance among the voters?[14] What would the composition of Congress be if voting men outnumbered women substantially in the United States, due to human enhancements, or vice versa [39]? The votes (and concerns) of the gender minority could be of less interest to the candidates for election, and to the elected representatives — or of more interest if they were tiebreakers in close races, as happened with newly-enfranchised members of certain tribes in Canada during the 1800s [8]. As will be suggested in Section 3, in such a situation, it would be particularly important to consider and be responsible for the humans in the minority. Before addressing that area, however, we will push the boundaries of "human" by considering voting implications of extreme human enhancements.

2.2. Extreme Human Enhancements and Voting

Under our rough categorization, humans with extreme enhancements would typically be socially perceived as "inhuman" or "non-human" by humans who did not have similar enhancements. In such a scenario, it is not clear that the concept of "humanity" would still be meaningful as a method of drawing boundaries between conscious beings who have a right to vote—or to be entitled to representation if unable to vote directly.

Humans on a wider scale might elect to have extreme enhancements if they were safe, available, and particularly if they were reversible. Consider "Muffe" on a Danish children's show, who had horns implanted under the skin of his bald head [40]. Or Dennis Avner (also known as "Stalking Cat") who held a world record for the most permanent transformations to look like an animal [41]. These may be outlier examples now but the popularity of the "animal" filters on social media networks suggests that humans do enjoy

---

[13] "The strictest voter ID laws require voters to present government-issued photo ID at the polls, and provide no alternative for voters who do not have one [35]." In addition, "each state handles gender and name changes differently. In Georgia, a judge has to approve a name change. To change a gender marker on your driver's license, you need a doctor's letter saying you've had gender reassignment surgery. The problem is that there are different kinds of transgender surgeries, and some trans people don't get any surgery at all [36]."

[14] In the United States, for example, the 116th Congress from the 2018 Midterms had more women than ever. And yet women still were only 25% of the Senate and 23% of the House, when they represented approximately 51% of the United States population as a whole [38].

pretending to be animals if they can still retain some element of humanity. In other words, one key part of such filters is that the person still be recognizable as the "cute" animal. If similar physical enhancements were reasonably reversible, they might be adopted more widely and they might be more extreme, such that the enhanced human was seen as more animal than human. In that case, unless the laws were changed, it is reasonably likely that they would be denied the right to vote in the United States since animals do not have the right to vote. A similar result would likely arise if an enhanced human had so many inorganic enhancements that they appeared to be a robot versus a human being.

Another example of an extreme enhancement might be a "cultured brain," such as that discussed by Professor Kevin Warwick:

"It is quite possible to culture networks of dissociated neurons grown in vitro in a chamber. The neurons are provided with suitable environmental conditions and nutrition. A flat microelectrode array is embedded in the base of the chamber, thereby providing a bi-directional electrical interface with the neuronal culture. The neurons in the culture rapidly reconnect, form a multitude of pathways and communicate with each other by both chemical and electrical means. Although for most research in the field thus far, the neurons are typically taken from rat embryos, it is quite possible to use human neurons instead once sufficient connections have been made between the neurons so that, in research, the cultured brain is given a robot body with the ability to sense the world and move around in it" [42]

Should a "cultured brain" grown from human neurons be treated as a person having a right to vote if it was implanted in a human body versus a robot body? Should it even matter which body a brain is in? If a human being, say, for example, Dr. Stephen Hawking, was only able to communicate via neural implants (versus only via a single cheek muscle as was the case toward the end of his lifetime), would he be stripped of his right to vote because he no longer had a body that functioned? Alternatively, a sentient and sapient non-human consciousness ("Artificial Intelligence Plus" or "AI+") could reside (permanently or even temporarily) in a human body, such as via neural implants that connected the body to the AI+, which could also extend globally via Internet connections and a public or private cloud-based network. Such AI+ might not need to be physically constrained to a particular body—and could move between host human bodies at will or occupy more than one human body at a time, in constant communication.[15] Another extreme enhancement example could be humans linked to form an extended consciousness, such as via Brain-to-Brain Interface (BBI) technology,[16] or even a "hivemind" of multiple enhanced humans (with both organic and inorganic based consciousness)[17] who shared an interconnected consciousness, which might be given no vote, one vote, or the number of votes of the humans who were linked together.

Would such enhanced humans be entitled to vote? If so, why, where (in which country or countries), and how many votes would they have? Such extreme enhancement scenarios put pressure on the relevance of the human body to "voting". As Dr. Chia Wei Fahn has articulated for disabled bodies as interrogating existing normative standards of humanity, "technological growth and innovative design are now seen as having a unique influence regarding disability and posthumanism; disabled bodies are a 'dynamic hybrid' that is focused 'not on borders but on conduits and pathways, not on containment but on leakages, not on stasis but on movements of bodies, information and particles' that transcend corporeal boundaries and join the biological to the technological in posthuman embodiment."[18]

---

[15] Such a scenario is described very powerfully in Ann Leckie's Imperial Radch science fiction trilogy, starting with *Ancillary Justice* [43].

[16] One specific form of BCI development, Brain-to-Brain Interface (BBI), may lead to particularly novel social and ethical concerns. BBI technology combines BCI with Computer-to-Brain Interfaces (CBI) and, in newer work, multi-brain-to-brain interfaces— ... in which "real-time transfer of information between two subjects to each other has been demonstrated [44]."

[17] Consider Professor Minoru Asada's hypothesis that a nervous system for pain sensation is necessary to shape the conscious minds of artificial (inorganic) systems [45].

[18] Ref. [46] (citing Nayar, P.K. *Posthumanism*; Polity: Oxford, UK, 2014).

For extreme enhancements, it is not clear whether "human DNA" should be required at all for voting rights, and if it is, in what percentage. Similarly, it is questionable whether the platform for the minds of such enhanced humans, whether neurons in the case of biological minds or silicon in the case of digital/software minds should be relevant to their voting rights.

Based on the history of discrimination in voting against those viewed as "other" as described above, a human who was enhanced to such an extreme level that they no longer appeared to be human at all—such as a full or hybrid inorganic or animal body, with varying transient embodiments, or with no embodiment whatsoever—would likely be denied the right to vote. It took Native Americans until 1924 to get the right to vote under federal law (and even until 1957 some states barred Native Americans from voting) [22]. Given this history, it seems unlikely that a cyborg with a human brain in a robot body (regardless of the source of the brain, birth brain versus cultured brain) or a "Frankenstein" with one person's brain implanted in another person's body would easily be given the legal right to vote, particularly if they were in the minority of the overall population and viewed as "freaks" or "less than human."[19] For example, in what might have been the first attack by non-enhanced humans on a human cyborg, Professor Steve Mann was wearing a system he called "EyeTap" physically connected to his skull. When he visited a restaurant in Paris, two employees allegedly tried to remove it from his head by force [47]. Given the prevalence of violent attacks (both in person and online) on those viewed as "other", such as transgender or gender-nonconforming individuals, as well as people who are physically or mentally disabled, it is likely that there would be efforts to prevent humans with extreme enhancements from voting, particularly if they were in the minority of the human population [48–50].

But what if certain enhancements, whether moderate or extreme, became the norm for the general population? In such a case, the non-enhanced might be (legally or practically) left behind when it came to voting.

2.3. Disenfranchisement of the Non-Enhanced or Less-Enhanced

In the case of either moderate or extreme enhancements, there is a risk that if such enhancements become prevalent among the existing population of voter, non-enhanced or less-enhanced humans could be legally and effectively denied the right to vote. For example, if moderate or extreme enhancements that allowed for a particularly convenient method of voting became ubiquitous for other reasons, then there could be effective denial of the right to vote for those who lacked the wealth or resources to have such physical enhancements.

2.3.1. Accessibility Challenges

What we can see from consideration of the physical methods of voting in the United States is that those who lack certain physical abilities that are considered "normal" (such as the ability to see or walk), or who lack access to certain technology or even the resources to physically get to a polling location, can be effectively denied the right to vote. We can expect the same result if human enhancements lead to a shift that allows votes to be cast by neural implants, or if other enhancements that facilitate voting become prevalent. Then the non-enhanced (or less-enhanced) humans may be perceived as disabled or less-abled and legally or practically denied the right to vote.

The United States Federal Americans with Disabilities Act of 1990 "provides protections to people with disabilities to ensure that they are treated equally in all aspects of life. Title II of the ADA requires state and local governments ("public entities") to ensure that people with disabilities have a full and equal opportunity to vote. The ADA's provisions apply to all aspects of voting, including polling places (or vote centers)" [51]. Similarly, under the Voting Rights Act of 1965 (VRA) election officials are required to allow a voter

---

[19] This is a subjective judgment of course. Humans from several hundred years ago might think enhanced humans of today are "beyond humanity".

who is blind or has another disability to receive assistance from a person of the voter's choice (other than the voter's employer or its agent or an officer or agent of the voter's union). But the reality is that despite these laws, such persons are less able to vote than the general population.[20] In an audit conducted in 2020, more than 40 states were found to have "absentee ballot applications that were not fully accessible to millions of visually impaired voters and those with other disabilities [53]."

As Dr. Fahn has explained in a similar context of prosthetics and other technologies: "At present, biotechnological marvels remain a form of class consumption. Advanced technology is only open to those who can afford the purchase, causing people in one country to experience the same impairment differently, based on their socioeconomic status. The "have nots" and the "haves" will lead directly to an "ability divide" that is not only reflected by individuals, but sets a division on a global scale [46]."

Consider enhancements that allow humans to "vote" by using a neural implant and a secure wireless internet connection with their unique digital and biological signature (for authenticity). This is not so far-fetched as Prof. Warwick, one of the world's first cyborgs, described in his experiment involving a "basic form of telegraphic communication" between two human nervous systems. As Prof. Warwick explained:

> "A volunteer assisted by having microneurography. Essentially, two very thin needles were pushed into the nervous system in their left arm. With this in place, we set up in the lab with a group of people around the volunteer and another group around me—we had a variety of different observers to oversee what we were doing. The volunteer and I were not able to see each other. We set the experiment up purely based on hand closures. When the volunteer closed their hand, I received a stimulating pulse on my nervous system, and the same happened vice versa. For me, it meant that my brain recognised the pulse. I shouted out "Yes" every time I felt a pulse, but only when I felt a pulse. Only the group around the volunteer could witness when they had closed their hand and when not. We achieved this with 100% success—the same being true in reverse. What I found exciting was that as the groups were splitting up, I felt a couple of quick pulses one after the other. Subsequently, the volunteer confirmed that they had done this. It was a 'secret' message between the two of us, a new means of communication". [42]

Similarly, entrepreneur Elon Musk has long been a proponent of neural implants, such as those pioneered by Jan Scheuermann as part of the DARPA initiatives [54]. In 2020, Elon Musk's company Neuralink showed a demonstration involving a pig who had a brain–computer interface implanted that allowed a display of her real-time brain activity [55]. While the initial stated goal was to assist humans with paralysis or a serious illness, the "augmentation" of people who are healthy and without disabilities "is an obvious result," according to the Director of Implant Systems at Neuralink [56]. "It's being able to enhance our ability to interact with the world." [56] Similarly, an extended mind, as discussed in the work of Professors Dunagan, Grove and Halbert in *"The Neuropolitics of Brain Science and its Implications for Human Enhancement and Intellectual Property Law"*, could also be considered an enhancement that could become the norm.[21]

---

[20] "According to a 2017 report by the U.S. Government Accountability, "Voters with Disabilities: Observations on Polling Place Accessibility and Related Federal Guidance," roughly two-thirds of the examined polling places had at least one potential barrier such as lack of accessible parking, poor paths to the building, steep ramps, or lack of a clear path to the voting area. Although most polling places had at least one accessible voting system, roughly one-third had a voting station that did not afford an opportunity for a private and independent vote. The report also noted that Department of Justice guidance does not clearly state the extent to which federal accessibility requirements apply to early in-person voting. People with disabilities also continue to report barriers including a lack of accessible election and registration materials prior to elections, lack of transportation to polling places, and problems securing specific forms of identification required by some states [52]."

[21] Ref. [57] Perhaps this "extended mind" is already becoming the norm in practice, even if not technically a "human enhancement" for purposes of this Special Issue, when one considers the expanded computational power and informational access available at any time 24/7 to any human with a smartphone with internet access. Consider also Professor Fiorella Battaglia's proposal that an "extended mind" offers the potential of a new way to view the "mind-body" problem as an opportunity rather than a problem, emphasizing the epistemic difference between being a mind and being an extended mind in her forthcoming article in this Special Issue—"Agency, Responsibility, Selves, and the Mechanical Mind" [58].

Such enhancements could become common because they offer more convenient lifestyle activities in addition to voting, such as electronic funds transfers, secure communications and hands-free use of all the apps and programs currently available on desktops, laptops, tablets and smartphones. If such enhancements became common for everyone who had the resources to purchase them, then access to voting could favor those who had such implants. In that case, unenhanced or less-enhanced humans could be effectively disenfranchised. We can see this happening already in the United States with respect to voting access. Traditionally, voting in the United States was conducted in person, at designated polling locations, for registered voters only. This has shifted to a greater percentage of mail-in ballots, and that shift was already occurring even prior to the pandemic of 2020. Voting patterns in the United States reveal that wealthy voters are taking greater advantage of the ability to cast votes by mail[22] [53]. It seems reasonable to anticipate that there will be a shift to computer voting in the future, and in that case voters who are poor and lack computer and internet access could be less able to vote. Voters with few resources may not have reliable addresses in order to be able to vote in person or by mail—or may not have access to printers or scanners to submit applications to vote by mail. Voters who are in poverty and/or have caregiver responsibility for friends or relatives may not be able to wait in long lines or to drive long distances to be able to cast ballots in person.

In addition to accessibility challenges, enhanced or non-enhanced/less-enhanced humans may also be effectively disenfranchised due to language barriers.

2.3.2. Language and Communication Barriers

With the click of a button, anyone with access to the Internet can easily translate a meaningful amount of the content into another language, even if the translation is imperfect. It is not a stretch to imagine that an enhanced human, such as via a neural implant or a genetic enhancement for linguistic ability, could readily speak and read a large number of languages, particularly those for which online translations are accurate. However, such rapid translations are not available for the vast majority of languages spoken in the world currently.[23] The importance of using a language that voters can understand, and the impacts of the failure to do so, cannot be underestimated.[24] Moreover, humans with extreme enhancements could "speak" new languages or even communicate in new ways that would not be intelligible to humans without similar enhancements. If such new languages became the "official" languages for voting, it could effectively disenfranchise the humans without such enhancements.

The choice to recognize a language as an "official language" is itself a decision that makes meaningful participation in government and voting more difficult for those who do not speak and read the official language (or languages) fluently. In many countries in Africa, the official languages may be French or English even though those are not the languages most widely spoken by the residents of such countries. In the European Union, for example, English, French and German are given preferential treatment as "procedural" languages although there are twenty-four official languages. Irish was only given the status of an official language in 2007 and was temporarily derogated through 31 December 2021 [62]. The fact that we face this challenge currently is an indicator that it should not be ignored in considering new languages that could be introduced via human enhancements.

---

22 "Among the obstacles for poor Philadelphians: Lack of stable housing makes it difficult to depend on the mail and know which address to provide when applying for a ballot to be mailed weeks or months later. Those with limited English proficiency have difficulty navigating the vote-by-mail process, and governmental voter outreach can miss them. Lack of internet service or home computers can complicate requesting ballots or finding key information about them [59]." In addition, a "2020 report by Native American Rights Fund determined that some members of the Navajo Nation must travel 140 miles roundtrip for postal services. Many do not have access to personal vehicles or public transportation to get them there [60]."

23 Google Translate purports to support over 100 languages, out of approximately 7000 languages spoken globally. "Odia, the official language of the Odisha state in India, with 38 million speakers, . . . has no presence in Google Translate [61]."

24 An example from the Canadian period of enfranchisement and disenfranchisement illustrates that the challenges of different languages is not unique to any particular country or region. In the 1800s, one chief grew so frustrated by the use of English at a particular Council of tribes that he "led a break away council with other northern bands" and was "only lured back with the promise of an interpreter [8]."

In enacting requirements to provide certain materials in additional languages as part of 1975 amendments to the Voting Rights Act, the United States Congress recognized that "Through the use of various practices and procedures, citizens of language minorities have been effectively excluded from participation in the electoral process. Among other factors, the denial of the right to vote of such minority group citizens is ordinarily directly related to the unequal educational opportunities afforded them resulting in high illiteracy and low voting participation. The Congress declares that, in order to enforce the guarantees of the fourteenth and fifteenth amendments to the United States Constitution, it is necessary to eliminate such discrimination by prohibiting these practices, and by prescribing other remedial devices."[25] Applying this goal to enhanced and non-enhanced/less-enhanced humans, if certain current or new languages became dominant, a meaningful right to vote would require, at a minimum, that translations be made easily available for the less dominant languages.

A similar additional potential cause of disenfranchisement might be if enhanced humans had such dramatically increased cognitive ability levels that it became practically impossible for the non-enhanced, or less-enhanced to "read" and comprehend the ballots even if they were reasonably fluent in the language itself.

2.3.3. Cognitive Ability Requirements

Cognitive enhancements, such as genetic cognitive enhancements via genetic engineering or embryo selection, pharmacological substances, Transcranial Magnetic Stimulations or Transcranial Direct Current Stimulation, or other techniques developed in the future,[26] could become a common method to increase the intelligence of humans before birth or other modifications could be made after birth to the same purpose. In such a situation, it might become normal to include extremely lengthy and complex voting information on the ballot,[27] or even ballot questions themselves (at least in states where ballot questions were permitted). Non-enhanced (or less-enhanced) humans could be unable to even decipher such questions at the polls. If they were not sufficiently enhanced due to lack of financial resources, then it would also be unlikely they would have access and time to peruse such ballot questions in advance.[28] Ballot questions can be challenging currently even for highly educated and intelligent non-enhanced voters. If the "normal" enhanced human had the ability to visually access extensive background on a ballot proposition by visually looking at the equivalent of a "QR code" or through the internet via neural implant, then the ballots might not include the written summary at all, for example to save printing costs and for efficiency. Anyone who lacked the necessary enhancement(s) might not have the ability to obtain such details easily. In an extreme case, they might not even know what they were voting about at all.

Parents in poverty might, at least hypothetically, be more willing than wealthy parents to undertake intelligence or other enhancements for their unborn children even when they were risky. For example, as discussed by Professor Marcelo de Araujo, in initial research by the SIENNA Project, it appeared that "Scientifically and technologically developed countries are not especially supportive of research on human enhancement technologies. . . .

---

[25] For a determination of which minority languages would be covered – see the Federal Register [63]. One goal of the amendments was to clarify coverage of certain racial and ethnic minorities, including members of the "Hispanic" or "Latino" population "who had suffered discrimination in the political process, but whose group status under the law remained uncertain" and "by self-designation or by ascription, often eluded clear racial categorization and transcended strict racial labels such as 'black' and 'white [64].'"

[26] Ref. [65] The work of Professor Nicole A. Vincent and others on neurointerventions and the law is also very relevant to any consideration of cognitive enhancements [66].

[27] For example, ranked-choice voting has been challenging for many voters to understand. "Although no federal constitutional arguments have prevailed in Maine or in any of the other litigation around the country, opponents have argued primarily that it violates equal protection (one person, one vote) and due process (too confusing). Opponents generally attack [ranked-choice voting] on the grounds that it is too confusing and too costly. It requires voters to understand how the votes will be cast and counted, and then to vote accordingly [67]." Future possible voting methods, particularly those developed with the input of a diverse group of enhanced and non-enhanced humans as outlined in Section 3, could, at least in theory, be more complicated by one or more orders of magnitude.

[28] An extremely lengthy and complex detailed "summary" of Ballot Question #1 was offered for voters in Massachusetts in 2018 [68].

Brazil, South Africa, and Poland were most supportive, while the Netherlands, Germany, and France were least supportive of cognitive enhancement technologies [65]." While no definitive conclusion can be reached without a deeper analysis, this disparity at least suggests that it is possible that wealth disparities (either individually or at a country-level) might potentially correlate with the level of support for cognitive enhancement technologies. As one possible example, a positive view of cognitive enhancement technologies, such as by genetic engineering or embryo selection, may, at least in part, be influenced by the desperation of poverty.

What this situation tells us about voting is that if, in the future, the "norm" for the general population becomes humans with physical and/or mental enhancements that make it easier for them to vote through certain processes, then people who lack such enhancements may be effectively deprived of the right to vote unless accommodations are made.

With these examples in mind, it is important to consider a possible ethical theory that could be used as a framework to protect both enhanced and non-enhanced humans from being disenfranchised, to the extent possible.

3. Discussion—Human Enhancements and Voting—Responding to the Appeal of the Other

"The will of all beings, and responsibility toward all beings (whether voting or non-voting), shall be the basis of the authority of government; this will, and responsibility, shall be expressed through universal and equal suffrage (directly and/or through representatives) or through such other process and method as may be developed and ratified with the universal and equitable participation of such beings."

—Draft, proposed Declaration of Rights and Responsibilities of Beings

In Section 2, we considered the phenomenon of voting and particularly how past experience might be relevant for enhanced and non-enhanced humans in a world where human enhancements, whether moderate ("still human") or extreme ("beyond human" or "non-human") were more common. It seemed likely that, based on the history of discrimination in voting rights and disenfranchisement of those deemed "other" or "lesser", long-established patterns of discrimination could impact both enhanced humans and, if those with enhancements became dominant, those without similar enhancements. In addition, distinctions based on ancestry, gender, and even embodiment itself would likely become meaningless artifacts, if they ever had meaning at all, particularly if such characteristics could be changed prior to birth and/or at will—either permanently or temporarily—at any time during the (potentially greatly extended) lifetime of an enhanced human.

In this Section 3, we will explore one possible philosophical framework for voting—an ethics of human rights as outlined by Loidolt that is extended to cover humans with extreme enhancements who may no longer be viewed as "human" by those without such enhancements—as well as to cover humans who may be viewed as so primitive and irrational that they are considered to be "less than" human by the future enhanced humans.[29] Such an extension would admittedly broaden Loidolt's already inclusive framework. However, the goal would be to preserve the core elements of the framework, including retaining the recognition of the impossibility of truly understanding and speaking for the "other." In *Citizen Cyborg*, Professor James Hughes predicted that "the emerging 'biopolitical' polarization between bioLuddites and transhumanists will define twenty-first century politics" [70]. For purposes of this analysis, it is not necessary to predict when, or if, such polarization will in fact occur, but it is important to recognize the possibility of different levels of power, and perspectives, between the enhanced and the less enhanced or non-enhanced.

After a brief introduction to Loidolt's theory of engagement, we will consider how it could translate to universal suffrage for *all* humans—enhanced and non-enhanced—

---

[29] Much as humans of the 21st century questioned whether Neanderthals were a separate species despite interbreeding with Homo sapiens [69].

regardless of whether they are even still viewed as being human at all. Such an extension could include, for example, all "living" beings and all "conscious" beings as having rights and responsibilities to other beings. In attempting to provide support for this possible extension, we will consider specific criteria that may no longer be appropriate as voting requirements, including "reason"/"rationality" (which potentially impacts restrictions based on gender and age, for example), and physical geographic location within certain political boundaries. We will also consider scenarios where certain groups have an indirect voice through representative voting, in lieu of—or in addition to—direct voting. Finally, applying this extended version of the engagement-based ethical model, we will propose that the entire framework and legal and social structure of what "voting" means should be itself designed with the broadest possible participation, rather than dictated "from the top down." In other words, the humans who have the power at a particular period in time, whether locally, nationally, regionally or globally, cannot legitimately dictate what is best for future enhanced or non-enhanced humans or even for humans who otherwise are disenfranchised or were excluded from participation in the design of the architecture of governance. Such "others" (current or future) should have a voice in the design of the governance structure, and whether and what voting should mean, in order for such structure to have legitimacy. If they were not able to do so originally, there should be periodic opportunities for re-design and re-ratification.

3.1. A Brief Introduction to an Extended Theory of Engagement

One potential theory of engagement that can be applied to a world of enhanced and non-enhanced humans is offered by Loidolt[30] [1]. By shifting the focus from a set of indispensable needs to the core *responsibility* of meeting such needs through an analysis of the structure of subjectivity, Loidolt outlined two themes that could serve as guidelines for an ethics of human rights in a theory of engagement, which we can extend to voting and human enhancements. The first theme emphasizes the transcendental importance of legitimization and justification in a meaningful world. The second theme seeks to "face the appeal of the other" and determine what is "right" "from a phenomenological first-person perspective instead of the classical 'objective' third-person perspective of reciprocity"[31]. Loidolt contends that the process of constituting the world is a priori occurring in a context of legitimizing intentionality. Specifically, "the meaning that something is 'right' or has 'a right', is not something that is perceived, but something that is achieved through passing a judgement: By judging something, by the means of implementing a norm or any measure, the meaning of 'right' or 'a right' originates as the formal expression of an 'accordingness.'"[32]

---

[30] Other possible ethical frameworks would be interesting to explore and likely would be fruitful for the analysis of voting rights and human enhancements, such as Professor Christine Korsgaard's Kantian ethical analysis of duties of humans to animals [71,72]. The work of Professor Simone Goyard-Fabre in combining an approach from Husserl's phenomenology with a Kantian analysis would also be relevant due to the deeply "legal" nature of the analysis of voting rights. Professor Maria Golebiewska provides a thought-provoking analysis in this area [73]. Another interesting framework to consider for voting rights could be Ernst Knapp's philosophy of technology, as proposed by Battaglia as a possible way to look at the "mind-body problem" in the context of a human-machine hybrid "extended mind" by using the "mechanical mind" as an explanatory device [58].

[31] Ref. [1] at 3.

[32] Ref. [1] at 7.

This theory of engagement is a much stronger version of the suggestion made by Professor Chris Hables Gray in *Cyborg Citizen*: "Along with the realization that no totalizing theory explains everything is the corollary that a number of different perspectives together are capable of creating a better model of reality than any one point of view" [74]. Under Loidolt's framework, there is no utilitarian analysis of what is "better" (and no objective third-person "reality" to be modeled). Instead, the participation of others and consideration of their views is itself *necessary* for legitimization and justification in a meaningful world, even if the end result might later be deemed "worse" under a particular utilitarian ranking model than the result that would have been reached without their participation (assuming, hypothetically, that an accurate empirical evaluation was possible).

Under such a theory of engagement, there is an "intersubjective community of communication" and "being receptive as such" makes humans, as conscious beings, "answer [the appeal of the "other"] in a category of legitimization." Extending this theory to the right to vote and human enhancements, we will ask how, in a world of enhanced and non-enhanced humans, an ethical enhanced or non-enhanced human should answer the appeal of the "other" in the context of the right to vote. In doing so, we will follow Loidolt's guidance and treat "consciousness" not as a "secluded sovereign entity" but instead as "openness as such."[33]

3.2. Universal and Equal Suffrage for Enhanced and Non-Enhanced Humans

As we look toward the future, we can draw inspiration from the response to the horrors of World War II, in which approximately 6 million Jews[34] [75] were systematically murdered by the Nazis. Victims included approximately one million children and teenagers. "Thousands of Roma children, disabled children, and Polish Catholic children were also among the victims. Like their parents, they were singled out not for anything they had done, but simply because the Nazis considered them inferior" and sub-human [76]. After World War II, the Universal Declaration of Human Rights ("UDHR" or "UN Declaration") was adopted. The UDHR was a collective shout of optimism that, by coming together, there would never be a repeat of such atrocities. The dream was that humans could live together in peace and harmony. While recognizing that such goals were not achieved,[35] we can nevertheless seek inspiration from that work in the context of voting rights and enhanced humans and hope for further progress, even if it is incremental and imperfect.

For the current analysis, we will focus on Article 21(3) of the UDHR [2], which states:

> *(3) The will of the people shall be the basis of the authority of government; this will shall be expressed in periodic and genuine elections which shall be by universal and equal suffrage and shall be held by secret vote or by equivalent free voting procedures.*

In extending the right to vote in a world of enhanced and non-enhanced humans, we will modify Article 21(3) to remove the term "people" and replace it with the more inclusive term "beings". Thus, for purposes of discussion of voting rights in such a context, one possible way to re-frame Article 21 is as follows:

> *"The will of all beings shall be the basis of the authority of government; this will shall be expressed through universal and equal suffrage."*

With the reference to "people" changed to "beings," this framework could be used for extreme human enhancements, where the enhanced human may no longer be recognized by non-enhanced humans as "human" at all, as well as for more moderate enhancements, where the broader community may be oblivious to important differences because they are not readily perceptible or detectable absent invasive testing. Similarly, it would apply to non-enhanced or less-enhanced humans who might be viewed by future enhanced humans

---

[33] [1] at 8–9.

[34] According to the testimony of Dr. Wilhelm Hoettl in the Nuremberg Trials in 1945 regarding a statement made to him by Adolf Eichmann, Chief of the Jewish Section of the Gestapo: "Approximately 4 million Jews had been killed in the various concentration camps, while an additional 2 million met death in other ways, the major part of which were shot by operational squads of the Security Police during the campaign against Russia" [75].

with extreme enhancements as "sub-human" or otherwise "inferior". In the ethical model we seek to extend to voting and human enhancements, "justice" is "not a self-assured calculation with symmetrical portions of free will, but an urgent conceptual reaction to an overwhelming appeal that can never be adequately responded to."[36] Extreme enhanced humans would still have a right to vote as long as they were viewed as "beings" meaning, in a very broad definition, that they were either "living" or they were "conscious". Similarly, non-enhanced/less-enhanced humans viewed as "other" or "lesser", would also have a right to vote as "beings".

In the following sections, we will seek to provide some support for such an expansion, including considering arguments that would retain the requirement of "humanity" for voting rights and therefore could exclude humans with extreme enhancements from voting. Finally, we will acknowledge the concern that an expansion of legal rights to votes broadly to all enhanced and non-enhanced humans could contribute to an unintended consequence that those who lacked power currently could be disenfranchised or otherwise harmed. Since such a harm is possible based on past experience with discrimination against the "other", we will suggest that, under the extended ethical framework being applied, such humans should be involved in any decisions regarding their enfranchisement and indeed the entire voting process and very framework of governance.

3.3. Casting a Broad Net for the Definition of "Human" in the Context of Voting

By intent and express language, the Universal Declaration of *Human* Rights covers the rights of "human beings." Under Article 1, "All *human beings* are born free and equal in dignity and rights. They are endowed with reason and conscience and should act towards one another in a spirit of brotherhood." However, in trying to meet the appeal of the "other" (and indeed the multitude of "others") in the context of voting and human enhancements, we propose to reject "humanity" as a litmus test and instead cast a wider net in defining the "other" as any living or conscious being.[37] We will "recognize the urgency that demands" the impossible and commit to the "rights of the others, . . . to equality, dignity etc., with the insight that [equality, dignity, etc.] will have been dependent on that commitment" and with "*awareness* of the imperfection and urgency of the case, thus *alertness* to the disturbing, responsivity to and responsibility for it, while being aware that [the disturbing demands of the others] can never be fully incorporated; and finally *openness* to that universality in progress (or universality to come) that keeps being constituted from the outside."[38] In this context, "disturbing" may perhaps be understood to mean both changing/disrupting the *status quo* as well as troubling/unsettling to the being who is called to respond to the appeal of the "other".

Under this extended ethical framework of engagement and under an expanded Heideggerian approach, enhanced humans and non-enhanced humans would all be "beings" entitled to certain rights but also with certain duties and responsibilities towards others.

"Man is not the lord of beings.

Man is the shepherd of being.

---

[36] Ref. [1] at 12 (referencing the work of Emmanuel Levinas, 1978).

[37] "Social robots should not be regarded as proper objects of moral concern unless and until they become capable of having conscious experience. While that does not entail that they should be excluded from our moral reasoning and decision-making altogether, it does suggest that humans do not owe direct moral duties to them." [78] Under an extended ethical theory of engagement and the new Declaration of Rights and Responsibilities of Beings toward which this Article on voting rights is intended to provide a stepping stone, the goal is to be as inclusive as possible. Ideally, all living beings, whether "conscious" or not (e.g., plant-based life or a human who is in a coma), and all "conscious" entities should be included in the definition of "Beings". Full articulation of the meaning, scope and boundaries for such an inclusive definition, as well as the philosophical basis (or bases) will obviously require significant work. A Kantian perspective of "personhood" and "identity" would be useful to consider, such as that presented by Professor Béatrice Longuenesse: "Kant's criticism of the paralogism of personhood opens the way to to substituting for the rationalist concept a rich and complex concept of a person as a spatiotemporal, living entity endowed with unity of apperception and with the capacity for automomous self-determination." [79].

[38] Ref. [1] at 13.

Man is the neighbor of being"[39] [80]

Heidegger struggled with anthropocentrism, eventually concluding that "until we arrive at a fuller understanding of the nature of world, we cannot pass judgement on the ultimate legitimacy of his conception of the animal as *poor in world*. Only then can it be determined whether there is "poverty in the animal's specific manner of being as such", or if such poverty obtains only in comparison to *Dasein*" [82,83]. In our extended ethical theory of engagement, we propose that it is not legitimate to deprive any enhanced or non-enhanced humans of voting rights while we await a "fuller understanding of the nature of the world." Instead, we will value *beings* in the lower-case versus the "Dasein" of Heidegger (with Dasein meaning roughly a *human* being "there"—totally immersed in the world of other things—a "Being-in-the-world"). An essential element of such an equitable approach is an awareness that there is an inherent risk of error any time one being represents another being or purports to understand what another being wants or needs. Indeed, Heidegger may have underestimated the depth of the "Abyss" between humans while overestimating the depth of the "Abyss" between humans and animals.[40] Thus, a sufficiently "full" understanding of the nature of the world, such as that contemplated by Heidegger, may be impossible to achieve. As Loidolt explained, we must try to "realize the intentional movement of reason towards complete legitimization: even if it cannot be achieved, legitimizing intentionality can not stop at an unjustified benchmark, but transgress it necessarily with critique"[41].

Such an extended ethical approach of course requires explicitly rejecting the anti-semitism and other arbitrary biases that permeated the philosophy of Heidegger—and Nietzsche—and seeking a commitment to respect for others. In his provocative concluding paragraph of his article in this Special Issue, Warwick draws from Nietzsche to speculate about a possible bleak future for non-enhanced humans: "The fundamental philosophy that underpins the concept of the future relationship between superhumans and humans comes straight from Nietzsche. We need to look no further than humans' present-day relationship with other animals. Humans cage them, destroy their habitat and treat them as captives, slaves or pets. Thus, if we look to the future, the best that a human could hope for might be that they become the pet of a superhuman." [42].

Surely we can *hope* for better than this. We can attempt to design an ethical structure of voting rights and responsibilities to reflect such aspirations while also considering the existing realities of imbalances in political power and other relevant factors. As an example, the brutal legacy of years of systematic discrimination has led to widely disparate income, education, health, medical treatment, access to housing, credit, incarceration, job opportunities and death (including killings by law enforcement agents) for people of color in the United States.[42] Voters (and their elected representatives) should consider not only their own interests but also their responsibility toward other beings, particularly those who cannot vote or whose votes may "count" for less than those in power. While it is theoretically possible that an enhanced human might have the ability (and will) to control other beings without voting or control others in voting, we should not automatically assume that is what would happen. An enhanced human might also have greater compassion and empathy, as well as improved emotional intelligence such that they would not wish to use their power in that manner. Any human, with any level of abilities or power, is capable

---

[39] Ref. [80] In applying any phenomenological approach that may have been influenced by the works of Heidegger, it is important to explicitly acknowledge the deep-rooted antisemitism in Heidegger's philosophical views. For a powerful rebuttal of the idea that Heidegger's antisemitism can be easily separated from his philosophical viewpoints, see Professor Donatella Di Cesare's work [81].

[40] "Although at one time exploring the proximity between the human and the animal, Heidegger later claimed that an *abyss* separates the human from the animal." [84] at 110 (citations omitted).

[41] Ref. [1] at 18.

[42] For example, in 2017, the median net worth of Black residents in Boston was *$8* versus *$247,500* for whites [85]. Also, "Black female restaurant workers in Massachusetts make on average $7.79 less per hour—including tips—than white men in the same positions, which amounts to 60 percent less" [86]. As just one example that discrimination and violence based on skin color is not limited to the United States, see an example from France [87].

of doing good or evil. Adolf Hitler was able to do great evil without enhancements, and Nelson Mandela was able to do great good without enhancements. Applying an ethical theory of engagement, the appropriate response to the possibility of a hypothetical future sociopath having tremendous power is not to abandon any path toward self-determination. We take that risk now—and likely that will be the case whatever form of government we establish for future enhanced or non-enhanced humans.

A more fundamental challenge to enfranchisement of "beings" than the spectre of a super-powered tyrant may be the view that humans are innately and metaphysically superior to all other beings. Such an objection could be based, for example, on the contention that there is inherently a hierarchy of species and that any broad approach that allows enhanced humans who are viewed as being more animal than human, more plant than human, or more computer than human to be treated as "human" upsets the proper hierarchy. The view that "pure" humans are of special and unique value higher than any other living being—whether conscious or non-sentient—is deeply entrenched in many widely-respected philosophical, scientific and religious frameworks.[43] Someone who holds that view may well feel that a human who has an extreme enhancement (either prior to birth or after birth) is not a human at all, or is *less-than* human. In that case, the extension of the right to vote to such an enhanced human likely would be rejected as the extension of the right to vote to non-human animals and to plants or other life forms would also be denied. Such a rejection need not be made callously or with indifference to others. Representation alone, as outlined in Section 3 below, could be viewed as sufficient to respond to the appeal of the "other" even under an ethical theory of engagement. However, in this analysis we are pushing the definition of "other" to reach beyond "humans" in the context of extreme human enhancements. This is a fundamental difference in perspective. In order to support this expansion, we will consider some of the reasons typically given for excluding animals[44] from having the same rights and responsibilities as humans, and express concern about their validity and the danger that they can be used as a proxy for improper discrimination (including against other humans). However, the core philosophical differences will not be reconciled in the analysis herein, and it is not clear that they are capable of reconciliation.

3.4. A Provisional Rejection of Ranking Systems/Hierarchies

One of the problems with any hierarchical approach, which is particularly acute in the context of enhanced humans, is that it is challenging to articulate meaningful ranking criteria that can withstand scrutiny and that do not naturally lead to individual humans being further ranked and excluded from voting. In addition to being pretexts for racism, sexism, classism, etc., ranking systems of the past have turned out to be factually false or absurd. For example, Plato felt that walking upright was a key distinguishing feature of humans. Darwin considered that bipedalism represented evolutionary advancement. However, it is difficult to see why walking upright is intrinsically a high value physical trait, or even that it is unique to humans.[45] From a value perspective, is walking upright better than flying, running fast like a Cheetah, or having an extremely precise sense of smell like a dog? Why is one physical trait inherently better than another? Second, once one establishes any form of ranking criteria, there is a natural tendency to further rank within each category. For example, if a human cannot walk upright at birth, or at some point during their lifetime due to a physical challenge, are they no longer human or less

---

43 Also, in certain religions. For example, Genesis 1:26 "Then God said, 'Let us make humankind in our image, according to our likeness; and let them have dominion over the fish of the sea, and over the birds of the air, and over the cattle, and over all the wild animals of the earth, and over every creeping thing that creeps upon the earth.'" For an alternative interpretation, see the work of Professor Ryan Patrick Mclaughlin, who proposes that, under one credible reading of Genesis, "human dominion" should be "peaceful and other-affirming" [88].

44 Although the present ranking discussion focuses on animals, a typical reason given for excluding computers (even ones that can communicate in a manner that is difficult to distinguish from an actual human) is that they lack sentience. In other words, they are neither conscious nor self-aware.

45 "If an ape who stands upright (even with the help of a mobility device) can be seen as more human, what happens to humans who do not or cannot stand upright? Monkey-like posture was one of many simian characteristics used to dehumanize people of color, particularly people of African descent, from the seventeenth century on." [3] at 86-87 (discussing illustration from 1699 of an ape standing erect with a walking stick).

than other humans at that point? If so, the famous physicist Dr. Stephen Hawking was not human (or was less than human) once he could no longer walk upright. Such a conclusion would not be acceptable even under the current version of the UDHR.

We will consider in more depth two other potential candidates as requirements for voting rights under the expanded version of UDHR and as potential justifications to retaining the requirement of "humanity" for voting rights for enhanced and non-enhanced humans—"reason" (also referred to herein interchangeably with "rationality") and "dignity". Both terms are used in Article 1 of the UDHR[46] and are frequently used in ethical theories relating to differences between humans and others.

3.4.1. Rejection of "Reason"/"Rationality" as a Voting Requirement

As discussed above in the context of a history of discrimination, the "ability to reason" and "rationality" have often been put forth as a purportedly neutral and objective requirement for voting rights. Aristotle viewed humans as the only rational beings. If only humans are rational, and if rationality is essential for voting, then perhaps enhanced humans who are more animal than human, more plant than human, and/or more machine than human should be excluded from voting?

One problem with this approach is that the definition of "reason" or "rationality" is key and it is not automatically clear what such terms mean in the context of voting rights and responsibilities of enhanced and non-enhanced humans. Does "rationality" mean the ability to think logically? If so, at an advanced level such as Aristotle demonstrated? If we have enhanced and non-enhanced humans, should all voters be tested on basic logic and, if they fail, be disenfranchised? If only some voters are tested, then the potential for discrimination based on improper categories is high. Also, why should logical thinking be of greater importance than creative thinking or compassionate thinking when it comes to voting? Should a human who excels at logical thinking or mathematical computational abilities be valued more than another human who cannot perform as well? When will the testing for such abilities be administered, by whom and on whom? Or, is "reason" only an ability to understand cause and effect in one's daily life? Many animals can understand cause and effect. Otherwise they could not be trained by humans.[47]

Returning briefly to the example of gender that was discussed in Section 2 above, we can perhaps uncover even more that may be relevant to the potential requirements of "rationality" or "reason" in voting in a world of enhanced and unenhanced humans in examining the argument that *"Women have the right to vote, for instance, because they are just as capable of making rational decisions as men are."*[48] This statement assumes that "rationality" is part of the act of voting, when observation of the phenomenon of voting reveals that "[t]he separateness of vote and decision decouples any pretence of rationality from the decision. It makes little sense to talk of rational democratic decisions, although we may approve of decisions for other reasons. Nor will you find support for rationality within the events leading to the decision. The naïve account of events is that the participants think about each specific decision with the penetration of Socrates and stand by their rationally derived conclusion with commendable integrity. Observe voting and you will see that the vast majority of decisions are preceded by inadequate and contradictory thought, much of it around matters that are peripheral to the specific content of the decision to be taken.

---

[46] "All human beings are born free and equal in dignity and rights. They are endowed with reason and conscience and should act towards one another in a spirit of brotherhood."—UDHR Art. 1 [2].

[47] And as many people with pet cats and dogs can attest, animals also can train their humans to behave in a manner desired by the animal at a particular time, in response to a particular prompt from the animal.

[48] See Singer's work in [33] at 148–162 (describing an initial possible response to Thomas Taylor's satire of Mary Wollstonecraft's "Vindication of the Rights of Women" in 1972—"A Vindication of the Rights of Brutes").

With equal claim to rationality politicians may determine to vote in accordance with their constituents' wishes, higher economic theory, or their perceived personal self-interest."[49]

Also, are all *men* who have the right to vote capable of making "rational" decisions? Should "rationality" even, in a naïve account of events, be an essential condition for having a right to vote, versus empathy for others, compassion, generosity or some other quality or quantity? If it were possible to enhance humans so that they were able to make better moral decisions, [90,91] should the votes of such enhanced humans be given greater weight? At a more basic level, what is the meaning of "male" or "sex" in the context of voting, given that one hopes that genitalia are neither involved in nor displayed when a being votes? Similarly, the argument that *"dogs, on the other hand, are incapable of understanding the significance of voting, so they cannot have the right to vote"* assumes that it is an essential requirement to "understand the significance of voting" in order to have a right to vote [33].

In addition, "rationality" has been used as the basis to deny the right to vote to those under a certain age. In the United States, the right to vote in federal elections was lowered through the Twenty-Sixth Amendment XXVI—which changed the minimum voting age from 21 to 18 years. Are humans under 18 not "rational"? One argument for why those under 18 do not have the right to vote is that their brains have not developed to the point that they can weigh risks "rationally" with an "appropriate" level of conservatism [92]. But people under 18 are part of the population and have good reason to be concerned about choices made by members of an elected body who do not reflect their views and interests. Also, as described above, based on the past and current use of testing for intelligence or knowledge, we should be very skeptical that such cognitive capacity requirements would be applied in a non-discriminatory manner when enhanced humans and non-enhanced humans are involved. If a human could change their physical age via enhancements—at will—and/or increase their cognitive capacity at will, then it is not clear that their actual date of birth would be particularly relevant to their right to vote (if it ever was relevant). Potentially, "age" requirements for voting could be eliminated, particularly in a world where enhanced humans were common.[50]

Fundamentally, is not clear that voters really need to "understand the significance" of voting in order to have a right to vote in a theory of engagement that requires responding to the appeal of the other. "A part of the myth of democracy is that it requires higher thought processes and rationality" [89]. To exclude an entire class of enhanced or non-enhanced humans on the ground that they could not "understand the significance of voting" could easily lead to other exclusions that have been used to discriminate in the past (and in the present), such as a requirement that voters have at least a certain level of formal education or that applicants for citizenship, and therefore the right to vote, pass certain tests.[51] For example, American suffragist Elizabeth Cady Stanton argued that (white) educated women were more deserving of the vote than "ignorant" men, including many formerly enslaved people, working-class people and immigrants [93]. Intentionally drawing on racist stereotypes, she explained: "Think of Patrick and Sambo and Hans Yung Tung, who do not know the difference between a monarchy and a republic, who can not read the Declaration of Independence or Webster's spelling-book, making laws for Lucretia Mott, Ernestine L. Rose, and Anna E. Dickinson [94]."

Similarly, by this argument, cognitively enhanced humans might perhaps deny the right to vote to those without enhancements. In rejecting artifactual and frequently discriminatory distinctions such as "rationality" and "reason", we propose that even enhanced

---

49 Ref. [89]. An alternative model of representation is offered by the Japanese Constitution, where Members of Congress are given the role of "representatives of the whole citizenry", which is understood to mean (1) "a congressman/woman is not a representative of each electoral zone, but a representative of all Japanese nationals" and (2) "he/she is not bound by voters in each electoral zone or by his/her supporters' organization." Ref. [9] at 35 (discussing Article 43(1)).

50 One of the obvious enhancements that is currently sought out via plastic surgery and cosmetics is to change one's age.

51 Should voters be required to pass a test for drug or alcohol impairment on the day they cast their vote? Screening for mental illness? If so, which types of mental illnesses? Depression? Who would decide and how would adminstration be executed in practice in a non-arbitrary and non-discriminatory manner?

humans who are seen as more animal than human, more plant than human, more machine than human or not even permanently embodied at all (or changing between such forms at will), should still have a right to vote.

In considering voting rights in a world of enhanced and non-enhanced humans, it is critical to strive to apply the rights and responsibilities of beings carefully, with due consideration of systematic injustice or harm that can result from stereotypes of from treating any beings or group of beings as "other" or "lesser." The way that a particular right is applied to a particular being in a particular context can and, in many cases, should vary based on the nature of the being but any variations should be equitable and grounded in respect for the fundamental autonomy of each being. Respect for "dignity" of humans (enhanced and non-enhanced) is a core part of the proposed extended ethical framework under consideration. We commit to the "rights of the others, commit to equality, dignity etc., with the insight that [equality, dignity, etc.] will have been dependent on that commitment."[52] However, that does not mean that some assessment of "dignity" of any individual human or a particular type of enhanced or non-enhanced human is a requirement for voting rights.[53]

3.4.2. Rejection of "Dignity" as a Voting Requirement

Moving beyond "rationality" or "reason", should "dignity" instead be a key ranking requirement? If so, what do we define as "dignity"? Who defines it? Why should we assume in advance that animals do not have any dignity or that if they do their dignity is less than that of a human? Or, in the context of human enhancements, why would we predict that there is some threshold at which an enhanced human is "more animal than human"[54] or "more machine than human" and then are no longer humans at all?[55] Who would determine when such a threshold was reached and how would they do so?

The definition of "dignity" should not be based on physical abilities or it is just an extension of bipedalism. A human being who is in a wheelchair and needs assistance to go to the bathroom, to dress and undress themselves, and to feed themselves is not less "dignified" than another human who can engage in such activities without obvious assistance. As Taylor convincingly explains, the "independence" and self-sufficiency of those viewed as "able-bodied" is a myth [3]. Interconnectedness and mutual dependency on other beings (humans and animals) is the true norm. Even people who are not viewed as "disabled" need assistance from others to engage in all of these activities. Unless someone is living "off the grid" and hunting for their own food with a weapon they made themselves, skinning it with a knife they made themselves, cooking it on a fire they made with wood they gathered, making their own fabric and clothes, and building and maintaining their own sewer system, they are benefiting from the extensive help of many other humans on a daily basis (and they are highly and obviously dependent on animals, plants and nature). Often the others upon whom we depend are an invisible population of those viewed as "lesser"—such as immigrant agricultural workers in the United States and the Dalits in India who perform the dirtiest jobs such as cleaning sewers and taking away dead animals [97]. Are they less "dignified" in a ranking or hierarchy? If so, is that because of something innate or inherent in the world or because the societal choices have been made systematically for hundreds of years, both intentionally and unintentionally (by people with the privilege of being allowed to be oblivious to the needs of others who are less powerful), to treat such people as "lesser" and to make their daily life more difficult than necessary, such as by designing curbs to be inaccessible to wheelchairs versus sloped, or by

---

[52] Ref. [1] at 13.

[53] As discussed in Section 3 below, representative voting is an available option, such as for a human that is mentally incapacitated to such a severe degree that they are incapable of expressing their wishes, or is unconscious (for example in a coma).

[54] This can be especially challenging when humans already have substantial overlap with animals, as outlined by Professor Jacques Derrida [95].

[55] For example, Professor Francis Fukuyama expresses concern that one of the potential effects of biomedical advances is that humans may be altered beyond recognition and this could have negative effects on the belief that human beings are equal by nature [96].

preferring stairs versus ramps in building designs? These choices were not necessary for any reasons related to the "natural" world.

Should "dignity" instead be based on caring for others and having compassion? If that is the case, many animals have demonstrated that they care deeply and intensely for others, including members of their own species and of other species. For anyone who has never seen a pet mourning the loss of an owner or companion, there are a multitude of examples of animals showing compassion and care for others, including, as just a few examples, dogs comforting humans who are struggling physically and/or emotionally, elephants visiting the home of a conservationist who had worked to help them after his death,[56] or Koko the gorilla who could communicate with humans through sign language and grieved when her companion cat died. "[W]hen [her companion kitten] died in 1985 after being struck by a car, Koko pretended she didn't hear her handlers for about 10 min after they told her the news. Then, she started whimpering. She signed 'sleep cat' by folding her hands and placing them by the side of her head. Researchers gave her a stuffed animal, but she wouldn't play with it and kept signing: 'Sad'" [99]. Recognizing that animals do have compassion and caring for others does not mean we should assume that they are the same as enhanced or non-enhanced humans.[57] Such a view would be inconsistent with the ethical framework of responding to the appeal of the "other" because it would deny that there was anything new or different that the "other" might tell us or include in their appeal. Instead, recognizing that animals have "dignity" (if that is defined as caring or compassion for others, for example) means that it is not ethical to deny the right to vote to enhanced humans even if they are considered to be "more animal than human" or to deny the right to vote to non-enhanced (or less-enhanced) humans if, in the future, humans with extreme enhancements considered such beings to be "sub-human."

Article 1 of the UDHR sets out the overarching principle that "All human beings are born free and equal in dignity and rights. They are endowed with reason and conscience and should act towards one another in a spirit of brotherhood." This should most properly be read as a normative or prescriptive, aspirational statement rather than a reflection of some objective reality. It is difficult to imagine any country or even any city or town in the world in 1948 where all humans were actually born with equal rights. The drafters of the UDHR were certainly aware of differences in rights of humans. We have seen from the examples in Section 2 above that pretending that every human is, in practice, equal (and is treated equally) in voting rights and ability to vote is to deny the existence of racism, classism, sexism, ageism, caste systems etc. Voting rights that are granted and designed in denial of the impacts of systemic racism, for example, will fail in practice to achieve anything close to "equal" votes.

In extending the right to vote in a world of enhanced and non-enhanced humans, we have modified Article 21(3) to remove the term "people" and replace it with "beings." For someone who views humans as intrinsically higher than all "others", or for whom only humans should be ethically entitled to vote, then they *might* be willing to accept this expansion if the term "beings" is defined to require, in the context of human enhancements, only that the candidate for enfranchisement was at some point in time a non-enhanced human or descended from a non-enhanced human in order to qualify as a "being". Or they might reject the change altogether and insist that the right be limited to enhanced humans who are still genetically predominantly *Homo Sapien* based on their DNA. In other words, more human than anything else. However, this is not a truly satisfactory solution in the extended ethical framework of engagement. Such a distinction seems uncomfortably close to gender-based and ancestry-based biases such as a law prohibiting children from inheriting citizenship (and voting rights) from their mothers. This was the law in Iran until 2019 and is still the law around the world, where "25 countries bar women from

---

56   "Nonhuman animals are amazing beings. Daily we're learning more and more about their fascinating cognitive abilities, emotional capacities and moral lives" [98].

57   For an additional approach, see the work of Professor Lori Gruen [100].

passing citizenship to children, while more than 50 have other discriminatory nationality laws such as those that, for example, permit only men to pass on citizenship to a foreign spouse," according to the United Nations High Commissioner for Refugees [101]. It also seems too much like a species-biased[58] (and organic/biologically-skewed) counterpart of the so-called "Grandfather Clauses" used to deny the right to vote to former slaves and their descendants. As mentioned in Section 2 above, voting restrictions based on ancestry are often explicitly designed and enforced to disenfranchise those deemed "lesser" or "other" and privilege those who were already "grandfathered" in as voters. "The phrase 'grandfather clause' originally referred to provisions adopted by some states after the Civil War in an effort to disenfranchise African-American voters by requiring voters to pass literacy tests or meet other significant qualifications, *while exempting from such requirements those who were descendants of men who were eligible to vote prior to 1867*" [30].

The intended definition of "beings" in the revised Article 21(3) should extend to animals, plants, inorganic intelligence (such as AI+) and even extraterrestrial "beings" (if any existed and for some reason wanted to vote) who never were human and who never descended from humans. The core minimum requirements for qualification as a "being" under Article 21(3) definitely warrant a deeper consideration if they are expanded as intended to respond to the appeal of the "other" (defined broadly). However, for purposes of discussion as an initial proposal, the minimal requirements for voting could be "life" or "consciousness." If a being is *either* "living", even if unconscious or largely plant-based, or "conscious," including animals or an AI+ where the being might have no organic embodiment and could, for example, be sentient and machine-based, such a being would be entitled to a right to vote under our extended ethical framework. However, due to lack of consciousness or lack of communications abilities, such voting might, in practice, need to occur via a representative as discussed in Section 3 below.

In the ethical framework of engagement and responding to the appeal of the other, there is no search for a fixed objective third-person reality where one would measure "dignity", "reason," or "conscience" in order to determine if the "other" had the right to be heard, and if the being who was attempting to respond had a responsibility to do so. The goal is to intentionally "see" those beings who are "other" and "hear" their voices as much as possible, recognizing that any being (enhanced or non-enhanced) will inevitably not fully understand the views and needs of other beings in the same category (because each individual being is unique), much less those in a different category, but must nevertheless try to do so as it is better than the alternative. The result will be flawed and imperfect, but it has the potential to be more legitimate, from an ethical perspective, than the other path where beings are ranked based on certain traits that are difficult to define, potentially false and/or artifactual, and that is particularly the case when the highest valued traits typically turn out to be traits the being doing the ranking possesses.

3.5. Potential Backlash against Humans Who Are Currently Disenfranchised

One important possible objection to the approach of universal and equal suffrage for all enhanced and non-enhanced humans is a concern that broadening the definition of "beings" who have rights (and to whom other beings have responsibilities) may, in practice, dilute or diminish the current rights of those without power (or with less power), such as minority groups. One of the arguments made by those who opposed giving women the right to vote in the United States was that women would, in practice, lose the benefits of their "exalted" positions if they were given the right to vote, and they would, as a result, get less protection.[59] The Nazis had strong animal protection laws and yet gave no such consideration to human beings they deemed as "lesser" or "other" [104]. Early Europeans

---

[58] As Professor Alice Crary proposes, "the plain fact of being human is morally significant but the plain fact of being an animal is so as well." Ref. [102] at 122.

[59] "To man, woman is the dearest creature on earth, and there is no extreme to which he would not go for his mother or sister. By keeping woman in her exalted position man can be induced to do more for her than he could by having her mix up in affairs that will cause him to lose respect and regard for her [103]."

used systems of classification to justify colonialism by "highlighting the human character of apes while emphasizing the purported simian qualities of Africans."[60] In Canada, male members of certain indigenous peoples who were deemed "civilized" (in part by virtue of owning land but also, depending on the time period and the law, by passing a test showing they could read and write in English) were given the right to vote in 1886 in pursuit of an express goal of assimilation. This was opposed by certain tribal leaders, who correctly predicted that it would lead to increasing political division within their own nations, on political party lines, and ultimately a loss of sovereignty and further loss of land. In just over a decade, "as the franchise expanded, settler politicians expressed the growing popular racism that presumed Indians could never be civilized—at least not without the radical process of removing children to assimilate them at infancy into white society."[61] As a result, in 1898 the right to vote was revoked for all "persons of Indian blood" (which was defined even more broadly than the definition of "Indian" men under the Indian Act).[62] "Over the twelve years from the creation of the Indian franchise in 1886 to its revocation in 1898, the early Victorian dream of the transformation and assimilation of Indigenous peoples as respectable Christian subjects gave way to a cynical idea of segregation under the permanent regime of the Indian Act. (Only children would be excepted under a radical new program of forced separation and assimilation in residential schools.)"[63]

It is impossible to guarantee—or even predict—a particular result at a particular time in a particular country or region, or the world as a whole, if enhanced humans (as broadly defined as possible) are given the right to vote on an equal basis with non-enhanced humans and with other enhanced humans. While the ethical framework of responding to the appeal of the other is not a utilitarian analysis, harm to the other is of course important to consider. There is a real possibility that broad enfranchisement could, in a practical effect, lead to an objectively "bad" result of disadvantaged groups of humans, whether enhanced or non-enhanced, being further harmed—and in an extreme scenario depriving them of any right to vote at all (as happened in Canada in 1898) and to attempted genocide (as in the Holocaust). The fact that there were other considerations at play, such as the desire to appropriate land or other property, and that the grant of the franchise right was not the *sole* cause of the harm would not make the damage any less real. Thus, the goal of responding to the appeal of the other should mean giving such being (enhanced or non-enhanced) the opportunity to participate in the choice of whether or not they should have the right to vote and what that will mean, as well as in the very structure of government itself and what "voting" even means. Even if the intention of granting enfranchisement broadly to all types of enhanced and non-enhanced humans is good, if it would have a detrimental effect on a particular group that harm (and concern) should not be ignored. We will outline the application of a participatory engagement model to the design of voting and governance structure further below in Section 3.7. First, however, we will offer an additional example of representative voting that may be useful to include in such a process as a way to give a form of vote to those enhanced or non-enhanced humans who are not able to vote directly, and as a way to promote/amplify the voice of those who have been disenfranchised historically or who are otherwise less powerful than the dominant group (whether enhanced or non-enhanced humans).

3.6. Representative Voting Instead of—or in Addition to—Direct Voting

To provide the option of representative voting in a world of enhanced and non-enhanced humans, one possible way to revise the new draft of Article 21 is as follows:

---

[60] Ref. [3]. But it is unlikely that the *cause* of the Nazi mistreatment of other humans was that they respected animals. Similarly, the *cause* of the English colonial mistreatment of the African population was not that they respected apes. There were more complex factors at play, including greed. Among other things, the Nazis wanted what they saw as the wealth of the Jews and they took it. The English wanted the wealth of Africa—and they took it.

[61] Ref. [8] at 32.

[62] Ref. [8] at 34.

[63] Ref. [8] at 35.

> *"The will of all beings shall be the basis of the authority of government; this will shall be expressed through universal and equal suffrage, directly and/or through representatives."*

Most current governments with some form of democratic component rely on elected representatives to vote on particular laws or regulations. This is, in part, due to practical considerations. A governance concept of democracy "emerges from a particular imperative: there is paraphernalia required to make practical those decisions rendered by democracy. Accordingly, there emerges the modern distinction between governance with management, which is a tolerable version of the distinction between those with the ability to vote and those with the ability to dictate" [89]. In such models, it is important not to let ostensible considerations of efficiency obscure some elements of elitism that may be present in the representative model,[64] particularly where significant wealth and privilege in a particular society is needed to even have a chance to become a viable candidate for election and then to serve in that role, with all the costs that may entail, often including maintaining two residences, one in the home state or jurisdiction and one in the capital or other location where the legislative body meets. Similarly, it is important that the elected representatives remain responsive to their constituents. "Where citizens participate in the conduct of public affairs through freely chosen representatives, it is implicit in article 25 that those representatives do in fact exercise governmental power and that they are accountable through the electoral process for their exercise of that power" [106]. For purposes of the analysis of human enhancements and voting however, we will consider a different type of representative voting—representative voting for an elected representative where the votes themselves are cast by representatives of the individual voters.

Representative voting adds an additional layer between the individual "being" for whom the vote is cast by a representative of such individual and the actual elected governmental representative. Such individual being is not actually casting a direct vote for the elected representative. Another being, or group, instead casts the vote on behalf of the individual voter. An example of one representative voting model is the Electoral College system for electing the President in the United States. A small number of electors for each state actually cast the vote for the President that will be counted, and the electors who cast such votes (who are selected by their political party) is determined based on the political party that "won" the direct votes in the state—in other words it is (generally)[65] a "winner take all" system where the individual votes for President for the political party that did not receive a majority of the individual votes overall are disregarded in the final electoral vote total for the state. Although it is rare,[66] this structure means the President that wins the election based on the Electoral College votes could have had the minority of the direct individual votes as a whole for the country. The system is intended to be strictly representative (by political party). Indeed, the United States Supreme Court has unanimously rejected the argument that by requiring that electors "vote" and "by ballot", the Framers of the Constitution intended that the electors' votes would reflect their own judgments versus those of their political party.[67]

Another option for representative voting for a particular individual being who is not able to communicate effectively could include appointing a representative who can vote on behalf of such being. In the United States, such a representative often plays a role in court proceedings involving young children, such as a custody dispute or parenting plan in the event of a divorce, or in cases involving abuse or neglect of a child. There are two different

---

[64] John Stuart Mill offered a classic argument in favor of representative government [105].

[65] Maine and Nebraska are exceptions and allocate one elector per district.

[66] In the elections of 1824, 1876, 1888, 2000 and 2016 the Electoral College winners lost the popular vote. In 1824, no candidate received the majority of the electoral votes and the election was decided by the United States House of Representatives.

[67] The *Chiafalo v. Washington* U.S. Supreme Court decision related to the so-called "faithless electors" who were fined by the State of Washington for not voting according to their party affiliation in the 2016 Election, as required by state law. The Court upheld the fine [107]. However, even the Electoral College does not attempt to represent the will of all United States residents. For example, residents of permanently inhabited United States territories such as American Samoa, Guam, the Northern Mariana Islands, Puerto Rico and the United States Virgin Islands cannot vote (either directly or indirectly) for the President and Vice President of the United States.

models for this representative role—"guardian ad litem" and "attorney for the child", also called guardian for the child. In the "guardian ad litem" model, the guardian is appointed "ad litem", roughly meaning, appointed "for the lawsuit" or "for the action", and typically is required to represent "the best interest of the child" [108]. In such cases, considerations of the "best interest of the child," may include, for example and without limitation: "(1) The temperament and developmental needs of the child; (2) the capacity and the disposition of the parents to understand and meet the needs of the child; (3) any relevant and material information obtained from the child, including the informed preferences of the child; (4) the wishes of the child's parents as to custody; (13) the child's cultural background."[68] In the second model, the attorney for the child has a different role, as they are assigned to represent the child's wishes and advocate for their behalf. For example, in an abuse case the child may wish to stay with the abusive parent or may wish to at least have visits with such parent (including if the parent is incarcerated). The attorney for the child shares those wishes with the court and advocates for those wishes, even if the attorney personally thinks they are *not* in the best interests of the child.[69] In an ethical model of engagement that involves attempting to respond to the appeal of the "other," the "best interests" model is less preferred than the model where a voting representative advocates for the wishes of the particular being they represent.[70]

For children and voting (versus lawsuits), in theory the parent or legal guardian could represent the child and cast a vote on their behalf. That is attractive for efficiency purposes but it is not clear that the parent or guardian would represent the child's wishes versus what they thought was best for the child. For example, a parent may be less concerned about climate change and voting for a candidate who is a strong environmentalist than their child (or children).[71] And, moving beyond the example of children to consider the large scale needed for voting in a world of enhanced and non-enhanced human voters (and "beings" broadly defined), there may be too many beings who need representatives for each such being to have a dedicated voting representative appointed individually (at least absent future human enhancements that could make that possible). Thus, a "guardian ad litem" type of voter representative speaking for the best interests of their assigned beings when casting votes could be appointed for *groups* of beings. In that role, they might have a responsibility to speak only for their particular group members, roughly along the lines of the electors for a particular political party in a state, as in the United States Electoral College framework. Alternatively, a framework could be used that was

---

68 Connecticut General Statutes Section 46b–54(f) includes numerous factors, many of which are specific to a custody dispute or an abuse or neglect proceeding, but the general structure of the provision may be helpful to see an example framework that could be applied in voting: "[Sec. 46b–54](f). When recommending the entry of any order as provided in subsections (a) and (b) of §46b–56, counsel or a guardian ad litem for the minor child shall consider the best interests of the child, and in doing so shall *consider, but not be limited to, one or more of the following factors*: (1) The temperament and developmental needs of the child; (2) the capacity and the disposition of the parents to understand and meet the needs of the child; (3) any relevant and material information obtained from the child, including the informed preferences of the child; (4) the wishes of the child's parents as to custody; … (13) the child's cultural background; Counsel or a guardian ad litem for the minor child *shall not be required to assign any weight to any of the factors considered.*" (emphasis added) [109]

69 Ultimately, the court determines what is in the best interests of the child, in both the guardian ad litem model and in the attorney for the child model. And, in the approach taken in the Connecticut statute, the attorney for the child should also consider the best interests of the child so the role is actually more nuanced than presented in the text above.

70 The guardian ad litem model and the advocate for the child model are still embedded in an "authorities framework" that "focuses too narrowly on state and parental control over children, reducing children's interests to those of dependency and the attainment of autonomy" [108]. "In place of this limited focus, we envision a "new law of the child" that promotes a broader range of children's present and future interests, including children's interests in parental relationships and nonparental relationships with children and other adults; exposure to new ideas; expressions of identity; personal integrity and privacy; and participation in civic life. Once articulated, these broader interests lay the foundation for a radical reconceptualization of the field of children and law. We propose a new tripartite framework of relationships, responsibilities, and rights that aims to transform how law treats children and their interactions with others" [110]. Based on the discussion in Section 3.7 below, a fresh look and potentially a new ratification of the governance framework itself would be appropriate where newly enfranchised voting beings such as children were not able to participate in the design of the original governance structure and what voting means.

71 "The UN Convention on the Rights of the Child (CRC) articulates children's rights to be heard and to participate in decisions that affect them. In spite of widespread ratification of the CRC, and the recognition of youth as a major stakeholder group with the right to participate in conversations related to climate change, children have not been adequately included in climate change decision-making at a global level, and attempts to facilitate their participation have rarely been more than tokenistic." Ref. [110] at 104–114.

similar to the Japanese Constitutional approach in which Members of Congress are given the role of "representatives of the whole citizenry", which is understood to mean (1) "a congressman/woman is not a representative of each electoral zone, but a representative of all Japanese nationals" and (2) "he/she is not bound by voters in each electoral zone or by his/her supporters' organization."[72]

Representative voting can be used to dull the impact of a particular group of humans (enhanced or non-enhanced) by giving them less of a proportional vote, and having a representative "speak" for another "being" inherently increases the risk of errors (intentional or unintentional) in even understanding what the original being wants. Thus, representative voting should be used with caution. For example, in a world of enhanced humans with extreme enhancements, such as the ability to clone/duplicate themselves at will (whether biologically, digitally or some combination thereof), representative voting might be an attractive approach proposed by other humans to avoid having such enhanced humans dominate the voting process by sheer numbers, and thereby deny effective participation to other humans. Before jumping to that "solution" to a hypothetical problem however, it would be important to consider the circumstances more carefully (and to get the views of the enhanced humans with such cloning/duplication abilities before making any ultimate decision, as will be discussed in Section 3.7 below).

It is risky to make assumptions about what an enhanced human would do or want. More importantly, it is contrary to an ethical framework of engagement that involves trying to respond to the appeal of the other. "[T]his right of the other, which should give the ultimate grounding and measure of every right, is exactly one that will always remain 'haunted' by the appeal of excessive immeasurable terms (and this always includes the danger to treat the other wrongly, especially in his otherness)."[73] Can we be confident that an enhanced human who had the ability to clone/duplicate themselves at will would do so in order to impact voting in a particular election or elections? They would be faced with the consequences of having a large number of clones/duplicates. Presumably, if each clone/duplicate was a being with rights, the clones/duplicates could not just be terminated/deleted at the will of the "original". The original source of the clones/duplicates would need to coexist with the others for the remainder of its lifetime (which could be quite extended). Moreover, even if the clones/duplicates were created with the exact memories and mind of the adult original source, over time they could presumably diverge individually in their wishes and views if they had different life experiences. A single individual non-enhanced human may change their views (and their voting patterns) over time. Should we not expect that the same might be possible for clones/duplicates of an enhanced human? Alternatively, if the clones/duplicates were all under the total control of a single mind or consciousness and did not have free will, then it might be more accurate to treat the entire group as a single voting being with a single direct vote. In that case, representative voting would not be appropriate.

Although there are dangers in using representative voting as a substitute for direct voting, it can also be beneficial for the less powerful. For example, representative voting can be used as a way to amplify the voice of those beings (whether enhanced or non-enhanced) who are in a minority or who are otherwise disadvantaged in their ability to participate fully and effectively in governance due to historical factors. A combination of representative and direct voting can also be employed where each individual member of a group (such as a particular political party, religion, culture, indigenous group, or caste under current examples—and potentially unlimited future groupings in a world with extreme human enhancements) wants the group as a whole to have a voice in addition to the voice of the individual member, who may differ from the group as a whole in a particular voting situation.

---

[72] Ref. [9] at 35 (discussing Article 43(1) of the Japanese Constitution).

[73] Ref. [1] at 15.

As a first step in considering whether representative voting is appropriate for a particular group, we must recognize and resist the temptation to declare that a group is "unable" to vote when the reality is that the group in power simply does not wish to make accommodations needed to allow such a group to vote. This has been the case for humans who were not able to physically have access to a polling location, who were not able to read or mark the ballots (or even register to vote), due to being physically different, in abilities, than those humans with political power, which could also be due to age, such as if they are very young or very old, or to a lack of formal education. In an ethical framework of engagement, it is essential to always question whether access to direct voting is being denied or limited for "practical" reasons that are only choices of convenience for the humans (enhanced or non-enhanced) in power. If that is the case, representative voting can be, at best, a provisional approach while the access issues are addressed. Representative voting in lieu of direct voting should never be assumed to be an adequate or equitable response in such a situation.

In addition, there is an inherent danger of stereotyping or assuming that a particular group will (or should) vote in a certain manner when in fact there may be many individuals or subgroups who may have very different views. This raises questions of identity and representation.[74] This can be seen for example in the construction of a new constitution when India became independent from Britain. "The mechanics of representation was a serious problem facing Indian democratization, and it proved to be most challenging in the cases of Muslims and the lower castes."[75] The colonial state mediated citizenship through community affiliation and therefore "embraced a static vision of participation where the interests of individuals were established in advance."[76] This resulted in a failure "to provide for a fair and enduring solution to the problem of group diversity." It was a denial of agency, that "necessarily proceeded on the assumption that individuals within certain groups would act collectively and in specific ways."[77] India's founders therefore rejected group-based or communal representation drawn on religious lines. However, the situation was different for caste—and particularly lower caste groups. "At India's founding, the Constitution permitted reserved quotas for such groups,"[78] which was not on its face consistent with the rejection of communal representation for religious groups. However, a theory that has been offered to reconcile the two approaches is to view communal representation as "suitable for societies where one was a subject rather than an agent."[79] As Dr. B. R. Ambedkar, chair of the drafting committee for the new Constitution, argued, the caste system was artificially imposed to create a division of laborers into compartments.[80] Thus, membership in a caste was different from a religious group in which members presumably chose to self-identify and were (at least theoretically) free to leave by their own choice when they reached an age of adulthood. Moreover, due to historical discrimination, lower-caste groups had been effectively excluded from voting. Reserved quotas were presented as a temporary option to facilitate the transition to universal and effective franchise, and communal electorates could facilitate "a new cycle of participation in which the representatives of various castes who were erstwhile isolated and therefore antisocial will be thrown into an associated life."[81] In this model,

---

74 "As articulated in international human rights instruments, the right to effective participation can be exercised by both individuals and collectives. While collective bargaining usually increases the strength of indigenous voices in negotiations and provides for more effective participation, it is also necessary to take into account the claims of each individual belonging to an indigenous group, as dissenting views cannot always be adequately represented in collective claims." Ref. [9] at 15.

75 Ref. [111] at 111.

76 Ref. [111] at 111.

77 Ref. [111] at 138.

78 Ref. [111] at 141.

79 Ref. [111] at 142.

80 Ref. [111] at 142—citing the writings of B.R. Ambedkar to advocate for the lower caste groups, such as the Dalits.

81 Ref. [111] at 145 (quoting Ambedkar).

a system of joint electorates combined with reserved seats might offer a path toward the extinction of caste while recognizing the current and near-term reality of the dominance of the higher castes, and allowing for individual self-determination by all through universal adult suffrage.[82]

Another improper path of exclusion that could occur in a scenario of representative voting is by making the ballot itself (and voter registration and information materials) only in a particular language or languages, as discussed above, or only in a certain media (such as print when the method of communication of the "other" beings were through sound, light, or gestures). Where an enhancement was so extreme that non-enhanced humans could not (yet) meaningfully communicate with the enhanced humans, representative voting might be appropriate for one—or both—groups until direct voting was possible (and even then representative voting might be appropriate as an addition to direct voting). Plants and animals provide an example of a situation where broad and effective cross-species communication is not currently available. However, there are strong indications that animals do communicate with each other through language and in a complex social structure [112–115]. Plants may communicate also [116,117]. If that is the case, human enhancements could facilitate such cross-species communication but could also make it difficult for the enhanced human to communicate with other humans who did not have the same enhancement. This problem would be especially acute if the enhancement was not reversible. For example, it is at least theoretically possible that an enhanced human could be physically modified to be proficient in communicating with a particular type of bird or octopus (or digitally embedded in a machine learning/artificial intelligence network), but then could no longer communicate with humans who lacked similar enhancements.

Examples of representative voting that could be applied where direct voting was not possible due to basic communication challenges between enhanced humans, non-enhanced humans and other beings include, for example, the "Party for the Animals". "On 22 November 2006 the Dutch Party for the Animals was elected to the House of Representatives with 2 of the 150 seats. A worldwide first! On 5 [M]arch 2017 the Party for the Animals won 5 seats, an increase of 150 per cent. The party now has 80 elected representatives at European, national, regional and local level" [118]. As the Party explains: "The main driver of our party is to protect the interest of the weakest against the alleged right of the strongest. In all this, the animals are the most vulnerable and often come in last in a world focused on short-term interests. The mistreatment of animals and the destruction of their home take place on a larger scale than ever seen before, including factory farming, animal testing, and in nature itself. After the liberation of slaves and women, and giving rights to children, the next logical step is to take the interests of animals seriously."[83]

Another example is a city in Costa Rica that gave citizenship to bees, plants and trees [121]. "Now known as *Ciudad Dulce*"—Sweet City—Curridabat's urban planning has been reimagined around its non-human inhabitants. Green spaces are treated as infrastructure with accompanying ecosystem services that can be harnessed by local government and offered to residents. Geolocation mapping is used to target reforestation projects at elderly residents and children to ensure they benefit from air pollution removal and the cooling effects that the trees provide. The widespread planting of native species underscores a network of green spaces and biocorridors across the municipality, which are designed to ensure pollinators thrive." In this situation, the bees, plants and trees are not voting directly, but they are being recognized as "citizens" to whom the elected representatives must respond to "the appeal of the other", while also considering the well-being of human constituents as well. This is an expansion of the Japanese Constitutional model, where Members of Congress represent the whole citizenry"—meaning all Japanese

---

[82] Ref. [111] at 146–147.

[83] Ref. [118] Sue Donaldson and Will Kymlicka also offer an applied ethics political theory of relationships between animals and humans in *Zoopolis: A Political Theory of Animal Rights*. [119]. The pioneering work of Professor Donna J. Haraway in this area also merits deeper consideration [120].

nationals, not just their direct constituents or political party.[84] In a world of enhanced and non-enhanced humans, such a scenario of respect and consideration for the well-being of those deemed "lesser" or "other"—or even "non-human"—is preferrable to the pet or slave options offered by Warwick as a prompt for debate.[85] It is also more consistent with the extended ethical framework of engagement applied herein.

Human enhancement is here—now—and may increase in the future. When it comes to a core right to vote, and responsibility to afford others that right and consider their wishes and needs, we should question whether it is ever ethical to treat living or conscious "others" as captives, slaves or pets, and if so, under what circumstances. We should reject— as likely unethical under an extended version of the framework suggested by Loidolt—a model that in intent or effect advantages enhanced humans—even superhumans—over non-enhanced humans. Although they will not be covered in the manner they deserve at this time, potential approaches from one or more "posthumanist" perspectives also offer exciting alternatives. For example, "Posthumanism offers a generative critique of the subject/object dualism not by prioritizing one instead of the other, or assimilating one to the other, but by embracing both relationally, as intra-connected actants in an open and respondent context, which is also constantly shifting."[86]

For enhanced or non-enhanced humans who are not legally entitled to vote directly, or unable to vote directly using current voting methods, alternative systems of representation may be needed with the express recognition that they are inherently inadequate. In order to mitigate systematic discrimination against and/or by the enhanced humans of the future, it may be appropriate to reconsider and adjust the voting system. "Voting" in such a situation, should not be viewed just a right, but also as a responsibility toward others who may not be able to vote (or whose votes may "count" less). In a fair negotiation forum with a power imbalance (such as for a minority group), "the right to effective participation implies a duty on the part of members of the majority which is more onerous than that of minority participants."[87] The more powerful group members have a greater responsibility to make efforts to listen to, learn from and respectfully understand (with the least possible prejudice) the interests and demands of the less powerful group in order to try to achieve equality. As Loidolt has persuasively articulated, the more powerful will still likely fail to achieve true equality due to the inherent impossibility of fully understanding the appeal of the "other", but the process itself will have greater legitimacy than if the greater burden is placed on the less powerful to articulate their interests in a manner the more powerful will respect and honor.

Humans are not just the bearers of rights, but also beings that seek to adjudge and constitute "right" [1]. Such a framework suggests that less emphasis should be placed on artifacts of physical appearance, mental capacity, geographical location[88], birthplace, ancestry (whether "natural" or "enhanced") and age of human bodies, or even species and organic versus inorganic or hybrid life forms. In addition, more weight should be allotted to the idea that all living beings are coexisting on a planet that has limited resources and that therefore all living beings may have responsibilities toward each other, regardless of

---

[84] Ref. [9] at 35.

[85] For example, "one new and emerging approach is the use of legal personality to protect water systems in law through the granting of legal rights to rivers" [122]. By contrast, Professor Sandra Seubert contends that the "animal-as-citizen approach overestimates the potentials of human-animal communication and underestimates the abuse of power asymmetries. Humans can (and should) take (what *they* interpret as) animals' interests into account but they cannot deliberate about interest with animals on an equitable basis." Ref. [123] at 63–69.

[86] Ref. [124] at 165.

[87] Ref. [9] at 22-23.

[88] Legal or practical disenfranchisement due to geographical location can occur because of lack of a permanent residence address due to poverty (such as in Japan, where homeless people are legally disenfranchised if they are not on the "resident register" [125] or due to membership in a nomadic tribe (such as for the Kuchis in Afghanistan [126]. There are also millions of "stateless" people who are not only disenfranchised but struggle in many other areas as well, including travel restrictions. "There are at least 4.2 million stateless people in the 79 countries that report them, but the U.N. agency believes that to be a severe undercount and that the problem affects many millions more" [127]. "Statelessness arises from a variety of situations, including redrawn borders, discriminatory laws that prevent women from passing on their nationality to a child, births that go unregistered, or the mass expulsion of an ethnic group" [127].

what they look like or how their minds work. If so, under a new Declaration of the Rights and Responsibilities of Beings, the legal (and practical) right to vote directly, or through indirect representation where such voting is not (yet) possible, could be distributed as broadly as possible, and voters would be ethically required to consider not only their own interests but also their responsibility toward other beings, particularly those who cannot vote or whose votes "count" for less.

3.7. Participation of the "Other" in the Structure, Process and Meaning of Voting and Governance

> *"it has been said that democracy is the worst form of Government except for all those other forms that have been tried from time to time."*[89]

In the context of "voting," giving a voice to the "other" in a world of enhanced and non-enhanced humans, defined as broadly as possible, including those with extreme enhancements and those with no enhancements, requires re-examining the basic framework of government with the benefit of inclusive participation. In an ethical theory of engagement, legitimization and justification is essential in a meaningful world. We must "face the appeal of the other" and determine what is "right" "from a phenomenological first-person perspective instead of the classical 'objective' third-person perspective of reciprocity."[90] If the process of constituting the world is a priori occurring in a context of legitimizing intentionality, and entire groups of enhanced and non-enhanced humans are (intentionally or unintentionally) excluded from participating in the very design of the governmental structure, then, in the ethical framework outlined by Loidolt,[91] the resulting structure would not be "right" even if the end result was the same as what would be established with the benefit of their voices and judgment. Only the *means* can justify the *ends*—not vice versa. This is particularly true because, in advance, we recognize that "this right of the other, which should give the ultimate grounding and measure of every right, is exactly one that will always remain 'haunted' by the appeal of excessive immeasurable terms (and this always includes the danger to treat the other wrongly, especially in his otherness)."

"The foundation of democracy is to be seen in the human being pressing into the future and the inherent comportment towards future circumstances that are different from those of the present. In other words, it is to be found in the pre-rational, bodying along of ourselves with our distinctive way of being. Human beings have always been communal and they have always displayed a range of involvements with others of their kind. Democracy itself has its ground in one of these involvements. Probably, the involvements are those which couple the manifest tensions around decision making in emergent communities with the aspirations of individuals who seek to produce something that is beyond their own competence."[92] Thus, any governmental structure and process of voting that is put in place should not only involve as many voices as possible (including both enhanced and non-enhanced humans for purposes of the present focus on human enhancements) but also must recognize it is inherently imperfect and provisional—and subject to a fresh re-examination and ratification process if it is later determined that groups were excluded from participation or new groups (or types of beings) come into existence or are recognized/discovered. There has never been an easy answer to questions about how a democratic form of government with diverse participation should be structured and operate, and it is likely to become even more difficult as we have more diverse types of

---

[89] Ref. [128] at 574.

[90] Ref. [1] at 3.

[91] Ref. [1] at 7.

[92] Ref. [89]. The importance of participation by the less powerful is emphasized by Kawashima in the context of indigenous peoples. Drawing from the Inter-American Commission on Human Rights, Organization of American States, Proposed American Declaration on the Rights of Indigenous Peoples (1997), (OAE/Ser/L/V/11.95), she notes a key point that "indigenous peoples have a right to participate not only in the implementation and evaluation of policies, plans and programmes for national or regional development, but also in their *formulation*." Ref. [9] at 31. In addition, "indigenous peoples have a right to maintain and develop their own indigenous decision-making institutions and to select their representatives according to their own rules and customs." Id.

humans, including those with extreme enhancements. Yet we must face such challenges, even knowing our response will be flawed.

Based on these considerations, our working draft Article could be modified as follows:

"The will of all beings, and responsibility toward all beings (whether voting or non-voting), shall be the basis of the authority of government; this will, and responsibility, shall be expressed through universal and equal suffrage (directly and/or through representatives) or through such other process and method as may be developed and ratified with the universal and equitable participation of such beings."

In this theory of engagement, an inclusive process is not just for the benefit of the "other" or "lesser" participants. It is also essential for the powerful if they desire to create an ethical form of government in a diverse world, even if the new form of government is admittedly imperfect. The right to "vote" may not be sufficient, or may even be harmful, if the very structure itself is flawed. For example, the United States (and Canadian) governments were built on a foundation that intentionally benefitted wealthy white male property owners and intentionally disadvantaged indigenous peoples and white women. In part due to the challenge of "catching up" after decades of disenfranchisement, the United States federal government still hews more closely to the composition that the founders intended than the composition of the country as a whole. In 2020, of the 100 Senators, only 26 were women, three were Black, three were of Asian ethnicity (one of whom was also Black), four were Hispanic, and none were indigenous [129]. Only three Native Americans have served in the United States Senate in the history of its existence. The fact that the United States Senate official website only selected four categories to measure diversity also is an indicator of a lack of diversity. Canada also continues to try to meet the challenges of a legacy of discrimination. As Coel Kirby noted in 2018, "In the past three years the federal government has passed two-fifths of all Indigenous-related legislation since confederation. These changes implement a particular and contested constitutional vision of Canada as a single multicultural society comprised of two founding European nations and many less equal Indigenous nations."[93] In considering alternative visions, we can see "there were and are a multiplicity of ways of constituting our collective selves in common."[94]

3.7.1. The Importance of Broad Engagement and Participation for Legitimacy

To address the challenges of a world of human enhancements in the ethical framework proposed, it is crucial to listen to alternative viewpoints. In ignoring the voices of different tribal leaders in Canada and unilaterally imposing (and then revoking) Indian enfranchisement, multiple "alternative visions of constitutional co-existence within confederation" were lost. "The Anishinaabe-dominated [c]ouncil proposed various forms of dual-nationality to reconcile membership in both their treaty-recognized nations and the Canadian state. In contrast, [another council with Haudenosaunee leadership] insisted on Haudenosaunee autonomy mediated by a special treaty relationship with the Canadian and imperial governments."[95]

Comparing the situation in Canada with that of India's founders, a profound difference was the leadership and participation of Dr. B. R. Ambedkar. Dr. Ambedkar was born a member of the Mahar caste, one of the lowest social groups in India—the Dalits, sometimes offensively referred to as the "untouchables". Nevertheless, he succeeded in earning his Ph.D. at Columbia University and a doctoral degree in economics from the London School of Economics. He was elected to chair the committee drafting India's constitution in 1947.[96] Likely as a result of his role and advocacy for the lower castes, the Constitution as adopted

---

93   Ref. [8] at 4.

94   Ref. [8] at 4. Also, "[f]ocusing on and placing weight on a fair negotiation forum (and the right to effective participation which guarantees it), which requires the mutual cooperation and support of participants, could be understood as an Asian way of thinking, as opposed to a Western way of thinking, stressing unilateral assertions of an individual's right or interest." Ref. [9] at 22, footnote 7.

95   Ref. [8] at 2.

96   For additional background on Dr. Ambedkar, see [130].

reflected (somewhat) more of Dr. Ambedkar's views, and not those of M.K. (Mahatma) Gandhi, who was born into a more privileged caste than that of Dr. Ambedkar.[97] Gandhi admired the caste system, although he thought there should be no *hierarchy* between castes. Dr. Ambedkar's response to Gandhi was "the outcaste is a byproduct of the caste system. There will be outcastes as long as there are castes. Nothing can emancipate the outcaste except the destruction of the caste system."[98]

The challenges faced by the Ainu indigenous minority in Japan provide an additional example of what is lost (in terms of legitimacy) when the powerful disregard the appeal of the "others" who are less powerful, and what can be gained when those voices are heard and respected. One of the questions raised when the assimilationist 1899 Hokkaido Ex-Aborigines Protection Act was finally repealed in 1997, was how the Ainu should be treated, including whether they should have special representation in local and national legislative assemblies as well as a standing consultative Ainu body (which were two of the six provisions drafted by the Ainu Association of Hokkaido for inclusion in the new law). However, likely due to the increasing restriction of the participation of Ainu in the higher-level policy making processes,[99] the only provision that was ultimately incorporated related to the promotion of Ainu culture.[100] One reason given for the rejection of the provision regarding representation was "the recognition of special electorates distinct from those of general citizens is very likely to violate the Constitution."[101] Assuming, hypothetically,[102] that the reason given was valid, then, in the participatory ethical framework being applied herein, the next question would be whether the Ainu had participated in the drafting and enactment of the Constitution that was being applied to deny them the right to special representation. If they had been able to meaningfully participate in such drafting, as was the case for the lower-caste and less powerful Dalits in the Indian Constitution where Dr. Ambedkar not only was able to participate, he was the elected chair of the drafting committee, its application to their rights and role would be legitimate. If they had not been able to participate meaningfully in the drafting of the Constitution and formulation of the framework of government, then they could reasonably demand a fresh review and ratification—with their participation—of the Constitution, which might warrant an entirely new framework or an amendment.

3.7.2. Selected Examples of Difficult Questions That Require Broad Engagement

In a world of enhanced and non-enhanced humans, including humans with extreme enhancements, it will be even more challenging to find an ethical approach that balances the right to vote with responsibilities toward other. Discarding any vision or alternative presented by a particular group out of hand would be itself a violation of the ethical obligation to respond to the appeal of the other. The questions raised when one becomes open to hearing the appeal of the other are typically not amenable to easy ethical solutions. Participation of diverse voices is essential for the structure enacted to be legitimate under

---

[97] Ref. [130] at 150-151.

[98] Ref. [130] at 26.

[99] "It is especially regrettable that the Experts Meeting [Concerning Ainu Affairs – established in 1995 in response to the request of the first and then sole Ainu member of the Japanese Congress] did not include any Ainu members. This was pointed out by a member of the Experts Meeting, who suggested that a new Meeting be established, including Ainu persons, to contribute to the drafting process after the issuance of the report by the Experts Meeting. Nevertheless, this proposal was never realized." Ref. [9] at 38–40.

[100] Ref. [9] at 32–33.

[101] Ref. [9] at 34 (quoting from Section 3 in Hokkaido Utari Mondai Konwa Kai Tosin (A Report of the Round Table on a Policy for the Ainu Culture), March 1998.

[102] Kawashima contends that the Japanese Constitution would, in fact, allow for special representation for the Ainu. Ref. [9] at 34–36.

our extended ethical theory of engagement.[103] As examples of the challenges involved, we will consider a few examples of the types of questions that would warrant careful review, with the benefit of broad engagement by enhanced and non-enhanced humans: (a) voting proportionality, (b) geo-political borders, and (c) temporality.

Reconsidering Voting Proportionality

How many "votes" should each being get, and should there be a difference based on country of citizenship, region or nationality—or based on whether the being is enhanced or non-enhanced (and the nature of the enhancement) or some other factors? In a direct model of "one being = one vote", an enhanced human might be able to clone themselves or, as noted above, they might distribute their consciousness in multiple human and/or silicon-based embodiments. If duplication of the identity and consciousness of an enhanced human was as simple (and unlimited) as making copies of digital works or computer programs, in theory such enhanced humans could "stack" the vote count in a manner that would not be viewed as fair to the other voters, whose votes could be rendered meaningless. A version of this problem has existed to some extent for every direct or representative democratic system of government with minorities whose votes would count "less" than the majority. It was the reason the U.S. Constitution established a hybrid approach (based on the House of Commons and House of Lords model of England) with the House of Representatives having a number of elected representatives for each state based roughly on the population of the state, and the Senate having a fixed number where each state had two Senators. But the model of rights and voting is a very particular model, and may not be the only appropriate model to consider, either historically or under our ethical framework of responding to the appeal of the other. We can also consider a model of responsibility toward others in voting, not just an individual (or communal) right to vote.

A group of humans operating as a hive mind or a single human with an extended consciousness that extended throughout a global network (both on land and in satellites) might well feel that communal voting was preferable for their well-being (and even for their very sanity), despite the fact that having one vote for each participant or "node" in the network might give them more votes. Operating by consensus might be tremendously valuable to them. This was the case in Canada during the period in the 1800s when Indians were briefly enfranchised before being disenfranchised. Certain tribal voters "were deeply split between the Conservative and Liberal parties that transformed the differences managed by chiefly consensus into factions competing in public electoral contests."[104] Thus, it is important not to make assumptions about what another person or group would want or would think is best. They need to be involved in such decisions, to the extent possible.

Reconsidering Political Geographical Borders as Determinants for Governance

While laudable and a tremendous advance for the time it was written, Article 21 is likely too narrow for a world of enhanced humans under the ethical framework we are applying. For one thing, it is very likely limited to citizens through the reference to "his country" and it unduly preferences votes of "people" in that country, which means humans with extreme enhancements likely would not qualify. They might not be citizens of any country. The issue of geographic borders played a major impact in the American

---

[103] The concept of "ancestry" itself, while frequently used as a proxy for race in discriminatory denial of enfranchisement as discussed above, may be legitimate when used by the group itself as a key cultural, social and/or religious requirement for membership in an indigenous group. See, for example, the struggle of the Hawaiian people or *Kānaka Maoli* to assert their sovereignty and right to self-determination, including the very right to define group membership at least in part based on ancestral terms. Ref. [131] at 2605 (the "OHA [Office of Hawaiian Affairs] constituted an attempt by the State of Hawaii to enable Kānaka Maoli self-determination. By rejecting this model, the [United States Supreme] Court in *Rice* demonstrated a troubling inability to understand indigenous self-governance as possible outside of federally recognized tribal governments—an oversight that continues to stifle indigenous self-governance in the U.S. territories to this day. Ultimately, as Hawaiian scholar Noelani Goodyear-Ka'ōpua writes, by invalidating Hawaiian-only voting for OHA trustees, *Rice* eliminated "the small measure of electoral control over resources Kānaka Maoli could collectively exercise within the settler state system.") (citations omitted).

[104] Ref. [8] at 35.

Revolution, as discussed above, where American Colonists felt they should be entitled to the right to vote in England as citizens. Such questions continue to be of concern.

For example, in 2018, the following question was submitted to the European Parliament, "Six EU Member States (Cyprus, Denmark, Germany, Ireland, Malta and the United Kingdom) deprive their nationals of the right to vote in national elections on account of residence abroad, on the basis of the assumption that expatriates are not affected by political decisions taken in their country of origin. These Member States also disenfranchise their nationals in European elections if they live permanently in a third country, and two of them (Ireland and the United Kingdom) even do so in respect of nationals who are resident in the EU. In some Member States (e.g., Spain and Portugal) there are insufficient or no facilities to vote from abroad, which is an obstacle to the exercise of the right to vote" [132].

Individual voting only by country or region on issues that could lead to mass extinction is problematic even in the current world because certain countries have a disproportionate impact on the global climate. In our extended ethical framework, responding to the appeal of the "other" requires consideration of the appeal of other beings regardless of where they are physically located. This is particularly evident in the case of extreme human enhancements that could allow an individual voter to "reside" in multiple countries at the same time or instantaneously change their geographic residence. We are facing catastrophic climate change impacts:

"'We're eroding the capabilities of the planet to maintain human life and life in general," said Gerardo Ceballos, an ecologist at the National Autonomous University of Mexico. The current rate of extinctions vastly exceeds those that would occur naturally. Scientists know of 543 species lost over the last 100 years, a tally that would normally take 10,000 years to accrue. 'In other words, every year over the last century we lost the same number of species typically lost in 100 years,' Dr. Ceballos said. If nothing changes, about 500 more terrestrial vertebrate species are likely to go extinct over the next two decades alone, bringing total losses equivalent to those that would have taken place naturally over 16,000 years" [133].

As it becomes clear that humans live in an increasingly interconnected and interdependent world, geographic borders could become largely irrelevant for votes that would impact global well-being (and potentially the very survival of humanity and other living beings).[105] As will be discussed below, in an ethical theory of engagement, the answers to the questions of who can vote and where they can vote, even if necessarily imperfect, require the participation of and discourse among all affected humans, enhanced and non-enhanced.

As a final example of difficult/disturbing questions that require a response in an extended ethical framework of engagement, we will consider the temporality and cadence of voting.

Reconsidering Temporality/Cadence of Voting and Ratification of the Framework of Government

The temporality of voting itself should also be subject to review and consideration by enhanced and non-enhanced humans. Voting annually, or after a variable number (n) years, could be ridiculous for an enhanced human who experienced time on a different scale, such as a human whose consciousness was embodied in an extended computer network. To such a human, a calendar year might be the subjective equivalent of millennia. If such humans were in the minority, the ethical theory of engagement being applied in this context would require that they have a voice in the design of the voting process itself and that other humans consider their needs and desires and not dismiss them. And if such humans became dominant in terms of wealth, numbers and political power, it would still be necessary to listen to and respect the wishes of humans without such

---

[105] An additional concern about using geographic borders to determine voting rights is that such borders too often reflect a legacy of violent appropriation of lands and divisions in which the dispossessed lacked a meaningful voice. Ref. [134] "'They make magic lines only they can see' noted a member of the Hupacasath First Nation of British Columbia, as colonial surveyors sliced up his ancestral lands into tidy parcels, a fraction of which would become the Hupacasath's reservation."

enhancements (whether due to poverty or self-determination that they did not wish to have such enhancements).

Looking to the future with hope that we have the opportunity to do better than the past, versus resignation that we are at or past the pinnacle of our achievements as the human race, we should periodically consider what form(s) of government are available for an inclusive and diverse population of enhanced and non-enhanced humans. We should take into account current and historical disparities and discrimination and future possibilities and what "voting" even means or should mean. Even if they were the best options available to date, which may or may not be true, existing approaches are inherently unreliable when we try to consider the diverse and potentially unlimited possibilities of a world of human enhancements—moderate to extreme—and current and future beings. The reason is that the approaches that were not designed with the input and participation of such diverse groups are fundamentally illegitimate under the extended ethical framework we are applying. They were largely designed by people of a certain wealth, gender, and religion in a particular country at a particular time. They may not represent the wide variety of perspectives that are needed in order to make self-determination meaningful in a design "from the ground up" versus "from the top down"—either currently or in the future, when there may be beings (enhanced humans or otherwise) who do not even exist today. The governed should be entitled to a voice in the design of the very system of governance, not just in its execution or in a retrofit—such as, in the United States example, by giving voting rights to male former slaves, then to women, then to indigenous peoples etc. while keeping the core (two-political party) system intact.[106] They may inherit an existing system of government as a practical matter but it should not be viewed as the best option available just by virtue of its age.[107] Thus, a foundational question that must be addressed in responding to the appeal of the "other" is when and how the very framework of government warrants re-examination and potentially a fresh ratification by the beings who are impacted by such framework. There is no easy answer to that question, but an answer that would be incorrect would be for the decision to be made only by one group of beings (whether enhanced humans or non-enhanced humans) without the input of the other.

The need to include a diverse group of "others" (whether enhanced or non-enhanced humans or other beings) in the design of government and what voting as a social and legal construct "means", is not a utilitarian argument or a prediction about what will result from diverse participation. The end result may not be "better" under some weighing method comparing costs and benefits. For one thing, any such weighing should be inherently suspect when the person or group doing the weighing does not represent the broader group because they tend to over-value their own interests and under-value or ignore the costs to others who are not participating in the weighing. But more fundamentally, the very premise of self-determination and enfranchisement means having the ability to make decisions that at the time or in hindsight will be seen as "bad" or "wrong." This is true at the individual level as well as at the broader citizen level.

We are responsible for the rights of others, but "at the same time we cannot point to evident legitimizing grounds—and if we do, we know that they are never enough for the ethical appeal that confronts us."[108] The "other" remains "radically impenetrable or inaccessible as other" and therefore "brings subjectivity into an anarchical and asymmetrical relation with his infinite and radical transcendence."[109] This will be true in a world of enhanced humans, as it is already true today. We deceive ourselves when we believe we

---

[106]   The two-party system in the United States has led to absurd extremes of partisan gerrymandering, such as the examples shown in *Rucho v. Common Cause*, in which the Supreme Court held involved "political questions" beyond the reach of the federal courts absent an equal protection (one person, one vote) violation or racial discrimination [135].

[107]   Dr. Ambedkar "thought of the [India] Constitution as a work in progress. Like Thomas Jefferson, he believed that unless every generation had the right to create a new constitution for itself, the earth would belong to 'the dead and not the living.'" Ref. [130] at 46 (citation omitted).

[108]   Ref. [1] at 18.

[109]   Ref. [1] at 19.

truly understand another human being (enhanced or non-enhanced) in the way that we understand ourselves.

4. Conclusions

"We're out to repair the future.

We are here for the storm that's storming because what's taken matters" [136]

Loidolt concludes that "this right of the other, which should give the ultimate grounding and measure of every right, is exactly one that will always remain 'haunted' by the appeal of excessive immeasurable terms (and this always includes the danger to treat the other wrongly, especially in his otherness). However, an ethics of human rights must not be paralyzed by such a situation. It must undertake the responsibility of an urgent judgment that proves its engagement by its openness for a universality to come."[110] In the context of extreme human enhancements, we will respond to the challenge of an ethics of the rights of living beings and of conscious beings—whether "still human" or "beyond human". Loidolt's "ethics of discourse" is offered as a commitment to "a critique and the ongoing process of legitimization (which would be a strategy to cope with historical and cultural relativism)" thereby opening a horizon, "where responding to the ethical appeal of the other becomes conceivable as an attitude of commitment which resists the totality of having everything at [one's] disposal."[111]

As the UDHR was built out of the ashes of World War II and the mass genocide whose roots could be traced to centuries of persecution (including brutal attacks on minorities and Jews who were blamed for the plague epidemic of 1348–1351 (the "Black Death"), [137] it seems appropriate to consider if we can do even better than the drafters of the UDHR as we come out of a global pandemic, mass extinctions of species and harms to humans due to climate change, and as we seek to address (and redress) the consequences of centuries of systematic legal and social discrimination, and violence, against certain people due to their skin color, gender, sexual preferences, religion, ethnicity and/or social class or caste. All of this is occurring as technology and human enhancements are advancing at a rapid rate. Just as the UDHR of 1948 was a massive step forward from the American Declaration of Independence of 1776, which enfranchised white men only and left the abomination of slavery intact, we can, and must, do better than the UDHR in years to come.

Funding: This research received no external funding.

Acknowledgments: The author greatly appreciates the guidance provided by Woodrow Barfield, a leader in the world of human enhancements—both as a technical expert and as a futurist.

Conflicts of Interest: The author declares no conflict of interest.

References

1. Loidolt, S. Notes on an Ethics of Human Rights: From the Question of Commitment to a Phenomenological Theory of Reason and Back. *IWM Jr. Fellows Conf.* **2005**, *20*, 9–24.
2. United Nations General Assembly. *Universal Declaration of Human Rights*; Department of State, United States of America: Minneapolis, MN, USA, 1948.
3. Taylor, S. *Beasts of Burden: Animal and Disability Liberation*; New Press: New York, NY, USA, 2017.
4. Brodeur, M. For composer Molly Joyce, disability is no barrier to virtuosity. *Washington Post.* 13 December 2020. Available online: https://www.washingtonpost.com/entertainment/molly-joyce-cant-use-her-left-hand-its-made-her-a-virtuosically-unique-composer/2020/12/11/c429012a-3899-11eb-98c4-25dc9f4987e8_story.html (accessed on 9 January 2021).
5. Christopher, A.D.C. Skin Bleaching, Self-Hate, and Black Identity in Jamaica. *J. Black Stud.* **2003**, *33*, 711–718.
6. Associated Press. California sues Cisco alleging discrimination based on India's caste system. *Los Angeles Times.* 2 July 2020. Available online: https://www.latimes.com/business/story/2020-07-02/california-sues-cisco-bias-indian-caste-system (accessed on 9 January 2021).

---

110   Ref. [1] at 15.

111   Ref. [1] at 15.

7. Korff, J. Voting rights for Aboriginal people. *Creative Spirits*. 14 August 2020. Available online: https://www.creativespirits.info/aboriginalculture/selfdetermination/voting-rights-for-aboriginal-people (accessed on 9 January 2021).
8. Kirkby, C. Reconstituting Canada: The Enfranchisement and Disenfranchisement of 'Indians', c. 1837–1900. *Univ. Tor. Law J.* **2019**, *69*, 497–539. [CrossRef]
9. Kawashima, S. The Right to Effective Participation and the Ainu People. *Int. J. Minority Group Rights* **2004**, *11*, 21–74. [CrossRef]
10. Zhao, T.; Zhu, Y.; Tang, H.; Zie, R.; Zhu, J.; Zhang, J. Consciousness: New Concepts and Neural Networks. *Front. Cell. Neurosci.* **2019**, *13*, 302. [CrossRef] [PubMed]
11. *Citizens United v. Federal Election Comm'n*, 558 U.S. 310 (2010) (5-4 decision) (Supreme Court of the United States) (Opinion for the majority written by Justice Anthony M. Kennedy, joined by Chief Justice Roberts and Justices Scalia, Alito and Thomas). Available online: https://supreme.justia.com/cases/federal/us/558/310/ (accessed on 9 January 2021).
12. U.S. Defense Advanced Research Projects Agency. DARPA and the Brain Initiative. Available online: https://www.darpa.mil/program/our-research/darpa-and-the-brain-initiative (accessed on 9 January 2021).
13. Strickland, E. DARPA Wants Brain Interfaces for Able-Bodied Warfighters. In *IEEE Spectrum: Technology, Engineering, and Science News*; IEEE: New York, NY, USA, 10 September 2018.
14. Fontelo, P.; Hawkings, D. Ranking the Net Work of the 115th. *Roll Call*. 2018. Available online: https://www.rollcall.com/wealth-of-congress/ (accessed on 9 January 2021).
15. Cornell, S.; Leonard, G. *The Partisan Republic: Democracy, Exclusion and the Fall of the Founder's Constitution, 1780s–1830s*; Cambridge University Press: Cambridge, UK, 2019; p. 2.
16. *Scott v. Sanford*, 60 U.S. 393, 404-405 (1857) (the "*Dred Scott* decision") (United States Supreme Court) (Opinion for the majority written by Chief Justice Roger B. Taney) (Holding that people born as slaves did not change their status and were not entitled to freedom despite having been held as a slave in a "free" state and then in a state where slavery was illegal). Available online: https://supreme.justia.com/cases/federal/us/60/393/ (accessed on 9 January 2021).
17. Haag, M. Rachel Dolezal, Who Pretended to Be Black, Is Charged with Welfare Fraud (Published 2018). *The New York Times*. 25 May 2018. Available online: https://www.nytimes.com/2018/05/25/us/rachel-dolezal-welfare-fraud.html (accessed on 9 January 2021).
18. Cook-Martin, D.; FitzGerald, D.S. How Legacies of Racism Persist in U.S. Immigration. *Sch. Strategy Netw.* 2014. Available online: https://scholars.org/contribution/how-legacies-racism-persist-us-immigration-policy (accessed on 9 January 2021).
19. Bridges, K. *The Poverty of Privacy Rights*; Stanford University Press: Stanford, CA, USA, 2017; pp. 48–49.
20. *Project Citizenship Inc. v. Dep't of Homeland Security* (D. Mass. Civ. 1:20-cv-11545 filed 8/17/20) (lawsuit filed in the United States Federal Court for the District of Massachusetts on 17 August 2020). Available online: https://projectcitizenship.org/wp-content/uploads/2020/08/Project-Citizenship_Complaint_AsFiled.pdf (accessed on 9 January 2021).
21. Doherty, J. Naturalization Fee Hike Is 'A Wealth Test,' Says New Suit. *Law 360*. 18 August 2020. Available online: https://www.law360.com/articles/1302067/naturalization-fee-hike-is-a-wealth-test-says-new-suit (accessed on 9 January 2021).
22. Cahill, C.D.; Deer, S. In 1920, Native Women Sought the Vote. Here's What's Next. *The New York Times*. 31 July 2020. (Updated 19 August 2020). Available online: https://www.nytimes.com/2020/07/31/style/19th-amendment-native-womens-suffrage.html (accessed on 9 January 2021).
23. United States Holocaust Memorial Museum. Nuremberg Race Laws. Available online: https://encyclopedia.ushmm.org/content/en/article/nuremberg-laws (accessed on 9 January 2021).
24. Chou, V. How Science and Genetics are Reshaping the Race Debate of the 21st Century. *Science in the News*. 17 April 2017. Harvard University Graduate School of Arts and Sciences blog. Available online: http://sitn.hms.harvard.edu/flash/2017/science-genetics-reshaping-race-debate-21st-century/ (accessed on 9 January 2021).
25. Bridges, K. *Critical Race Theory: A Primer*; Foundation Press: St. Paul, MN, USA, 2019.
26. Wines, M. Illegal Voting Gets Texas Woman 8 Years in Prison, and Certain Deportation (Published 2017). *The New York Times*. 10 February 2017. Available online: https://www.nytimes.com/2017/02/10/us/illegal-voting-gets-texas-woman-8-years-in-prison-and-certain-deportation.html (accessed on 9 January 2021).
27. Gonzalez-Barrera, A. Mexican Lawful Immigrants Among the Least Likely to Become U.S. Citizens. *Pew Res. Cent.* 29 June 2017. Available online: https://www.pewresearch.org/hispanic/2017/06/29/mexican-lawful-immigrants-among-least-likely-to-become-u-s-citizens/ (accessed on 9 January 2021).
28. U.S. Citizenship and Immigration Service. Civics (History and Government) Questions for the Naturalization Test. (rev. January 2019). Available online: https://www.uscis.gov/sites/default/files/document/questions-and-answers/100q.pdf (accessed on 9 January 2020).
29. Onion, R. Take the Impossible "Literacy" Test Louisiana Gave Black Voters in the 1960s. *Slate*. 23 July 2013. Available online: https://slate.com/human-interest/2013/06/voting-rights-and-the-supreme-court-the-impossible-literacy-test-louisiana-used-to-give-black-voters.html (accessed on 9 January 2021).
30. *Comstock v. Zoning Board of Appeals of Cloucester*, Slip Op. 19-P-1163 at 8, fn. 11 (Opinion by Justice Milkey) (Supreme Judicial Court of Massachusetts—3 August 2020). Available online: https://www.massmunilaw.org/wp-content/uploads/2020/08/HENRY-W.-COMSTOCK-JR.-v.-ZBA-OF-GLOUCESTER.pdf (accessed on 9 January 2021).

31. Lorber, J. Paradoxes of Gender Redoux: Multiple Genders and the Persistence of the Binary. In *Gender Reckonings: New Social Theory and Research*; James, W.M., Patricia, Y.M., Michael, A.M., Raewyn, C., Eds.; New York University Press: New York, NY, USA, 2018; p. 298.

32. Jones, P. Roosters, Hawks and Dawgs: Toward an Inclusive, Embodied Eco/Feminist Psychology. *Fem. Psychol.* **2010**, *20*, 365–380. [CrossRef]

33. Singer, P. All Animals Are Equal. In *Animal Rights and Human Obligations*; Regan, T., Singer, P., Eds.; Prentice Hall: Englewood Cliffs, NJ, USA, 1989; pp. 148–162.

34. Jefferson, T. Declaration of Independence: A Transcription. Available online: https://www.archives.gov/founding-docs/declaration-transcript (accessed on 9 January 2021).

35. O'Neill, K.; Herman, J. The Potential Impact of Voter Identification Laws on Transgender Voters in the 2020 General Election. *UCLA School of Law Williams Inst.* 2020. Available online: https://escholarship.org/uc/item/1qx199j7 (accessed on 9 January 2021).

36. Hurt, E. Voter ID Laws May Disproportionately Affect Transgender Community. *NPR.* 29 October 2018. Available online: https://www.npr.org/2018/10/29/661676054/voter-id-laws-may-disproportionately-affect-transgender-community (accessed on 9 January 2021).

37. Denyer, S.; Gowen, A. There's Too Many Men: What Happens When Women Are Outnumbered on a Massive Scale. *The Washington Post.* 18 April 2018. Available online: https://www.washingtonpost.com/graphics/2018/world/too-many-men/ (accessed on 9 January 2021).

38. Hansen, C. 116th Congress by Party, Race, Gender, and Religion. *U.S. News.* 19 December 2019. Available online: https://www.usnews.com/news/politics/slideshows/116th-congress-by-party-race-gender-and-religion (accessed on 9 January 2021).

39. Breeden, A. City of Paris Fined Nearly $110,000 for Appointing Too Many Women. *The New York Times.* 22 December 2020. Available online: https://www.nytimes.com/2020/12/16/world/europe/paris-too-many-women-fine.html (accessed on 9 January 2021).

40. Erdbrink, T.; Sorensen, M.S. A Danish Children's TV Show Has This Message: Normal Bodies Look Like This. *The New York Times.* 7 October 2020. Available online: https://www.nytimes.com/2020/09/18/world/europe/denmark-children-nudity-sex-education.html?action=click&algo=als_engaged_control_desk_filter&block=editors_picks_recirc&fellback=false&imp_id=666102250&impression_id=88bb3550-fae0-11ea-8f6c-292671e378eb&index=0&pgtype=Article®ion=footer&req_id=28074295&surface=home-featured (accessed on 9 January 2021).

41. BBC—Science & Nature—Human Body and Mind—Obsessions—Gallery. Available online: https://www.bbc.co.uk/science/humanbody/mind/articles/disorders/gallery/gallery_case2.shtml (accessed on 9 January 2021).

42. Warwick, K. Superhuman Enhancements via Implants: Beyond the Human Mind. *Philosophies* **2020**, *5*, 14. [CrossRef]

43. Leckie, A. *Ancillary Justice*; Orbit Books Hachette Book Group: New York, NY, USA, 2013.

44. Coin, A.; Mulder, M.; Dubljević, V. Ethical Aspects of BCI Technology: What Is the State of the Art? *Philosophie* **2020**, *5*, 31. [CrossRef]

45. Asada, M. Artificial Pain May Induce Empathy, Morality, and Ethics in the Conscious Mind of Robots. *Philosophies* **2019**, *4*, 38. [CrossRef]

46. Fahn, C.W. Marketing the Prosthesis: Supercrip and Superhuman Narratives in Contemporary Cultural Representations. *Philosophies* **2020**, *5*, 11. [CrossRef]

47. Biggs, K. Augmented Reality Explorer Steve Mann Assaulted at Parisian McDonald's. *TechCrunch.* 16 July 2012. Available online: https://techcrunch.com/2012/07/16/augmented-reality-explorer-steve-mann-assaulted-at-parisian-mcdonalds/ (accessed on 9 January 2021).

48. Avery, D. Killings of transgender Americans reach all-time high, rights group says. *NBC News.* 7 October 2020. Available online: https://www.nbcnews.com/feature/nbc-out/killings-transgender-americans-reach-all-time-high-rights-group-says-n1242417 (accessed on 9 January 2021).

49. Violent hate crime against disabled has risen by 41 per cent in the last year, figures suggest. *Independent.* 9 October 2019. Available online: https://www.independent.co.uk/news/uk/home-news/disabled-hate-crime-rise-41-cent-last-year-leonard-cheshire-a9148301.html (accessed on 9 January 2021).

50. Samuel, S. Humans keep directing abuse—Even racism—at robots. *Vox.* 2 August 2019. Available online: https://www.vox.com/future-perfect/2019/8/2/20746236/ai-robot-empathy-ethics-racism-gender-bias (accessed on 9 January 2021).

51. U.S. Department of Justice, Civil Rights Division, Disability Rights Section. ADA CHECKLIST FOR POLLING PLACES (includes checklist from 2016). Available online: https://www.ada.gov/votingchecklist.htm (accessed on 9 January 2021).

52. Pendo, E. Blocked from the Ballot Box: People with Disabilities. *Am. Bar Assoc.* 6 June 2020. Available online: https://www.americanbar.org/groups/crsj/publications/human_rights_magazine_home/voting-in-2020/blocked-from-the-ballot-box/ (accessed on 9 January 2021).

53. Abrams, A. Absentee Ballot Applications Are Not Accessible to Voters with Disabilities in 43 States. *TIME.* 30 September 2020. Available online: https://time.com/5894405/election-2020-absentee-ballot-applications-disability-rights/ (accessed on 9 January 2021).

54. Phillip, A. A paralyzed woman flew an F-35 simulator—Using only her mind. *The Washington Post.* 3 March 2015. Available online: https://www.washingtonpost.com/news/speaking-of-science/wp/2015/03/03/a-paralyzed-woman-flew-a-f-35-fighter-jet-in-a-simulator-using-only-her-mind/ (accessed on 9 January 2021).

55. Neuralink: Elon Musk unveils pig with chip in its brain. *BBC News*. 29 August 2020. Available online: https://www.bbc.com/news/world-us-canada-53956683 (accessed on 9 January 2021).
56. Virginia, C. Innovators under 35: Visionaries. *MIT Technology Review*. 2020. Available online: https://www.technologyreview.com/innovator/dongjin-seo/ (accessed on 9 January 2021).
57. Dunagan, J.; Grove, J.; Halbert, D. The Neuropolitics of Brain Science and Its Implications for Human Enhancement and Intellectual Property Law. *Philosophies* **2020**, *5*, 33. [CrossRef]
58. Battaglia, F. Agency, Responsibility, Selves, and the Mechanical Mind. *Philosophies*. (forthcoming).
59. Lai, J.; Melmed, S.; Bond, M. Pa. Now Lets Everyone Vote by Mail. But Poor People in Philly Remain Forgotten. *The Philadelphia Inquirer*. 22 October 2020. Available online: https://www.inquirer.com/politics/election/a/pennsylvania-mail-ballots-election-2020-poor-voters-20201022.html (accessed on 9 January 2021).
60. Norwood, C. How the pandemic has complicated voting access for millions of Native Americans. *PBS News*. 6 October 2020. Available online: https://www.pbs.org/newshour/politics/how-the-pandemic-has-complicated-voting-access-for-millions-of-native-americans (accessed on 9 January 2021).
61. McCulloch, G. The Widely-Spoken Languages We Still Can't Translate Online. *Wired*. 28 November 2018. Available online: https://www.wired.com/story/google-translate-wikipedia-siri-widely-spoken-languages-cant-translate/ (accessed on 9 January 2021).
62. Communication Department of the European Commission. Official Website of the European Union, Europa.com, Publications Office—Interinstitutional style guide—7.2.4. Rules governing the languages of the institutions.*OP/B.3/CRI*. 2019. Available online: http://publications.europa.eu/code/en/en-370204.htm (accessed on 9 January 2021).
63. Federal Register, Vol 81, No. 233. Available online: https://www.justice.gov/crt/file/927231/download (accessed on 28 December 2020).
64. Ancheta, A.N. Language Accommodation and the Voting Rights Act. In *Voting Rights Act Reauthorization of 2006: Perspective on Democracy, Participation, and Power*; Ana, H., Ed.; Berkley Public Policy Press: Berkley, CA, USA, 2007; p. 296.
65. De Araujo, M. The Ethics of Genetic Cognitive Enhancement: Gene Editing or Embryo Selection? *Philosophies* **2020**, *5*, 20. [CrossRef]
66. Santoni de Sio, F.; Faulmuller, N.; Vincent, N. How cognitive enhancement can change our duties. *Front. Syst. Neurosci.* **2014**, *8*, 1031. [CrossRef] [PubMed]
67. Neal, J. Ranked-choice voting, explained. *Harvard Law Today*. 26 October 2020. Available online: https://today.law.harvard.edu/ranked-choice-voting-explained/ (accessed on 9 January 2021).
68. Massachusetts Question 1, Nurse-Patient Assignment Limits Initiative. *Ballotpedia*. 2018. Available online: https://ballotpedia.org/Massachusetts_Question_1,_Nurse-Patient_Assignment_Limits_Initiative_(2018) (accessed on 9 January 2021).
69. Glausiusz, J. Were Neanderthals More Than Cousins to Homo Sapiens. *Anthropology Magazine SAPIENS*. 2020. Available online: https://www.sapiens.org/biology/hominin-species-neanderthals/ (accessed on 9 January 2021).
70. Hughes, J. *Citizen Cyborg: Why Democratic Societies Must Respond to the Redesigned Human of the Future*; Westview Press: Boulder, CO USA, 2004.
71. Korsgaard, C.M. Fellow Creatures: Kantian Ethics and Our to Animals. *Tanner Lect. Hum. Values* **2004**, *24*, 77–110. Available online: https://dash.harvard.edu/handle/1/3198692 (accessed on 9 January 2021).
72. Korsgaard, C.M. Kantian Ethics, Animals, and the Law. *Oxf. J. Leg. Stud.* **2013**, *33*, 629–648. [CrossRef]
73. Maria, G. The Phenomenological Critique of Law according to Simone Goyard-Fabre. In *Archive of the History of Philosophy and Social Thought*; Oxford University Press: New York, NY, USA, 2016; Volume 61.
74. Gray, C.H. *Cyborg Citizen: Politics in the Posthuman Age*; Routledge: New York, NY, USA, 2002.
75. The trial of German major war criminals: Proceedings of the International Military Tribunal sitting at Nuremberg Germany. Available online through the Avalon Project of the Yale Law School. Available online: https://avalon.law.yale.edu/imt/12-14-45.asp (accessed on 9 January 2021).
76. Bachrach, S.D. Introduction by Jeshajahu Weinberg, Director (1989–1995) United States Holocaust Museum. In *Tell Them We Remember: The Story of the Holocaust*; United States Holocaust Memorial Museum. Little, Brown: Boston, MA, USA, 1994.
77. Stone, L. Quantifying the Holocaust: Hyperintense kill rates during the Nazi genocide. *Sci. Adv.* **2019**, *5*, eaau7292. [CrossRef]
78. Mosakas, K. On the moral status of social robots: Considering the consciousness criterion. In *AI & SOCIETY*; Springer: Berlin/Heidelberg, Germany, 2020; Available online: https://www.researchgate.net/publication/342192401_On_the_moral_status_of_social_robots_considering_the_consciousness_criterion (accessed on 9 January 2021).
79. Longuenesse, B. *I, Me, Mine: Back to Kant and Back Again*; Oxford University Press: New York, NY, USA, 2017; p. 163.
80. Heidegger, M. ; Capuzzi, F., Translator; Letter on "Humanism" (1949). Available online: https://warwick.ac.uk/fac/arts/english/currentstudents/undergraduate/modules/en354/syllabus/seminars/letteronhumanism1949.pdf (accessed on 9 January 2021).
81. Di Cesare, D. *Heidegger and the Jews: The Black Notebooks*, English edition; Polity: Cambridge, UK, 2018.
82. Tanzer, M. Heidegger on Animality and Anthropocentrism. *J. Br. Soc. Phenomenol.* **2016**, *47*, 18–32. [CrossRef]
83. Lindberg, S. Heidegger's Animal. *Phänomenologische Forschungen* **2004**, 57–81. Available online: www.jstor.org/stable/24360638 (accessed on 9 January 2021). [CrossRef]
84. Zimmerman, M. Heidegger on Techno-Posthumanism: Revolt against Finitude, or Doing What Comes "Naturally". In *Perfecting Human Futures: Transhuman Visions and Technological Imaginations*; Hurlbut, J., Tirosh-Samuelson, H., Eds.; Springer VS: Hesse, Germany, 2016. [CrossRef]

85. Johnson, A. That Was No Typo: The Median Net Worth of Black Bostonians Really Is $8. *The Boston Globe*. 11 December 2020. Available online: https://www.bostonglobe.com/metro/2017/12/11/that-was-typo-the-median-net-worth-black-bostonians-really/ze5kxC1jJelx24M3pugFFN/story.html (accessed on 9 January 2021).
86. Sorensen, M. Report Outlines 'Shocking' Wage Gap between Black Women and White Men Working in Mass. Restaurants. *Boston Globe*. 14 August 2020. Available online: https://www.bostonglobe.com/2020/08/14/business/report-outlines-shocking-wage-gap-between-black-women-white-men-working-mass-restaurants/ (accessed on 9 January 2021).
87. Onishi, N. Black Man Is Beaten on Camera, Thrusting French Police Into Spotlight. *The New York Times*. 4 December 2020. Available online: https://www.nytimes.com/2020/12/04/world/europe/Michel-Zecler-france-police-video-beating.html (accessed on 9 January 2021).
88. Mclaughlin, R. A Meatless Dominion: Genesis 1 and the Ideal of Vegetarianism. *Biblical Theol. Bull.* **2017**, *47*, 144–154. [CrossRef]
89. Shaw, R. The Phenomenology of Democracy. *Policy Futures Educ.* **2009**, *7*, 340–348. [CrossRef]
90. Latheef, S.; Henschke, A. Can a Soldier Say No to an Enhancing Intervention? *Philosophies* **2020**, *5*, 13. [CrossRef]
91. Pugh, J.; Hannah, M. Drugs That Make You Feel Bad? Remorse-Based Mitigation and Neurointerventions. *Crim. Law Philos.* **2017**, *11*, 499–522. [CrossRef] [PubMed]
92. Allen, A. Why Do Teens Do Such Crazy Stuff? It's All in Their Brains and How They Grow. *Washington Post*. 1 September 2014. Available online: https://www.washingtonpost.com/national/health-science/risky-behavior-by-teens-can-be-explained-in-part-by-how-their-brains-change/2014/08/29/28405df0-27d2-11e4-8593-da634b334390_story.html (accessed on 9 January 2021).
93. Lemay, K.C. Winning the Right to Vote Was the Work of Many Lifetimes. *The New York Times*. 13 August 2020. Available online: https://www.nytimes.com/2020/08/13/us/suffrage-generations-vote.html (accessed on 9 January 2021).
94. McDaneld, J. White Suffragist Dis/Entitlement: The *Revolution* and the Rhetoric of Racism. *Legacy* **2013**, *30*, 243–264. [CrossRef]
95. Derrida, J. *The Animal That Therefore I Am*; Marie-Lous, M., David, W., Eds.; Fordham University Press: New York, NY USA, 15 April 2008.
96. Fukuyama, F. *Our Posthuman Future: Consequences of the Biotechnology Revolution*; Picador: New York, NY, USA, 2003.
97. Martin, P. Even with a Harvard Pedigree, Caste Follows 'like a Shadow.'. *The World from PRX*. 5 March 2019. Available online: https://www.pri.org/stories/2019-03-05/even-harvard-pedigree-caste-follows-shadow (accessed on 9 January 2021).
98. Do Animals Have Feelings? Examining Empathy in Animals. *UWA Online*. 3 April 2019. (discussing research of Professor Marc Bekoff). Available online: https://online.uwa.edu/news/empathy-in-animals/ (accessed on 9 January 2021).
99. Hailey, B.-P. Koko, the gorilla whose sign language abilities changed our view of animal intelligence, dies at 46. *Baltimore Sun*. 21 June 2018. Available online: https://www.baltimoresun.com/la-sci-sn-koko-gorilla-dies-20180621-story.html (accessed on 9 January 2021).
100. Gruen, L. *Entangled Empathy: An Alternative Ethic for Our Relationships with Animals*; Lantern Books: New York, NY, USA, 2015.
101. Berger, M. Breaking with Some Mideast Neighbors, Iran Now Lets Mothers Give Their Citizenship to Their Children. *Washington Post*. 26 December 2020. Available online: https://www.washingtonpost.com/world/middle_east/iran-women-refugees-rights-citizenship/2020/12/24/0b5f74b0-445d-11eb-ac2a-3ac0f2b8ceeb_story.html (accessed on 9 January 2021).
102. Crary, A. *Inside Ethics: On the Demands of Moral Thought*; Harvard University Press: Cambridge, MA, USA, 2016.
103. Sanford, J.B. Senator, 4th District, Letter Received by Secretary of State Frank Jordan on June 26th, 1911, for publication as part of a voters' information manual. Document is currently filed in the California State Archives under: Secretary of State Elections Papers, 1911 Special Election. Available online: https://sfpl.org/pdf/libraries/main/sfhistory/suffrageagainst.pdf (accessed on 9 January 2021).
104. Arnold, A.; Boria, S. Understanding Nazi Animal Protection and the Holocaust. *Anthrozoos Multidiscip. J. Interact. People Anim.* **1992**, *5*, 6–31. [CrossRef]
105. Mill, J.S. *Considerations on Representative Government*; Cambridge University Press: Cambridge, UK, 1861. [CrossRef]
106. United Nations Office of the High Commissioner for Human Rights. General Comment No. 25: The right to participate in public affairs, voting rights and the right of equal access to public service (Art. 25): 12/07/96. CCPR/C/21/Rev.1/Add.7, General Comment No. 25 regarding the United Nations International Covenant on Civil and Political Rights. Available online: https://treaties.un.org/doc/publication/unts/volume%20999/volume-999-i-14668-english.pdf (accessed on 9 January 2021).
107. *Chiafalo v. Washington*, No. 19-465 (U.S. Supreme Court decision) (9-0) (Justice Kagan delivered the opinion of the Court) (6 July 2020). Available online: https://www.supremecourt.gov/opinions/19pdf/19-465_i425.pdf (accessed on 9 January 2021).
108. Dailey, A.C.; Rosenbury, L.A. The New Law of the Child. *Yale. L. J.* **2018**, *127*, 1448–1741.
109. State of Connecticut, Connecticut General Statutes Section 46b-54. Available online: https://www.cga.ct.gov/current/pub/chap_815j.htm#sec_46b-54 (accessed on 9 January 2021).
110. Vaghri, A. Climate Change, an Unwelcome Legacy: The Need to Support Children's Rights to Participate in Global Conversations. *Child. Youth Environ.* **2018**, *28*, 104–114. [CrossRef]
111. Khosla, M. Identity and Representation. In *India's Founding Moment: The Constitution of a Most Surprising Democracy*; Harvard University Press: Cambridge, MA, USA, 2020.
112. Meijer, E. *When Animals Speak: Toward an Interspecies Democracy*; Part 1 of Animals in Context; New York University Press: New York, NY, USA, 2019.
113. Meijer, E.; Watkinson, L. *Animal Languages*; MIT Press: Cambridge, MA, USA, 2020.

114. McElligott, A.G.; O'Keeffe, K.H.; Green, A.C. Kangaroos Display Gazing and Gaze Alternations during an Unsolvable Problem Task. *Biol. Lett.* **2020**, *16*, 20200607. [CrossRef]
115. Godfrey-Smith, P. *Other Minds: The Octopus, the Sea, and the Deep Origins of Consciousness*, 1st ed.; Farrar, Straus and Giroux: New York, NY, USA, 2016.
116. Gorzelak, M.A.; Asay, A.K.; Pickles, B.J.; Simard, S.W. Inter-Plant Communication through Mycorrhizal Networks Mediates Complex Adaptive Behaviour in Plant Communities. *AoB Plants* **2015**, *7*, plv050. [CrossRef] [PubMed]
117. Ferris, J. The Social Life of Forests. *New York Times*. 12 December 2020. Available online: https://www.nytimes.com/interactive/2020/12/02/magazine/tree-communication-mycorrhiza.html (accessed on 9 January 2021).
118. Party for the Animals. *Party for the Animals*. Available online: https://www.partyfortheanimals.com/en/who-we-are (accessed on 9 January 2021).
119. Donaldson, S.; Kymlicka, W. *Zoopolis: A Political Theory of Animal Rights*; Oxford University Press: Oxford, UK, 2013.
120. Haraway, D.J. "*A Cyborg Manifesto*" and "*The Companion Species Manifesto*". In *Manifestly Haraway*; University of Minnesota Press: Minneapolis, MN, USA, 2016.
121. Greenfield, P. Sweet City: The Costa Rica Suburb That Gave Citizenship to Bees, Plants and Trees. *The Guardian*. 29 April 2020. Available online: https://www.theguardian.com/environment/2020/apr/29/sweet-city-the-costa-rica-suburb-that-gave-citizenship-to-bees-plants-and-trees-aoe (accessed on 9 January 2021).
122. O'Donnell, E.L.; Talbot-Jones, J. Creating legal rights for rivers: Lessons from Australia, New Zealand and India. *Ecol. Soc.* **2018**, *23*, 1. [CrossRef]
123. Seubert, S. Politics of Inclusion: Which Conception of Citizenship for Animals? *Hist. Soc. Res.* **2015**, *40*, 154. Available online: http://www.jstor.org/stable/24583241 (accessed on 9 January 2021).
124. Ferrando, F. *Philosophical Posthumanism*; Bloomsbury, Academic—Paperback; Bloomsbury Academic: London, UK, 2020.
125. *Voting by Homeless People* question on the ACE Electoral College Network online community from an ACE Project user in Japan. 13 October 2008. Responses facilitated by Stina Larserud. Available online: https://aceproject.org/electoral-advice/archive/questions/replies/536533290 (accessed on 9 January 2021).
126. Afghan Nomads Cite Poor Representation in Elections. *Gandhara RFE/RL*. 11 October 2018. Available online: https://gandhara.rferl.org/a/afghan-nomads-cite-poor-representation-in-elections/29538652.html (accessed on 9 January 2021).
127. Londoño, E. Stateless, She Became the Face of a Largely Invisible Plight. *The New York Times*. 25 December 2020. Available online: https://www.nytimes.com/2020/12/25/world/americas/stateless-brazil.html (accessed on 9 January 2021).
128. Churchill, W.; Richard, M.L. *Churchill by Himself: The Definitive Collection of Quotations*, 1st ed; PublicAffairs: New York, NY, USA, 2008.
129. U.S. Senate. Ethnic Diversity in the Senate. (as of 2020). Available online: https://www.senate.gov/senators/EthnicDiversityintheSenate.htm (accessed on 9 January 2021).
130. Roy, A. *The Doctor and the Saint: Caste, Race, and Annihilation of Caste, the Debate between b.r. Ambedkar and m.k. Gandhi*; Haymarket Books: Chicago, IL, USA, 2017.
131. Pino, L.M. Colonizing History: *Rice v. Cayetano* and the Fight for Native Hawaiian Self-Determination. *Yale L. J.* **2020**, *8*, 2574–2605.
132. European Commission, Official Website of the European Union. European Parliament, Parliamentary Questions 21 June 2018—Question for oral answer O-000069/2018 to the Council. Available online: https://www.europarl.europa.eu/doceo/document/O-8-2018-000069_EN.html (accessed on 9 January 2021).
133. Nuwer, R. Mass Extinctions Are Accelerating, Scientists Report. *The New York Times*. 1 June 2020. Available online: https://www.nytimes.com/2020/06/01/science/mass-extinctions-are-accelerating-scientists-report.html (accessed on 9 January 2021).
134. Whyte, M. At the Addison, Contemporary Artists Push Back on Traditional Mapmaking. *Boston Globe*. 17 December 2020. Available online: https://www.bostonglobe.com/2020/12/17/arts/marking-their-spot-contemporary-artists-push-back-traditional-mapmaking/ (accessed on 9 January 2021).
135. *Rucho v. Common Cause*, No. 18-422, 588 U.S. ___, 139 S. Ct. 2484 (United States Supreme Court) (5-4) (2019) (Chief Justice Roberts delivered the opinion of the Court). Available online: https://supreme.justia.com/cases/federal/us/588/18-422/ (accessed on 9 January 2021).
136. Rankine, C. Weather. *The New York Times*. 15 June 2020. Available online: https://www.nytimes.com/2020/06/15/books/review/claudia-rankine-weather-poem-coronavirus.html (accessed on 9 January 2021).
137. Colet, A.; Santiveri, J.X.M.; Ventura, J.R.; Saula, O.; de Galdàcano, M.E.S.; Jáuregui, C. *The Black Death and Its Consequences for the Jewish Community in Tarrega: Lessons from History and Archeology*; Pandemic Disease in the Medieval World: Rethinking the Black Death; *The Medieval Globe Books*. 1.; Arc Medieval Press: York, UK, 2014.

MDPI
St. Alban-Anlage 66
4052 Basel
Switzerland
Tel. +41 61 683 77 34
Fax +41 61 302 89 18
www.mdpi.com

Philosophies Editorial Office
E-mail: philosophies@mdpi.com
www.mdpi.com/journal/philosophies